Advances in Agroforestry

Volume 15

Advances in Agroforestry publishes critical reviews and state-of-the-art research results in all areas related to the science and practice of agroforestry worldwide. Its scope covers any aspects of agroforestry research, development, and application, from basic knowledge to the latest findings and innovations. This book series brings together exciting contributions in agroforestry as an increasingly prominent land-use that can harmonize productivity and ecosystem conservation at the farmer and landscape levels. Subjects include, but are not restricted to, aspects of agroforestry establishment, management, and design, financial and social considerations, as well as functions such as the enhancement of environmental services and values.

Contributions are welcome, as well as solicited from prominent experts. Volumes in the series consist of Contributed Volumes, collecting focused reviews written by leading experts, and single-authored or multi-authored monographs providing a comprehensive overview of their research topic.

Bhupendra Singh • Munesh Kumar
Deepa Rawat • Vinod Prasad Khanduri
Editors

Himalayan Agroforestry Systems

Challenges and Opportunity to Sustainable Development

Springer

Editors
Bhupendra Singh
Department of Silviculture and Agroforestry
College of Forestry
Veer Chandra Singh Garhwali Uttarakhand
University of Horticulture and Forestry
Tehri Garhwal, Uttarakhand, India

Munesh Kumar
Department of Forestry and Natural
Resources
Hemvati Nandan Bahuguna Garhwal
University (A Central University)
Srinagar Garhwal, Uttarakhand, India

Deepa Rawat
Department of Forest Resource
Management
College of Forestry
Veer Chandra Singh Garhwali Uttarakhand
University of Horticulture and Forestry
Tehri Garhwal, Uttarakhand, India

Vinod Prasad Khanduri
Department of Silviculture and Agroforestry
Veer Chandra Singh Garhwali Uttarakhand
University of Horticulture and Forestry
Tehri Garhwal, Uttarakhand, India

ISSN 1875-1199 ISSN 1875-1202 (electronic)
Advances in Agroforestry
ISBN 978-3-032-25095-7 ISBN 978-3-032-25096-4 (eBook)
https://doi.org/10.1007/978-3-032-25096-4

This Springer imprint is published by the registered company Springer Nature Switzerland AG
The registered company address is: Gewerbestrasse 11, 6330 Cham, Switzerland

Preface

The editors are pleased to bring out this edited book titled *Himalayan Agroforestry Systems: Challenges and Opportunities for Sustainable Development*. The north-western Himalayan region of India is unique for its agrobiodiversity, due to its rich variety of flora and fauna and traditional knowledge systems. In this region, farmers have been traditionally growing different multipurpose tree species (MPTs) in the bunds of agricultural fields to meet their essential requirements. The existing agroforestry practices in the Indian Himalayan region support sustainable land-use systems and deliver numerous environmental and economically viable benefits through the spatial and temporal integration of agricultural crops and trees. The Indian Himalaya is diverse in its agro-climatic conditions and suitable for cultivating different types of fruits, vegetables, medicinal plants, and traditional crops. In the region, home gardens (HGs) serve as a unique example of sustainably managed systems within small landholdings.

There are a number of obstacles to agroforestry promotion, aside from the valuable ecosystem services it provides. There are also many opportunities arising from international climate change negotiations and national commitments to mitigate the negative effects of climate change. Governance structures and ecological, economic, social, and cultural factors all influence the promotion of agroforestry systems in the current context. The well-known ecological benefits of agroforestry include reduced soil erosion (73%), enhanced biodiversity conservation (65%), and climate change adaptation and mitigation (58%).

This book comprises 14 chapters that cover a wide range of topics related to agroforestry in the Himalayan region. It discusses the concept of various agroforestry practices and their challenges and opportunities to mitigate climate change, enhance biodiversity, and increase farmers' income. Comprehensive information on various agroforestry systems in the Himalayan region, as well as the functions and changes needed to adopt agroforestry in the region, is included in the book.

The editors and authors hope that readers from diverse research fields, professions, and backgrounds will respond favorably and provide input to promote constructive criticism and professional dialogue that will advance understanding of the science, innovations, and policies related to agroforestry. This will help identify new research topics that can fill the existing knowledge gaps and support the implementation of agroforestry systems efficiently across the varied Himalayan region.

The editors would like to express their gratitude to all of the contributors for accepting our invitation, sharing their research and expertise, and contributing their efforts to chapter development and revision to create this book. This volume will be valuable for researchers, academicians, and scholars interested in agroforestry.

We would also like to thank the team from Springer International Publishing for their continuous support throughout the book production process.

Tehri Garhwal, Uttarakhand, India — Bhupendra Singh
Srinagar Garhwal, Uttarakhand, India — Munesh Kumar
Tehri Garhwal, Uttarakhand, India — Deepa Rawat
Tehri Garhwal, Uttarakhand, India — Vinod Parsad Khanduri

Contents

Editors and Contributors

About the Editors

Bhupendra Singh is working as an Assistant Professor (Silviculture and Agroforestry) in the Veer Chandra Singh Garhwali Uttarakhand University of Horticulture and Forestry, Ranichauri, Uttarakhand, India, and did his MSc and PhD degrees in Forestry at H.N.B. Garhwal University, Srinagar Garhwal. He has over 20 years of research experience in forest resource assessment, tree crop interactions, vegetative propagation, and nutrient analysis of fodder trees in the Garhwal Himalaya. He has published more than 95 research papers in different peer-reviewed national and international scientific journals. He is serving as an editorial board member of the *Discover Forests* journal.

Munesh Kumar is a Professor in the Department of Forestry and Natural Resources, H.N.B. Garhwal University, Srinagar Garhwal, Uttarakhand, India. His areas of research interest are Forest Ecology and Biodiversity, Agroforestry, and Ethnobotany. He obtained his MSc and PhD degrees in Forestry from H.N.B. Garhwal University (A Central University), India. He has more than 20 years of teaching and research experience. He has published more than 180 research papers in international and national journals. He has published an edited book as lead editor from CRC Press titled *Diversity and Dynamics in Forest Ecosystems* (ISBN: 9781771889797), co-editor *Sustainable Management and Conservation of Environmental Resources in India* (ISBN: 9781774915929), and co-editor *Climate Change in the Himalayas* (Elsevier, ISBN: 9780443194153), and a few books are in the process of publication in Springer and Elsevier. He also worked as a Guest Editor in several journals of MDPI, *Frontiers*, *Journal of Land Degradation and Development*, *Trees, Forests and People*, etc.

Deepa Rawat is an Assistant Professor in the Department of Forest Resource Management at the College of Forestry, Ranichauri, VCSG Uttarakhand University of Horticulture & Forestry, Bharsar, Uttarakhand, India. She holds MSc and PhD degrees in Soil Science from GB Pant University of Agriculture & Technology, Pantnagar, Uttarakhand, and has teaching and research experience of over 10 years. Her area of research focuses on soil-plant interactions, nutrient management, and sustainable land-use systems in the Himalayan region. She has published 7 book chapters and 20+ research articles in peer-reviewed national and international journals with an impact factor. She has also served as the Principal Investigator for three externally funded research projects. Her research focuses on improving soil fertility, assessing soil health, and evaluating soil carbon sequestration potential to enhance productivity and resilience in mountain agroforestry and other land-use systems.

Vinod Prasad Khanduri is Professor of Forestry (Silviculture) and Head of the Department of Silviculture and Agroforestry at the College of Forestry, VCSG Uttarakhand University of Horticulture and Forestry, Uttarakhand, India. He holds MSc and PhD degrees in Forestry from Hemwati Nandan Bahuguna Garhwal University (a central university), Srinagar Garhwal, Uttarakhand. He has expertise in silviculture, forest ecology, forest biology, tree improvement, and agroforestry. He served as Assistant Professor and Associate Professor in the Department of Forestry, Mizoram University, Aizawl, and subsequently joined the College of Forestry, Ranichauri, as Associate Professor. With more than 24 years of teaching and 28 years of research experience, he has supervised 6 doctoral scholars, more than 40 master's students, and handled 2 major research projects. He has also held key academic and administrative positions, including Head, Department of Forestry (2012–2019); Dean, College of Forestry, Ranichauri (2019–2023); and Registrar (2016, 2017–2018) at Veer Chandra Singh Garhwali Uttarakhand University of Horticulture and Forestry, Bharsar, Uttarakhand. At present, he serves as Head, Department of Silviculture and Agroforestry, and also holds the additional charge of Chief Personnel Officer of the University. He has authored 2 books, 8 book chapters, and over 124 research papers in reputed national and international journals. He also serves as a potential reviewer for various journals.

Contributors

Taufiq Ahmad Department of Extension Education, College of Forestry, Veer Chandra Singh Garhwali Uttarakhand University of Horticulture and Forestry, Ranichauri, Tehri Garhwal, Uttarakhand, India

Benatul Behar Division of Silviculture and Agroforestry, Faculty of Forestry, SK University of Agricultural Sciences & Technology of Kashmir, Jammu, Jammu and Kashmir, India

D. R. Bhardwaj Dr. Yashwant Singh Parmar University of Horticulture and Forestry, Solan, Himachal Pradesh, India

G. M. Bhat Division of Silviculture and Agroforestry, Faculty of Forestry, SK University of Agricultural Sciences & Technology of Kashmir, Jammu, Jammu and Kashmir, India

Garima Bhatt Department of Silviculture and Agroforestry, College of Forestry, Veer Chandra Singh Garhwali Uttarakhand University of Horticulture and Forestry, Ranichauri, Tehri Garhwal, Uttarakhand, India

Sumit Chakravarty Department of Forestry, Uttar Banga Krishi Viswavidyalaya, Cooch Behar, West Bengal, India

Aman Dabral Department of Forest Biology and Tree Improvement, College of Forestry, Ranichauri, Veer Chandra Singh Garhwali Uttarakhand University of Horticulture and Forestry, Bharsar, Uttarakhand, India

M. V. Durai ICFRE – Institute of Wood Science and Technology, Bengaluru, Karnataka, India

Vaishnu Dutt Division of Silviculture and Agroforestry, Faculty of Forestry, SK University of Agricultural Sciences & Technology of Kashmir, Jammu, Jammu and Kashmir, India

Etalesh Goutam Faculty of Agriculture, Guru Kashi University, Talwandi Sabo, Bathinda, Punjab, India

M. A. Islam Division of Silviculture and Agroforestry, Faculty of Forestry, SK University of Agricultural Sciences & Technology of Kashmir, Jammu, Jammu and Kashmir, India

Manish Jangid Department of Silviculture and Agroforestry, College of Forestry, Veer Chandra Singh Garhwali Uttarakhand University of Horticulture and Forestry, Ranichauri, Tehri Garhwal, Uttarakhand, India

Alisha Keprate Dr. Yashwant Singh Parmar University of Horticulture and Forestry, Solan, Himachal Pradesh, India

Vinod Prasad Khanduri Department of Silviculture and Agroforestry, College of Forestry, Veer Chandra Singh Garhwali Uttarakhand University of Horticulture and Forestry, Ranichauri, Tehri Garhwal, Uttarakhand, India

Atul Kumar Department of Forest Products and Utilization, College of Forestry, Veer Chandra Singh Garhwali Uttarakhand University of Horticulture and Forestry, Ranichauri, Tehri Garhwal, Uttarakhand, India

Jatin Kumar Dr. Yashwant Singh Parmar University of Horticulture and Forestry, Nauni, Solan, Himachal Pradesh, India

Munesh Kumar Department of Forestry and Natural Resources, Hemvati Nandan Bahuguna Garhwal University (A Central University), Srinagar Garhwal, Uttarakhand, India

Pankaj Kumar Department of Genetics and Plant Breeding, College of Hill Agriculture, Chirbatiya, at Ranichauri, Veer Chandra Singh Garhwali Uttarakhand University of Horticulture and Forestry, Ranichauri, Tehri Garhwal, Uttarakhand, India

Sandeep Kumar Department of Silviculture and Agroforestry, Veer Chandra Singh Garhwali Uttarakhand University of Horticulture and Forestry, Ranichauri, Tehri Garhwal, Uttarakhand, India

Department of Basic and Social Sciences, Veer Chandra Singh Garhwali Uttarakhand University of Horticulture and Forestry, Ranichauri, Tehri Garhwal, Uttarakhand, India

Tarun Kumar Department of Environmental Science, Indira Gandhi National Tribal University, Amarkantak, Madhya Pradesh, India

Bijendra Lal Department of Forest Products & Utilization, College of Forestry, Ranichauri, Veer Chandra Singh Garhwali Uttarakhand University, Bharsar, Uttarakhand, India

Nazir A. Pala Division of Silviculture and Agroforestry, Faculty of Forestry, SK University of Agricultural Sciences & Technology of Kashmir, Jammu, Jammu and Kashmir, India

Prabhat Kumar Pal Department of Agricultural Extension, Uttar Banga Krishi Viswavidyalaya, Cooch Behar, West Bengal, India

Vipin Panwar Department of Botany, Bharat Singh Rawat, Govt. Degree College, Pauri Garhwal, Uttarakhand, India

Pooja Department of Silviculture and Agroforestry, Dr YS Parmar University of Horticulture and Forestry, Solan, Himachal Pradesh, India

Uday Bhanu Pratap Department of Forest Products & Utilization, College of Forestry, Ranichauri, Veer Chandra Singh Garhwali Uttarakhand University, Bharsar, Uttarakhand, India

Modala Rakesh ICFRE – Forest Research Institute, Dehradun, Uttarakhand, India

T. A. Rather Division of Silviculture and Agroforestry, Faculty of Forestry, SK University of Agricultural Sciences & Technology of Kashmir, Jammu, Jammu and Kashmir, India

Deepa Rawat Department of Forest Resource Management, College of Forestry, V.S.C.G. Uttarakhand University of Horticulture and Forestry, Ranichauri, Tehri Garhwal, Uttarakhand, India

J. M. S. Rawat Department of Agriculture, Government Degree College, Rudraprayag, Uttarakhand, India

Vidyawati Rawat Department of Agriculture, Government Degree College, Rudraprayag, Uttarakhand, India

Gaddam Sudharshan Reddy ICFRE – Institute of Wood Science and Technology, Bengaluru, Karnataka, India

Manoj Kumar Riyal Department of Basic and Social Sciences, Veer Chandra Singh Garhwali Uttarakhand University of Horticulture and Forestry, Ranichauri, Tehri Garhwal, Uttarakhand, India

Sarswati Prakash Sati Department of Basic and Social Sciences, Veer Chandra Singh Garhwali Uttarakhand University of Horticulture and Forestry, Ranichauri, Tehri Garhwal, Uttarakhand, India

Prashant Sharma ICAR-Indian Institute of Soil and Water Conservation, Research Centre, Datia, India

Vatika Sharma Department of Silviculture and Agroforestry, College of Forestry, Veer Chandra Singh Garhwali Uttarakhand University of Horticulture and Forestry, Ranichauri, Tehri Garhwal, Uttarakhand, India

Gopal Shukla Department of Forestry, North Eastern Hill University, Tura, Meghalaya, India

Abhishek Pratap Singh Department of Silviculture and Agroforestry, Acharya Narendra Deva University of Agriculture and Technology, Ayodhya, Uttar Pradesh, India

Bhupendra Singh Department of Silviculture and Agroforestry, College of Forestry, Veer Chandra Singh Garhwali Uttarakhand University of Horticulture and Forestry, Tehri Garhwal, Uttarakhand, India

Chandramohan Singh Department of Silviculture and Agroforestry, College of Forestry, Veer Chandra Singh Garhwali Uttarakhand University of Horticulture and Forestry, Ranichauri, Tehri Garhwal, Uttarakhand, India

Indra Singh Department of Silviculture and Agroforestry, College of Forestry, Veer Chandra Singh Garhwali Uttarakhand University of Horticulture and Forestry, Tehri Garhwal, Uttarakhand, India

Krishan Kumar Singh Faculty of Agriculture, Guru Kashi University, Talwandi Sabo, Bathinda, Punjab, India

Naresh Singh Department of Forest Biology and Tree Improvement, College of Forestry, Ranichauri, Veer Chandra Singh Garhwali Uttarakhand University of Horticulture and Forestry Uttarakhand, Bharsar, Uttarakhand, India

Bisleshna Tamang Department of Forestry, Uttar Banga Krishi Viswavidyalaya, Cooch Behar, West Bengal, India

Sumit Tewari Department of Silviculture and Agroforestry, College of Forestry, Veer Chandra Singh Garhwali Uttarakhand University of Horticulture and Forestry, Ranichauri, Tehri Garhwal, Uttarakhand, India

Arjun Tomar Department of Forestry and Natural Resources, Hemvati Nandan Bahuguna Garhwal University (A Central University), Srinagar-Garhwal, Uttarakhand, India

Pooja Uniyal Department of Silviculture and Agroforestry, College of Forestry, Veer Chandra Singh Garhwali Uttarakhand University of Horticulture and Forestry, Ranichauri, Tehri Garhwal, Uttarakhand, India

S. K. Verma Department of Silviculture and Agroforestry, Acharya Narendra Deva University of Agriculture and Technology, Ayodhya, Uttar Pradesh, India

Vineeta Department of Forestry, Uttar Banga Krishi Viswavidyalaya, Cooch Behar, West Bengal, India

Chapter 1
Multipurpose Trees in Agroforestry Systems of the Northwestern Himalaya

Chandramohan Singh, Hitesh Gupta, Vinod Prasad Khanduri, Bhupendra Singh, Deepa Rawat, and Munesh Kumar

Abstract The Northwestern Himalayan region of India encompasses the territories of Ladakh, Jammu and Kashmir, Himachal Pradesh, and Uttarakhand, covering approximately 10% of the country's geographical area. The country exhibits significant ecological and biological diversity, sustaining the livelihoods of a substantial proportion of rural populations in Himachal Pradesh (89.90%), Uttarakhand (69.40%), Jammu and Kashmir (72.62%), and Ladakh (61.33%), respectively. The Northwestern Himalayas are rich in natural resources, including forests, water, and a diverse range of flora and fauna. This region is characterized by an abundance of multipurpose tree species (MPTs), which are traditionally grown in agricultural fields by local farmers and managed through agroforestry practices to meet their fundamental requirements. The term MPT is used predominantly in tropical agroforestry to describe woody plants that are intentionally cultivated and managed to deliver a range of functions, products, and ecosystem services. Although relatively underutilized in commercial forestry and conventional agriculture, they constitute essential components of most agroforestry systems (AFS). While MPTs are recognized as vital fodder resources for livestock, their use is limited due to an insufficient understanding of their nutritional properties. Agroforestry represents a sustainable land-use system that delivers multiple ecological and economic benefits through the spatial and temporal integration of agricultural crops and trees. The AFS prevalent in the region play a crucial role in sustaining the livelihoods of millions of people by ensuring both food security and financial income. Furthermore,

C. Singh · H. Gupta · V. P. Khanduri · B. Singh (✉)
Department of Silviculture and Agroforestry, College of Forestry, Veer Chandra Singh Garhwali Uttarakhand University of Horticulture and Forestry, Tehri Garhwal, Uttarakhand, India

D. Rawat
Department of Forests Resource Management, College of Forestry, Veer Chandra Singh Garhwali Uttarakhand University of Horticulture and Forestry, Tehri Garhwal, Uttarakhand, India

M. Kumar
Department of Forestry and Natural Resources, Hemvati Nandan Bahuguna Garhwal University (A Central University), Srinagar Garhwal, Uttarakhand, India

B. Singh et al. (eds.), *Himalayan Agroforestry Systems*, Advances in Agroforestry 15,
https://doi.org/10.1007/978-3-032-25096-4_1

they contribute significantly to forest and biodiversity conservation, enhance soil quality and water retention, and provide opportunities for cultural, recreational, and spiritual activities. The chapter presents evidence on the significance of MPTs in the agroforestry landscapes of the Northwestern Himalayas, emphasizing their roles and contributions to sustainability for livelihood.

Keywords Multipurpose tree species · Agroforestry systems · Biodiversity conservation · Sustainable practice · Northwestern Himalaya

1 Introduction

In recent years, the world has experienced a significant increase in population, leading to an unprecedented demand for food and resources (Singh et al. 2014). According to FAO (2009), sustaining the global population by 2050 will require a 70% increase in the current level of worldwide food production. However, addressing this escalating demand in the context of climate change and increasing frequency of extreme weather events will likely intensify reliance on chemical inputs and nutrient supplementation, thereby exerting adverse impacts on ecosystem health (Sharma et al. 2022; Verma et al. 2023). The intensification of poverty, population pressures, agricultural expansion and amplification, and infrastructure development has increased the tremendous pressure on biodiversity and natural resources (Baboo et al. 2017). The overexploitation of natural resources has created a substantial gap between the demand and supply of goods and services (Kaushal et al. 2024). In response to this challenge, agroforestry appears to be a promising agricultural approach that supports the eco-friendly production of both edible and nonedible products (Sharma et al. 2017a, b, c; Bhardwaj et al. 2024), while also contributing to the implementation of several sustainable development goals (SDGs) (Panwar et al. 2022) and the sequestration and storage of carbon at the farm level (Verma et al. 2021). The Indian National Agroforestry Policy (2014) defines agroforestry as the practice of integrating trees into croplands and rural landscapes, either with or without livestock, to improve ecosystem productivity, profitability, diversity, and sustainability. A pan-European study by Rois-Díaz et al. (2017) states that "agroforestry is a modern term for an ancient and widespread practice." Recently, "agroforestry" has been defined as a land-use system that includes productivity, additional ecosystem services, and environmental benefits more comprehensively (Jose 2019).

The Indian Himalayan Region (IHR) is located between the latitudes of 23.23° and 34.08° N and longitudes of 73.17° and 95.51° E, covering about 12% of the total geographical area of the country (ISFR 2023), and is inhabited by around 51 million people, who engage in agricultural practices in fragile and diverse ecological environments. The region extends across the Eastern and Western Himalayas, encompassing 11 states and 2 union territories. Due to its rich biological and sociocultural diversity, the region has been identified as one of the 34 biological hotspots (Tewari et al. 2017). The Western Indian Himalayas harbor nearly one-tenth of the world's recorded high-altitude plant and animal species and account for approximately half of the country's native flora. This region is notably rich in biodiversity and also includes the Indian states of Jammu and Kashmir, Himachal Pradesh, and Uttarakhand, where elevations range from 300 meters to over 6000 meters (Padma 2014). In the Himalayan states of the country, indigenous AFS are vital for local communities, and planting trees on farmland allows farmers to fulfill their various needs, which increases tree density and reduces the pressure on existing forest resources (Phondani et al. 2020). Additionally, agroforestry, as an integrated farming system, plays a crucial role in sustaining the fragile ecosystems of the region (Kaler et al. 2017) and reduces farmers' investment risks by diversifying their crops and income sources, which in turn mitigates economic and social risks (Mey and Gore 2021). Various workers (Toky et al. 1989; Khosla and Toky 1996; Thakur et al. 2004) identified and studied several AFS in the Northwestern Himalayan region, focusing on their floristic diversity, biological productivity, carbon sequestration potential, and the impact on soil physicochemical properties. Traditional agroforestry encompasses centuries-old practices used worldwide, varying in structure, function, socioeconomic characteristics, and ecological services (Ghale et al. 2022). These systems were designed as multifunctional land-use practices that sustainably supplied the six Fs (food, fiber, fuel, fruits, floss, fertilizer, and fodder), thereby meeting the diverse subsistence and livelihood needs of farming families (Chavan et al. 2015). In the Northwestern Himalayan region, a variety of traditional agroforestry practices have been reported, including silvopastoral, agrisilviculture, agrisilvihorticulture, pastoral-silviculture, hortisilviculture, agri-horticulture, agrihortisilviculture, agri-silvopastoral, and pastoral silvihorticulture. These systems exhibit diverse structural and functional compositions, tailored to address the daily needs and preferences of local farmers while being well-adapted to the ecological conditions of the region.

The term "MPT" is a concept that is used almost exclusively in tropical agroforestry (Nair et al. 2021). Generally, the taller and long-lived plant components of AFS are typically categorized as trees. However, MPT refers to woody species that are intentionally cultivated and managed to provide multiple purposes, products, or services. These species are still underutilized in commercial forestry and agriculture, yet they are vital components of AFS. Most of the MPTs are designated by single-purpose-sounding terms such as fodder trees, fruit trees, fuelwood species, and other specialized trees; however, they offer multiple products and services (Chavan et al. 2021; Nair et al. 2021). They also play a crucial role in carbon capture, water conservation, and the rehabilitation of degraded lands (Semwal et al. 2013). MPTs not only meet the basic needs of rural people but also provide cash benefits and raw materials to local artisans and forest-based industries (Tewari 1995), along with essential ecosystem services (Sharma et al. 2017a, b, c). MPTs are recognized as a significant source of fodder for livestock, but their potential remains untapped due to a lack of knowledge about their nutritive value (Navale et al. 2022). However, in India, natural pastures are either degraded or have a carrying capacity of less than one livestock unit per hectare per year (PCI 2011). Simultaneously, growing forage crops on farms has become increasingly challenging because of the rapid rise in the human population and the demand for food grains (Sharma 2004). This is the case of leaf fodder from MPTs that is notable, as these trees are commonly regarded as emergency feedstuffs (Ganai et al. 2009), particularly in the hilly regions (Khatta et al. 1999), due to the high protein, soluble carbohydrate, mineral, and vitamin contents of their leaves (Azim et al. 2011). Therefore, MPTs are intentionally planted or left to grow along farm bunds (Khosla et al. 1992) because they offer a variety of products and can significantly contribute to rural economies (Sharma et al. 2017a, b, c; Sharma et al. 2022). As a result, MPTs can enhance livestock productivity by reducing the gap between demand and the availability of green fodder (Datt et al. 2008). NAP (2014) identified 20 significant MPTs, including commercial trees such as poplar and eucalyptus, at the national level as being exempt from all limitations on harvesting, transporting, and marketing cultivated under agroforestry systems. The nutritional value of some commercial MPTs used for forage has been studied extensively (Khosla et al. 1992; Datt et al. 2008; Navale et al. 2022); however, limited data are available on less commercially important but common MPTs, particularly regarding the changes in forage quality due to seasonal variation and its impact on palatability in the Northwestern Himalayas. In this chapter, we synthesize information on MPT in AFS practices, particularly focusing on the Northwestern Himalayan region, also highlighting their potential to meet society's growing demands for fuel, food, fodder, and wood, as well as their contribution to ecosystem services and sustainability.

1.1 Agroforestry Systems and Livelihoods in the Northwestern Himalayas: An Overview

In the Western Himalayas, land-use practices consist of agroecosystems, with 90% of the population residing in villages where agriculture, horticulture, and animal husbandry serve as the primary sources of livelihood (Atul et al. 1994). Agroforestry is traditionally practiced in tropical, subtropical, and temperate regions (Kumar and Srivatava 2018). It combines trees, crops, and livestock to create a multidisciplinary land-use system that meets both productive and protective goals (Singh et al. 2015). In the Himalayan region, traditional agroforestry is mainly practiced in an area of 6.51 million hectares, covering the Western Himalayas, which include Ladakh, Jammu and Kashmir, Himachal Pradesh, and Uttarakhand. It also extends to the Eastern Himalayas, encompassing Sikkim, Arunachal Pradesh, the hills of Darjeeling and Assam, as well as Manipur, Mizoram, Assam, Meghalaya, Tripura, and Nagaland. In this region, agroforestry has been identified as a productive land-use approach; nonetheless, the regional determinants of its adoption, the factors driving changes in traditional practices, and the socioeconomic implications associated with agroforestry must be examined thoroughly. Agroforestry meets the diverse needs of humans while providing economic benefits and environmental services, including carbon sequestration, watershed protection, and climate change mitigation and adaptation. The dramatic changes in the altitudinal ranges of the Western Himalayan region also affect the vegetation patterns (Tewari et al. 2017) and the composition of agroforestry systems. Over the years, farmers have accustomed several multipurpose tree species on their farmlands, which have evolved into extant agroforestry practices. In the IHR, the adoption of agroforestry technologies relies on edaphic and climatic conditions, the socioeconomic status and needs of farmers, while management is influenced by physical, demographic, and institutional factors (Nath et al. 2022; Sharma et al. 2022).

The major AFS in the Northwestern Himalayas are shown in Table 1.1. The major AGFs practiced in the region included agrisilvihorticulture, agrihorticulture, agrihortisilviculture, hortipastoral, agrisilviculture, hortiagriculture, hortisilviculture, hortiagrisilviculture, silvopastoral, hortisilvopastoral, pastoralsilviculture, agrisilvipastoral, and silvihorticulture. In Jammu and Kashmir, important tree species are *Populus* spp., *Ulmus wallichiana*, *Ailanthus altissima*, *Salix* spp., *Morus alba*, and *Aesculus indica*. In contrast, the Ladakh region is dominated by *Populus* spp., *Ulmus wallichiana*, *Salix* spp., and *Juniperus* spp. Several tree species, such as *Grewia* spp., *Celtis australis*, *Quercus leucotrichophora*, *Toona ciliata*, *Cedrus deodara*, *Morus* spp., and *Melia* spp., grow commonly in both Himachal Pradesh and Uttarakhand. Moreover, due to the diverse landscape, climate, and environmental conditions, AFS varies significantly across the Himalayan region. For deeper insights, see Kashyap et al. (2014) on the Northwestern Himalayas and Singh et al. (2014) on the Northeastern Himalayas. Both studies provide a thorough overview of AFS and their importance in soil conservation, ecosystem stability, and livelihood security. However, all these systems support livestock in various ways to fulfill the domestic needs of the people, as each household in the Himalayan region traditionally maintains 5–8 livestock units (Banyal et al. 2016).

India is home to the largest livestock population in the world, with 538.8 million livestock in 2019, which increased by 4.6% in comparison with 2012 (DAHD and F 2021) and contributes significantly to the economy, accounting for 5.1% of the total gross value added (GVA) and 17.1% of the GVA in the agriculture and allied sectors, respectively (NAS 2020). The increasing livestock population has created challenges in the fulfillment of the fodder requirement and impacted the availability of fodder resources needed to meet nutritional requirements (PCI 2011). The availability of fodder must be addressed, as feed accounts for approximately 70% of the cost of milk production and is a significant factor contributing to the 20–60% lower productivity of livestock in Indian conditions. According to a report by ICAR-IGFRI (2021), there is a 49.17% shortage of fodder in Jammu and Kashmir, 40–45% in Ladakh (Tewari et al. 2016), and about 33% in Himachal Pradesh (NITI Aayog 2018). By adopting alternative land-use systems such as silvopastoral and hortipasture systems, it is possible to enhance land productivity, meet fodder requirements,

Table 1.1 Functional composition of various agroforestry systems in the Northwestern Himalayan region of India

State/union territory	Agroforestry practice	Agroforestry trees	Horticulture trees	Agriculture crops	Major grasses	References
Uttarakhand	Agrisilviculture, agrisilvihorticulture, homegardens, agrisilvopastural, silvipastural, agrihorticulture, agrihortisilviculture, silvihorticulture, hortipastoral, medicinal agroforestry, block and boundary plantation	*Albizia procera*, *Eucalyptus globulus*, *Populus deltoides*, *Alnus nepalensis*, *Quercus leucotrichophora*, *Grewia oppositifolia*, *Celtis australis*, *Pinus roxburghii*, *Holoptelea integrifolia*, *Leucaena leucocephala*, *Prunus cerasoides*, *Rhododendron arboretum*, *Pyrus pashia*, *Cedrus deodara*, *Ficus auriculata*, *Ficus palmata*, *Morus serrata*, *Melia azedarach*, *Myrica esculenta*, *Toona ciliata*	*Artocarpus heterophyllus*, *Carica papaya*, *Citrus lemon*, *Citrus limetta*, *Citrus medica*, *Citrus paradise*, *Citrus sinensis*, *Litchi chinensis*, *Juglans regia*, *Prunus armeniaca*, *Prunus domestica*, *Malus domestica*, *Pyrus pyrifolia*, *Musa paradisiacal*, *Punica granatum*, *Carica papaya*, *Psidium guajava*, *Prunus* spp., *Phyllanthus emblica*, *Mangifera indica*	*Allium sativum*, *Coriandrum sativum*, *Zea mays*, *Triticum aestivum*, *Oryza sativa*, *Phaseolus vulgaris*, *Eleusine coracana*, *Amaranthus caudatus*, *Amaranthus spinulosa*, *Vigna umbellate*, *Vigna mungo*, *Cajanus cajan*, *Glycine max*, *Hordeum vulgare*, *Lens culinaris*, *Brassica campestris*, *Echinochloa frumentacea and Sorghum bicolor*, *Pisum sativum*, *Lycopersicon esculentum*, *Abelmoschus esculentus*, *Spinacea oleracea*, *Brassica juncea*, *Brassica oleracea*, *Coriandrum sativum*, *Curcuma longa*, *Zingiber officinale*, *Capsicum annuum*, *Allium cepa*	*Cynodon dactylon*, *Andropogon munroi*, *Apluda mutica*, *Euphorbia hirta*, *Avena fetua*, *Setaria* spp., *Paspalum* spp.	Yadav and Bisht (2013), Kashyap et al. (2014), Himshikha (2015), Mahato et al. (2016), Himshikha and Singh (2017), Kumar and Srivatava (2018), Vikrant et al. (2018), Yadav et al. (2019), Bhatt and Parihaar (2020), Gariya et al. (2020), Himshikha et al. (2020), and Kumar et al. (2021), Maurya et al. (2025)

Jammu and Kashmir	Agrisilviculture, agrihorticulture, hortiagriculture, hortisilviculture, homegardens, hortisilvopastoral, hortiagrisilviculture, silvopastoral, hortipastoral, agrisilvipastoral	*Grewia optiva*, *Celtis australis*, *Populus deltoids*, *Ulmus wallichiana*, *Aesculus indica*, *Salix alba*, *Robinia pseudoacacia*, *Morus alba*, *Ailanthus altissima*, *Cedrus deodara*, *Albizzia lebbeck*, *Acacia catechu*, *Dalbergia sissoo*, *Quercus leucotricophora*, *Toona ciliata*, *Bauhinia retusa*, *Anogeissus latifolia*, *Butea monosperma*	*Mangifera indica*, *Psidium guajava*, *Malus* spp., *Prunus amygdalus*, *Juglans regia*, *Prunus persica*, *Prunus avium*, *Cydonia oblonga*, *Punica granatum*, *Vitis vinifera*, *Morus laevigata*	*Oryza sativa*, *Zea mays*, *Brassica juncea*, *Daucas carota*, *Raphanus sativus*, *Brassica rapa*, *Abelmoschus esculentus*, *Glycine max*, *Triticum aestivum*, *Pisum sativum*, *Brassica oleracea*, *capitata*, *Brassica oleracea var. botrytis*, *Solanum tuberosum*, *Allium sativum*, *Cucumis sativus*, *Lagenaria siceraria*, *Momordica charantia*, *Phaseolus vulgaris*, *Vigna radiata*	*Trifolium repens.*, *Dactylis glomerata*, *Festuca pretense*, *Aegilops tauschii*, *Echinochola crus-galli*, *Lolium perenne*, *Lolium multiflorum*, *Bromus japonicas*, *Bromus inermis*, *Cynodon dactylon*, *Chrysopogon fulvus*, *Dicanthiumspp*	Choudhary et al. (2012), Kashyap et al. (2014), Gupta and Arora (2015), Banyal et al. (2016), Khaki et al. (2016), Ahmad et al. (2017), Islam et al. (2017), Dar et al. (2018), and Zahnoor et al. (2022)
Ladakh	Agrisilviculture, agrihortisilviculture, shelterbelts, windbreaks, agrihorticulture, hortisilvopastoral, hortipasture, silvopastoral, homegardens	*Betula utilis*, *Populus* spp., *Hippophae rhamnoides*, *Salix* spp., *Ulmus wallichiana*, *Juniperus* spp.	*Pyrus communis* *Malus domestica.*, *Prunus armeniaca*, *Juglans regia*, *Morus alba*	*Raphanus sativus*, *Triticum aestivum*, *Hordeum vulgare*, *Avena sativa*, *Fagopyrum esculentum*, *Fagopyrum tataricum*, *Panicum miliaceum*, *Pisum sativum*, *Allium cepa*, *Coriandrum sativum*, *Carum carvii*, *Chenopodium album*, *Setaria italic*, *capitata*, *Brassica oleracea*, *Phaseolus vulgaris*	*Avena sativa*, *Dicanthium* spp., *Trifolium* spp., *Medicago sativa*, *Iris lactea*	Pateria et al. (2003), Butola et al. (2012), Banyal et al. (2016), Handa et al. (2019), and Namgial et al. (2020)

(continued)

Table 1.1 (continued)

State/union territory	Agroforestry practice	Agroforestry trees	Horticulture trees	Agriculture crops	Major grasses	References
Himachal Pradesh	Agrisilviculture, agrihorticulture, agrisilvihorticulture, agrihortisilviculture, homegardens, hortiagriculture, silvopastoral, hortipastoral, pastoralsilviculture, agrisilvopastoral	*Grewia optiva*, *Celtis australis*, *Melia azedarach*, *Pistacea integrima*, *Dalbergia sissoo*, *Toona ciliata*, *Morus alba*, *Ficus roxburghii*, *Ficus palmata*, *Bauhinia variegata*, *Bauhinia vahlii*, *Bombax ceiba*, *Melia composita*, *Albizia chinensis*, *Crataegus songarica*, *Acacia catechu*, *Quercus spp*, *Cedrus deodara*, *Pinus roxburghii*, *Populus ciliata*, *Robinia pseudoacacia*, *Ulmus villosa*, *Salix alba*	*Morus alba*, *Mangifera indica*, *Citrus* spp., *Prunus domestica*, *Corylus jacquemontii*, *Litchi chinensis*, *Psidium guajava*, *Carica papaya*, *Phyllanthus emblica*, *Malus domestica*, *Prunus persica*, *Pyrus communis*, *Prunus armeniaca*, *Diospyros kaki*, *Punica granatum*, *Juglans regia*, *Prunus amygdalus*, *Pistacia vera*, *Prunus dulcis*	*Zingiber officinale*, *Zea mays*, *Oryza sativa*, *Vigna mungo*, *Phaseolus vulgaris*, *Solanum lycopersicum*, *Abelmoschus esculentus*, *Glycine max*, *Triticum aestivum*, *Lens culinaris*, *Hordeum vulgare*, *Brassica juncea*, *Cicer arietinum*, *Pisum sativum*, *Brassica oleracea var. capitata*, *Brassica oleracea*, *Solanum tuberosum*, *Allium sativum*, *Allium cepa*, *Avena sativa*, *Amaranthus hypochondriacus*, *Echinochloa frumentacea*, *Sorghum bicolor*	*Arundo donex*, *Imperata cylindrica*, *Chrysopogon montanus*, *Seteria glauca*, *Cymbopogon martinii*, *Apluda mutica*, *Heteropogon contortus*, *Dicanthium annulatum*, *Themeda anathera*, *Paspalum notatum*, *Apluda mutica*, *Andropogon nardus*, *Pennisetum clandestinum*, *Dactylis gloemerata*, *Arundinella nepalensis*, *Agrostis* spp., *Poa annua*, *Trifolium repens*, *Cynodon dactylon*	Toky et al. (1989), Kumari et al. (2008), Chisanga et al. (2013), Goswami et al. (2014), Kashyap et al. (2014), Gupta et al. (2017), Rajput et al. (2017), Tiwari et al. (2018), Kumar et al. (2018a, b), Salve and Bhardwaj (2020), Sharma et al. (2023a, b), Sharma et al. (2024), and Dogra et al. (2024)

reduce grazing pressure, and achieve positive environmental outcomes (Roy et al. 2019).

Additionally, the wide altitudinal variations in the Himalayan region lead to significant differences in climatic conditions, with some areas being covered in snow during the winter months. Under such conditions, fuelwood serves as a significant source of energy, with people primarily relying on forest resources (Kumar et al. 2020). Numerous studies indicate that 93% of the population in Himachal Pradesh relies on fuelwood for energy (Parikh 2011; TERI 2015), with 94% of those users depending on forests as their primary source. Fuelwood consumption in Jammu and Kashmir ranges from 0.05 to 5.50 kg per capita per day; in Himachal Pradesh, it varies from 0.91 to 5.13 kg per capita per day, while in Uttarakhand, it is reported to range from 1.13 to 8.75 kg per capita per day (Kumar et al. 2020), indicating that the inhabitants rely on fuelwood. Government initiatives like the Pradhan Mantri Ujjwala Yojana aim to promote the use of LPG as a cleaner alternative to traditional cooking fuels. Additionally, integrated farming practices play a crucial role in meeting the demand for fuelwood, which helps alleviate pressure on and the overexploitation of natural resources. The annual availability of fuelwood from trees outside forests (TOF) in Jammu and Kashmir, including Ladakh, Himachal Pradesh, and Uttarakhand, reported 0.365, 0.290, and 0.297 million tonnes, respectively. In comparison, the fuelwood available from forests in these respective union territories and states reported 0.02 million tonnes (Jammu & Kashmir), 0.00005 (Himachal Pradesh), and 0.005 million tonnes (Uttarakhand) (Dar et al. 2018). Moreover, the availability of fuelwood on farmland will support the use of cow dung as organic manure rather than burning it as an energy source. Farmers rely on natural resources for fodder, fuelwood, timber, and small wood, but this dependence can lead to exploitation when necessity turns into greed. Agroforestry provides both ecological services and economic benefits, as 65% of the country's timber demand is fulfilled by trees outside forests (Government of India 2016). In the current scenario, where there is a lack of data for both the demand and supply of tree-based products and natural forests are closed for protection and conservation purposes, there is a greater opportunity for promoting agroforestry practices (Parthiban et al. 2021). Furthermore, a dedicated agroforestry policy that supports the selection of species appropriate for specific regions makes quality planting material available, allows flexible harvesting and transit regulations, and provides marketing facilities can encourage wider adoption of scientific agroforestry practices.

2 Ecosystem Services Provided by MPTs in AFS

The "benefits that people derive from ecological systems" are referred to as ecosystem services (Millennium Ecosystem Assessment MEA 2005). It is also described as an "ecosystem that offers both direct and indirect benefits to human well-being" (The Economics of Ecosystems and Biodiversity TEEB 2010). Ecosystem services

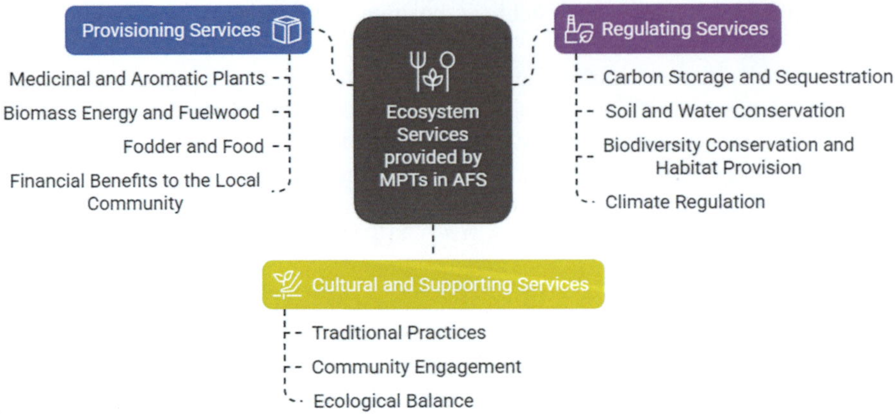

Fig. 1.1 Ecosystem services provided by multipurpose tree species

that benefit society are air quality, climate regulation, carbon sequestration, water purification, pollination, and erosion prevention, and they have been valued at 16–54 trillion US dollars per year (Costanza et al. 1997). The ecosystem service approach offers a promising strategy for preserving biodiversity-rich ecosystems given the willingness of society to safeguard natural biodiversity. Important services provided by MPTs in the Western Himalayan region are discussed as follows (Fig. 1.1).

2.1 *Provisioning Services*

The MEA provisioning service encompasses processes that provide food, fiber, fuel, water, biochemicals, medicinal plants, pharmaceuticals, and genetic resources. Provisioning services from MTPs primarily focus on food and raw materials supplied to rural households that directly rely on them for their livelihoods and increase food security for both humans and livestock. Various studies showed that farmers in the Northwestern Himalayan region practice agroforestry, but they often do so in an unstructured and unplanned manner. A general survey conducted across the hills of the Northwestern Himalayas identified various MPTs, shrubs, and grasses, along with their uses such as fodder, fuel, timber, medicine, construction, fruit/leaves, fiber, religious, and other uses (Table 1.2). Some provisioning services of the MTPs are briefly described as follows.

Table 1.2 Multidimensional uses of major plants in agroforestry systems of the Northwestern Himalayas

Botanical name	Common name	Multipurpose uses								
		Fodder	Fuel	Timber	Medicine	Construction	Fruit/ leaves	Fiber	Religious	Others
Acer caesium	Kainjal	✓	✓	X	X	✓	X	X	X	X
Acer oblongum	Putali	✓	✓	X	X	X	X	X	X	X
Adina cordifolia	Haldu	✓	✓	X	X	✓	X	X	X	X
Aesculus indica	Pangar	✓	X	X	✓	X	✓	X	X	X
Alangium salviifolium	Ankol	✓	X	X	✓	X	X	X	X	X
Albizia chinensis	Kala siras	✓	✓	✓	X	X	X	X	X	X
Albizia lebbeck	Siris	✓	✓	✓	✓	X	X	X	X	✓
Albizia procera	Safed siris	✓	✓	✓	✓	X	X	X	X	✓
Alnus nepalensis	Uteesh	✓	✓	✓	X	X	X	X	X	✓
Anogeissus latifolia	Dhauda	✓	✓	✓	✓	X	X	X	X	X
Bauhinia purpurea	Kaniyar	✓	✓	✓	✓	X	X	X	X	✓
Bauhinia racemosa	Jhingora/Amli	✓	✓	✓	✓	X	X	✓	X	X
Bauhinia retusa	Kachnar	✓	✓	✓	✓	X	X	X	X	✓
Bauhinia vahlii	Malu	✓	✓	X	X	X	X	✓	X	✓
Betula alnoides	Katbhoj	✓	✓	✓	X	X	X	X	X	✓
Boehmeria rugulosa	Genthi	✓	✓	X	✓	X	✓	X	X	X
Bridelia Montana	Gondni	✓	✓	X	✓	✓	X	X	X	X
Bridelia retusa	Kaja	✓	✓	X	✓	X	X	X	X	X
Bischoffia javanica	Kotsemla	✓	✓	X	X	X	X	✓	X	✓
Bombax ceiba	Semal	✓	✓	✓	✓	X	X	✓	X	X
Butea monosperma	Dhak	✓	✓	✓	✓	X	X	X	X	X
Casearia glomerata	Gilchi	✓	✓	X	X	X	X	X	X	✓
Callicarpa arborea	Ghiwala	✓	X	X	X	X	X	X	X	✓

(continued)

Table 1.2 (continued)

Botanical name	Common name	Multipurpose uses								
		Fodder	Fuel	Timber	Medicine	Construction	Fruit/ leaves	Fiber	Religious	Others
Celtis australis	Kharik	✓	✓	✓	✓	X	✓	X	X	X
Celtis eriocarpa	Kharik	✓	✓	X	X	X	✓	X	X	X
Celtis tetrandra	Khari	✓	✓	X	X	X	✓	X	X	X
Cordia vestita	Lassuri	✓	✓	X	✓	X	✓	X	X	X
Cordia obliqua	Lassura	✓	✓	X	✓	X	✓	X	X	X
Dalbergia sissoo	Kala-sheesham/ sisau	X	✓	✓	X	✓	X	X	X	✓
Dalbergia latifolia	Sheesham/sisau	X	✓	✓	X	✓	X	X	X	✓
Debregeasia salicifolia	Siharu	✓	✓	X	✓	X	✓	X	X	X
Dendrocalamus strictus	Bans	✓		X	X	X	X	X	X	✓
Diploknem abutyracea	Pahadi Mahua	✓	✓	X	✓	X	X	X	X	✓
Ehretia acuminata	Nalsura	✓	✓	X	X	X	✓	X	X	X
Emblica officinalis	Amla	✓	✓	X	✓	X	✓	X	✓	X
Ficus glomerata	Umra	✓	✓	X	✓	✓	✓	X	X	X
Ficus hispida	Bhumra	✓	✓	X	✓	X	✓	✓	X	X
Ficus nemoralis	Chil	✓	✓	X	X	X	✓	X	X	X
Ficus palmata	Bedu	✓	✓	X	✓	X	✓	X	X	X
Ficus religiosa	Pipal	✓	X	X	X	X	✓	X	✓	X
Ficus roxburghii	Timal	✓	X	X	X	X	✓	X	X	X
Ficus semicordata	Khaina	✓	✓	X	✓	X	✓	✓	X	X
Ficus subincisa	Chanchri	✓	✓	X	✓	X	✓	X	X	X
Grewia optiva	Bhimal/bihul	✓	✓	X	✓	X	✓	✓	X	X
Grewia oppositifolia	Bhimal	✓	✓	✓	✓	X	✓	✓	X	X

Leucomeris spectabilis	Phusara /Phusiari	✓	✓		X	X	X	X	X	X
Litsea monopetala	Kadwai	✓	✓	✓	✓	X	X	X	X	X
Madhuca indica	Mahwa	✓		✓	✓	X	X	X	X	X
Mallotus philippensis	Kamela	✓	✓	✓	✓	X	X	X	X	X
Melia azedarach	Batain/Drek	✓	✓	✓	✓	X	X	X	X	X
Morus alba	Sahtoot	✓	X	X	X	X	✓	X	X	✓
Morus laevigata	Sahtoot	✓	✓	X	✓	X	✓	✓	X	X
Morus serrata	Keemu	✓	✓	X	✓	X	✓	✓	X	X
Myrica esculenta	Kaphal	✓	✓	✓	✓	X	✓	X	X	X
Ougeinia oojeinensis	Sandan	✓	✓	✓	✓	X	X	X	X	X
Pistacea integerrima	Kakadshringi	✓	✓	✓	✓	X	X	X	X	X
Kydia calycina	Pulia	✓	X	X	✓	X	X	X	X	✓
Pinus roxburghii	Chir	X	✓	✓	X	✓	X	X	X	✓
Populus ciliata	Pahari pipal	✓	X	X	✓	✓	X	✓	X	X
Populus deltoides	Necklace poplar	✓	X	X	✓	✓	X	✓	X	X
Prunus cerasoides	Padam	✓	✓	✓	✓	X	X	X	✓	X
Pyrus pashia	Mehal	✓	✓	✓	✓	X	✓	X	X	X
Quercus glauca	Phalyant	✓	✓	✓	✓	X	X	X	X	X
Quercus leucotrichophora	Banj	✓	✓	✓	✓	X	X	X	X	X
Rhus parviflora	Murthi	✓	✓	X	✓	X	✓	X	X	X
Robinia pseudoacasia	Robinia	✓		✓	X	X	X	X	X	✓
Salix acmophylla	Willow	✓	✓		X	X	X	X	X	X
Salix babylonica	Bains/weeping willow	✓	X	✓	✓	X	✓	X	X	X
Saurauiana paulensis	Gogan	✓	X	X	✓	X	✓	X	X	X
Sterculia pallens	Udaal	✓	X	X		X	✓	X	X	X

(continued)

Table 1.2 (continued)

Botanical name	Common name	Multipurpose uses								
		Fodder	Fuel	Timber	Medicine	Construction	Fruit/ leaves	Fiber	Religious	Others
Syzygium cuminii	Jamun	✓	✓	✓	✓	X	✓	X	X	X
Terminalia bellerica	Bahera	✓	✓	✓	✓	X	✓	X	X	X
Terminalia chebula	Haira	✓	✓	✓	✓	X	✓	X	X	X
Toona ciliata	Toon	✓	✓	✓	X	X	X	X	X	X
Trema orientalis	Gio	✓	X	X	✓	X	X	X	X	✓
Ulmus wallichiana	Mahun	✓	X	X	✓	✓	✓	X	X	X
Zanthoxylum armatum	Timru	✓	X	X	✓	X	✓	X	X	X

✓ represents the use of the tree in the region
X denotes not in use

2.1.1 Medicinal and Aromatic Plants (MAPs)

MPTs are another source of livelihoods for mountain communities (Hoermann et al. 2010; Shah et al. 2018) and also play a crucial role in cultivating and sustaining MAPs within AFS. These trees provide shade, improve soil fertility, and produce medicinal products. Many medicinal and aromatic plants flourish under the canopy of MPTs, which create a favorable microclimate by moderating temperature extremes and reducing evapotranspiration. The use of medicinal plants in traditional medicine has been officially endorsed by several developing countries such as Sri Lanka, India, Cuba, China, and Thailand, where approximately 80% of the population relies on this system (Rao et al. 2004). The traditional AFS of the Himalayas conserve many medicinal plants, including Tulsi (*Ocimum sanctum*), drumstick (*Moringa oleifera*), curry leaf (*Murraya koenigii*), and *Terminalia arjuna*, which are renowned for their medicinal properties. An increase in household income through the farming of *Ocimum sanctum* in AFS has been reported in Uttarakhand, a state in the Western Himalayas (Shah et al. 2019). Other aromatic plants commonly cultivated in the Western Himalayan AFS include *Withania somnifera*, *Andrographis paniculata*, *Mentha* sp., *Aconitum* spp., *Rauvolfia* sp., *Matricaria chamomilla*, and *Ocimum basilicum*. Interestingly, cultivating peach (*Prunus persica*) alongside other crops increased the yield of the fruit compared to peach grown without these crops (Kashyap et al. 2014).

2.1.2 Biomass Energy and Fuelwood

Biomass energy, particularly fuelwood, has long been the primary energy source for the people of the Himalayan mountains for ages (Rawat et al. 2009), as 90% of the energy is derived from the burning of the fuelwood (Khuman et al. 2011). The lack of alternative energy sources and the extremely cold climate in high-altitude regions have increased the demand for fuelwood for cooking, heating (both room and water), lighting, and livestock rearing, leading to pressure on forests and ultimately causing deforestation (Kumar and Sharma 2009). The high demand for fuelwood can be reduced and offset using various MPTs within Himalayan AFS, particularly in agrisilviculture and agrihortisilvopastoral practices. Additionally, crop residues and cattle dung incorporated into agroforestry practices provide alternative fuel sources during extreme events (Kumar and Sharma 2009). A recent study found that fuelwood consumption is higher in the Eastern Himalayan region than in the Western Himalayas, possibly due to greater fuelwood availability resulting from a decline in shifting and Jhum cultivation and the expansion of AFS in the region (Bhatt et al. 2016).

2.1.3 Fodder and Food

MPTs play a crucial role in AFS by providing diverse products and services that improve food security and livestock nutrition. These trees make a substantial contribution to sustainable agriculture by providing edible components for humans and fodder for animals. MPTs provide a variety of edible products, including fruits, nuts, leaves, and seeds, which are essential food sources for rural communities. For instance, species such as *Moringa oleifera* and *Erythrina edulis* are well-known for their nutrient-rich leaves and seeds, respectively, enhancing dietary diversity and improving nutritional intake. Moreover, the deep-rooted systems of many MPTs access subsoil nutrients, improving soil fertility and supporting the growth of understory crops. Fodder is an essential provisioning service provided by MPTs, providing a reliable source in regions where seasonal variations affect the availability of herbaceous forage. Leaf fodder from MPTs is notable, as these trees are commonly regarded as emergency feedstuffs (Ganai et al. 2009), particularly in hilly regions (Khatta et al. 1999), due to the high protein, soluble carbohydrate, mineral, and vitamin content of their leaves (Azim et al. 2011). Therefore, MPTs are intentionally planted or left to grow along farm bunds (Khosla et al. 1992), because they offer a variety of products and can significantly contribute to rural economies (Sharma et al. 2017a, b, c; Sharma et al. 2022). As a result, MPTs can enhance livestock productivity by reducing the gap between demand and the availability of green fodder (Datt et al. 2008). Similarly, Yadav et al. (2023) evaluated the seasonal nutritive composition of leaves from two *Bauhinia* spp. (*Bauhinia retusa* and *Bauhinia variegata*) and found both to be valuable sources of green fodder for cattle. Specifically, *B. retusa* exhibited higher crude protein and ash content during the rainy season, while *B. variegata* showed elevated crude protein levels in the summer. These seasonal variations suggest that incorporating these species into cattle diets can help address protein deficiencies in livestock, particularly in the hill regions of Uttarakhand.

2.1.4 Financial Benefits to the Local Community

MPTs are vital for improving the financial well-being of local communities by offering a variety of products and services. These trees provide various benefits such as food, timber, fodder, and nontimber forest products (NTFPs) for both subsistence and commercial use. The integration and management of MPTs within local communities offer significant financial benefits by diversifying income streams, employment opportunities, and broader economic advantages. As a result, MPTs contribute significantly to sustainable development and the long-term economic resilience of these communities. Various MPTs, including *Tectona grandis*, *Melia azedarach*, *Populus deltoides*, *Casuarina equisetifolia*, *Grevillea robusta*, Acacia spp., and *Eucalyptus* spp., are grown for their valuable timber and fuelwood. Timber production offers significant income to local communities through direct sales, employment in forestry-related activities, and value-added processing industries (FAO 2020). Small-scale industries that utilize MPTs, such as beekeeping in *Acacia* forests and essential oil extraction from *Eucalyptus*, offer alternative income sources for local entrepreneurs (Jamnadass et al. 2013). Moreover, the commercialization of tree products promotes the development of local businesses and strengthens rural economies.

2.2 *Regulating Services*

MPTs provide a wide range of regulatory services, such as carbon storage and sequestration, soil fertility enhancement, biodiversity conservation, climate regulation, water conservation, and watershed management. It also enhances soil properties and processes, which are crucial for climate regulation through carbon sequestration and the reduction of greenhouse gas emissions, as well as for water management by regulating soil characteristics. Some regulating services of the MTPs are briefly described as follows.

2.2.1 Carbon Storage and Sequestration

A contemporary approach to addressing the challenges of climate change has drawn substantial interest is carbon sequestration (Hepburn et al. 2019). The AFS has the potential to mitigate global warming and climate change. It is currently estimated that approximately 26 million metric tons of carbon are sequestered, with this figure expected to increase of 45 million metric tons by 2040 as it incorporates fast-growing woody species (IPCC 2001). Incorporating MPTs into land-use systems provides a comprehensive approach to carbon sequestration, combining environmental, economic, and social benefits. The selection of suitable species, consideration of biodiversity, and an understanding of ecosystem interactions are essential for maximizing carbon storage potential. The ability of MPTs to capture carbon differs among species and is affected by growth rate, biomass accumulation, and environmental conditions. Various MPTs such as poplar (*Populus deltoides*), Nilgiri (*Eucalyptus* spp.), subabul (*Leucaena leucocephala*), tree of heaven (*Ailanthus excelsa*), willow (*Salix* spp.), malabar neem (*Melia dubia*), cadamba (*Neolamarckia cadamba*), and white teak (*Gmelina arborea*) are the suitable tree species for carbon sequestration under agroforestry (Chavan et al. 2023). Knowledge of the growth rates and ecological impacts of these tree species remains limited (Maikhuri et al. 2000). However, agroforestry development on degraded lands, beyond offering tangible benefits to local communities, could also contribute significantly to carbon sequestration, a global environmental advantage. This potential has been demonstrated in several case studies. For instance, Chambial (2016) observed the carbon sequestration potential of various MPTs in Himachal Pradesh, viz., *Grewia optiva*, *Melia azedarach*, *Bauhinia variegata*, *Celtis australis*, *Acer oblongum*, *Acacia catechu*, and *Morus alba*. The findings revealed that *Acer oblongum* and *Bauhinia variegata* sequestered the highest amounts of carbon, whereas *Morus alba*, *Grewia optiva*, and *Celtis australis* exhibited lower sequestration levels. Similarly, another study conducted by Kumar et al. (2012) in the villages of the Pauri Garhwal district of Uttarakhand aimed to estimate the structure and carbon sequestration potential of six traditional agroforestry systems. Their findings indicated that the average total carbon stock of trees in traditional agroforestry systems was 32.56 t ha^{-1} and the soil organic carbon was 56.74 t ha^{-1}. In hill-based agroforestry systems, farmers typically meet their fodder and fuelwood requirements through the practice of lopping, or selective pruning, of the trees. Among the MPTs in the Northwestern Himalayas, *Grewia optiva* is considered one of the most important. According to Verma et al. (2014), a 23-year-old *Grewia optiva* tree has an estimated biomass of 74.62 kg $tree^{-1}$, a carbon stock of 24.43 kg $tree^{-1}$, and a carbon sequestration potential of 1.06 kg $tree^{-1}$ $year^{-1}$. In comparison, *Morus alba* exhibits higher values, with a 22-year-old tree producing a biomass of 286.16 kg $tree^{-1}$, a carbon stock of 143.09 kg $tree^{-1}$, and a carbon sequestration potential of 6.50 kg $tree^{-1}$ $year^{-1}$ (Chambial 2016) (Fig. 1.2).

Promoting MPTs in sensitive mountain ecosystems vulnerable to deforestation and land-use changes can help reduce the negative impacts of global warming and climate change while supporting local communities. The removal of most biomass for human consumption in agriculture AFS leads to a lower amount of vegetation carbon in contrast to horticulture due to the limitation of the extraction of plant parts. However, numerous studies have found that variations in carbon stock occur in both vegetation and soil due to differences in elevation, agroforestry components, and management practices (Rajput et al. 2017; Chisanga et al. 2018; Singh et al. 2024; Vikrant et al. 2020). Besides these factors, plant growth habits, ages, and tree–crop interactions influence biomass accumulation and carbon stock in AFS (Jana et al. 2009; Kanime et al. 2013), while these interactions enhance carbon sequestration and contribute to CO_2 mitigation (Vikrant et al. 2018).

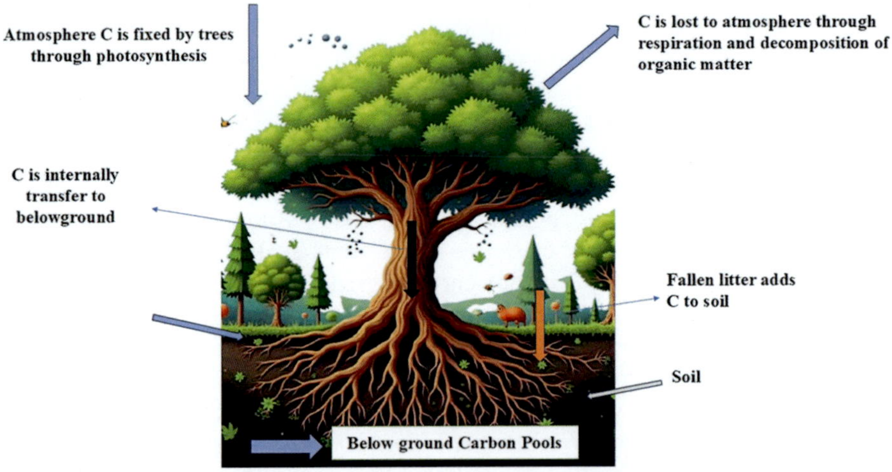

Fig. 1.2 The process of carbon sequestration through trees

2.2.2 Soil Fertility Enrichment and Water Conservation

The nutrient status of soil in an ecosystem is influenced by various factors, including edaphic conditions, topography, elevation, species composition, geological factors, and microbial activity (Singh et al. 2023; Furtak and Gałązka 2019). MPTs are essential for improving soil fertility by contributing organic matter, nitrogen fixation, and nutrient cycling and preventing erosion while their incorporation into agroforestry and sustainable farming practices significantly improves soil health and structure, reduces reliance on synthetic inputs, and strengthens overall ecosystem stability (Osei-Tutu et al. 2018) (Table 1.3). The deep root systems of MPTs enhance groundwater quality by trapping nutrients and metals (Young 1989) while absorbing nutrients from deeper soil layers beyond the reach of crop roots and recycling them to the top 15 cm of soil through fallen leaves and organic matter (Jadhav et al. 2006), thereby increasing nutrient capture and recycling efficiency. The nitrogen-fixing capability of MPTs reduces the need for synthetic fertilizers, promoting sustainable agricultural practices. Additionally, incorporating organic matter from MPTs promotes diverse soil microbial communities, which are vital for decomposing organic material and enabling nutrient transformations essential for plant uptake (Singavarapu et al. 2023).

2.2.3 Biodiversity Conservation and Habitat Provision

Biodiversity underpins all ecosystem services and plays a crucial role in enhancing the resilience of ecosystems and their services to both current and future environmental changes (Mace et al. 2012). The incorporation of trees within AFS increases habitat heterogeneity, creating ecological niches for a diverse array of organisms both above and below ground. This diversification initiates agroecological succession, leading to the development of functional ecosystems that support sustainability. MPTs play a vital role in supporting biodiversity by supporting a wide array of flora and fauna. It can also function as connective pathways between fragmented habitats, enabling the dispersal of area-sensitive species and consequently enhancing genetic diversity. By preserving and augmenting tree biodiversity, these corridors exert a beneficial and reinforcing effect on the overall biodiversity of ecosystems.

2.3 Climate Regulation

MPTs play a crucial role in climate regulation through the process of carbon sequestration. Their ability to capture and store atmospheric carbon within their biomass and soil contributes to mitigating the effects of climate change. AFS incorporating MPTs can augment carbon sequestration, with trees capturing carbon lost from deforestation while supplying supplementary resources such as food and timber.

Table 1.3 Total nitrogen stock and nitrogen fixation as indicators of soil fertility in various Himalayan agroforestry systems incorporating multipurpose tree species

Agroforestry type	Components	Location	Total N kg ha^{-1}	Nitrogen fixation kg ha^{-1} $year^{-1}$	References
Agrisilviculture	Multipurpose tree + crops	Himachal Pradesh, Western Himalaya	298.30	–	Keprate et al. (2024)
	Multipurpose tree + crops	Uttarakhand, Western Himalaya	224	–	Kaushal et al. (2024)
	Multipurpose tree + crops	Himachal Pradesh, Western Himalaya	394	–	Singh et al. (2015)
	Pecan nut + crops	Almora, Western Himalaya	231	–	Yadav et al. (2017)
	Alnus+ vegetables	Nagaland, eastern Himalaya	–	117	Kehie and Khamu (2018)
Silvopastoral	Multipurpose tree + grasses	Himachal Pradesh, Western Himalaya	405	–	Singh et al. (2015)
Agrosilvopastoral	Multipurpose fodder and timber trees+ bamboo	Meghalaya, eastern Himalaya	4360	59	Sharma et al. (2016)
Agrihortisilvicultural	Trees + fruits + crops	Ladakh, Western Himalayas	1	–	Kumar et al. (2009)
	Trees + fruits + crops	Himachal Pradesh, Western Himalayas	355	–	Singh et al. (2015)
Hortisilvicultural	Alnus+ cardamom	Sikkim, eastern Himalayas	7650	95	Sharma et al. (2016)
	Forest + cardamom		5440	9.5	Sharma et al. (2016)

The rotation age and utilization of tree products are pivotal factors that influence the quantity of carbon sequestered. Additionally, MPTs improve microclimatic conditions by providing shade, stabilizing soil temperature, and enhancing moisture retention. These effects are particularly beneficial in arid and semi-arid regions, where MPTs contribute to preserving suitable conditions for agricultural cultivation and maintaining overall ecosystem stability (Kisangau et al. 2021).

2.4 Cultural and Supporting Services

The Western Himalayas, renowned for their rich biodiversity and cultural heritage, host diverse agroforestry systems where multipurpose trees play a vital role in maintaining cultural traditions, providing essential ecosystem services, and sustaining local livelihoods (Khosla and Toky 1999). MPTs act as keystone species in agroforestry systems, supporting diverse flora and fauna by providing food, shelter, and breeding sites (Devi et al. 2023). While species such as *Quercus leucotrichophora* (ban oak) and *Morus alba* (white mulberry) contribute to biodiversity by enriching soil fertility, preventing erosion, and sustaining habitats for pollinators and seed dispersers. They help sustain native plant communities, which, in turn, support ecosystem stability. In the Western Himalayas, MPTs play a crucial role in the hydrological cycle by enhancing groundwater recharge, reducing surface runoff, and improving soil moisture retention. The deep-rooted species such as *Pinus roxburghii* (chir pine), *Quercus leucotrichophora* (ban oak), and *Alnus nepalensis* (Nepal alder) play a vital role in water conservation by enhancing infiltration rates and mitigating drought impacts in hill ecosystems (Jana et al. 2021). Additionally, these trees stabilize riparian zones, ensuring a steady water supply for agriculture and domestic use.

In the Himalayan region, steep slopes and erratic precipitation patterns make agroforestry highly vulnerable to landslides and soil degradation (Pant et al. 2022). MPTs contribute to slope stabilization by reinforcing soil structure with extensive root systems. For instance, *Bauhinia variegata* (mountain ebony) and *Salix species* (willows) effectively mitigate soil erosion and landslides by binding soil particles and regulating water flow. MPTs hold immense cultural significance in the Western Himalayas, deeply intertwined with folklore, religious practices, and traditional medicine (Singh and Rawat 2012). Sacred groves, home to species such as *Ficus religiosa* (Peepal) and *Cedrus deodara* (Himalayan cedar), play a vital role in local spiritual traditions and conservation ethics. These trees also enrich ethnobotanical knowledge, serving as essential resources for traditional healing and ritualistic practices within communities. Integrating MPTs into agroforestry creates educational opportunities for students, researchers, and conservationists while fostering awareness of sustainable land-use practices. Community-led initiatives promoting tree-based agroforestry function as living laboratories for sustainable land management. Ecotourism activities, such as guided nature trails through oak and rhododendron forests, attract researchers and tourists, generating income for local communities while fostering conservation awareness. Additionally, integrating MPTs into agroforestry practices in the Western Himalayas is essential for preserving cultural traditions and ensuring ecological stability. Their conservation supports sustainable livelihoods for local communities while enhancing environmental resilience. Acknowledging the multifunctional benefits of these trees can help policymakers and farmers develop more effective agroforestry strategies that harmonize cultural heritage with ecological sustainability.

3 Socioeconomic Contribution of MPTs in the Agroforestry System

Agroforestry and socioeconomic factors are closely interconnected, influencing one another. On one hand, improved socioeconomic status encourages farmers to integrate trees into their farmland. On the other, adopting agroforestry practices enhances the socioeconomic well-being of farming. MPTs primarily benefit rural communities, particularly those with limited resources. These techniques are suitable for small-scale land use, requiring minimal capital and energy while providing essential goods and services to meet immediate human needs, rather than prioritizing commercial gains (Nair 1993). Therefore, assessing social acceptability is crucial for determining the viability of MPTs and ensuring their social sustainability. The socioeconomic condition of a society reflects its combined social and economic status relative to others. It significantly influences resource accessibility, livelihood patterns, and food security (Roy et al. 2013). Additionally, it influences farm-based enterprises by affecting their organization, management, production, and marketing practices. MPTs in the Northwestern Himalayas have significantly enhanced local livelihoods and income generation for rural communities by providing multiple products, including timber, fuelwood, fodder, medicinal plants, and NTFPs. Their impact is particularly evident in improving food security, strengthening climate change resilience, and promoting community gender empowerment. Farmers extensively cultivate species such as *Quercus leucotrichophora* (ban oak) and *Grewia optiva* for fodder and fuelwood, which provide essential resources for livestock-dependent households (Singh and Rawat 2012). Additionally, the collection and trade of nontimber forest products (NTFPs), including wild edibles and medicinal plants like *Aconitum heterophyllum* and *Berberis aristata*, play a crucial role in enhancing household incomes, particularly in areas with limited alternative employment opportunities (Timmermann and Smith-Hall 2019). Many MPT species produce edible fruits, seeds, and leaves rich in essential nutrients. For example, *Morus alba* (white mulberry) and *Prunus armeniaca* (wild apricot) serve as valuable sources of vitamins and minerals, enhancing dietary diversity in regional communities (Madhu et al. 2023). Additionally, AFS integrating MPTs significantly boost farm productivity and diversification, thereby reducing the economic risks faced by small-scale farmers. Farmers reported lower daily farm expenses due to MPT cultivation, with the following preference trend: *Quercus leucotrichophora* (85.00%) >*Grewia optiva* (80.00%) >*Celtis australis* (65.00%) >*Bauhinia variegata* (60.00%) >*Albizia lebbeck* (33.33%) (Dogra et al. 2024). Local communities depend on MPTs to meet essential subsistence needs, including firewood for cooking and heating, particularly in regions with limited access to alternative energy sources (Singh and Rawat 2012). Among these, *Quercus leucotrichophora* serves as the primary species for charcoal production, with 71.67% of farmers using it for this purpose (Dogra et al. 2024).

Studies across the country highlight the beneficial effects of agroforestry on the socioeconomic well-being of farmers. Agroforestry supports women's welfare, uplifts marginalized communities, strengthens food security, enhances financial resilience, reduces crop failure, provides steady employment and income, improves land productivity, and generates economic benefits from diverse outputs, both annually and periodically. Implementing agroforestry on 75,500 hectares in the Indian Himalayan region has the potential to create over 5.76 million man-days of employment each year. This emphasizes agroforestry's importance as a sustainable land-use practice and its significant contribution to rural development, especially in the challenging terrains of the Himalayas (Arunachalam et al. 2020). Despite extensive research on the identification, productivity, and carbon sequestration potential of agroforestry systems, a significant gap remains in studies assessing their impact on farmers' socioeconomic conditions. This research gap poses a challenge in formulating effective policies for the betterment of the farming community. However, it also presents an opportunity for future studies to explore socioeconomic aspects and support more informed policy development.

4 Sustainable Development Goals (SDGs) Through MPTs in AFS

Integrating MPTs into AFS offers a multifunctional alternative to traditional agricultural practices, which often harm natural resources and the environment. By providing various ecological and provisioning services, AFS promotes sustainable land management and supports key United Nations SDGs. In 2015, the United Nations established 17 SDGs to promote global well-being. MPT play a crucial role in advancing several of these goals, particularly: SDG 1—Ending poverty in all its forms; SDG 2—Alleviating hunger and ensuring food security; SDG 3—Promoting good health and well-being; SDG 5—Achieving gender equality; SDG 6—Ensuring clean water and sanitation; SDG 7—Providing affordable, sustainable, and reliable modern energy services; and SDG 12—Promoting the efficient use of natural resources. Through their diverse benefits, MPTs support a more sustainable and resilient future, aligning with the United Nations' 2030 Agenda (Fig. 1.3). MPT provide extensive environmental benefits, directly supporting key SDGs, that is, SDG 13 (Climate Action) by enhancing crop resilience to climate change by improving water infiltration, mitigating extreme temperatures, and preventing soil erosion and SDG 15 (Life on Land) through ecosystem restoration, promote the sustainable use of wild resources, and contribute to biodiversity conservation and land rehabilitation. This science with MPTs enhances ecosystem productivity, improves landscape connectivity, and increases the carbon sequestration potential of farmlands. By incorporating diverse tree species, these systems also reduce pressure on natural forests, promoting sustainable land use and climate resilience (Mbow et al. 2014; Salvini et al. 2016).

Agroforestry enhances soil fertility and increasing agricultural productivity. Integrating MPT such as *Celtis australis* enriches soil organic matter, promotes efficient nutrient cycling, and boosts crop yields. These benefits directly support SDG 2 (Zero Hunger) by fostering sustainable farming practices that strengthen

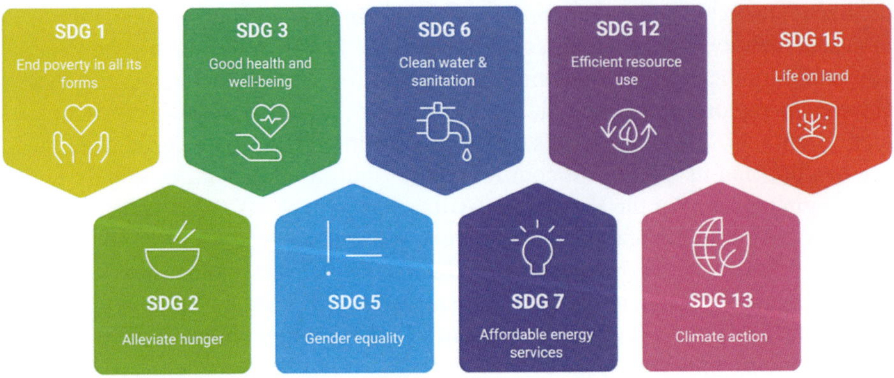

Fig. 1.3 Sustainable development goals through multipurpose trees

long-term food security (Purohit 2010). MPTs provide significant ecological benefits as well as play a crucial role in economic development by creating diverse income sources. Farmers gain economically from products such as timber, fodder, fuelwood, and nontimber forest products (NTFPs), which in turn strengthens rural livelihoods. This directly supports SDG 1 (No Poverty) by enhancing household incomes and SDG 12 by promoting the efficient and sustainable use of natural resources. Research indicates that adopting agroforestry in the Northwestern Himalayan region has led to increased agricultural yields and improved household incomes (MHWDP 2023). However, despite these clear benefits, several challenges still hinder its wider adoption. Key challenges to the wider adoption of agroforestry include limited awareness of native MPTs, a lack of quality planting materials, and inadequate technical knowledge among farmers. To address these barriers, targeted research is needed on the propagation of local MPTs, the establishment of nurseries to ensure a consistent supply of planting materials, and comprehensive training programs for farmers to promote sustainable agroforestry practices (Bhatt et al. 2010). Moreover, expanding agroforestry in mountainous regions enhances sustainability by contributing to food security, improving financial and livelihood conditions, enriching soil quality, and providing various environmental and economic benefits. These multifaceted advantages make agroforestry a more acceptable and favorable land-use system for long-term rural development.

5 Opportunities and Challenges

5.1 Opportunities

The Northwestern Himalayas are home to a rich diversity of multipurpose tree species MPTs that form a crucial component of the region's AFS. These species provide essential resources such as fodder, fuelwood, timber, and NTFPs, which support local livelihoods (Bhatt et al. 2010; Yadav and Bisht 2015). Beyond their economic importance, MPTs play a vital role in ecological functions, including enhancing soil fertility, controlling erosion, and sequestering carbon. They also create income opportunities and supply raw materials for local artisans and forest-based industries. However, their cultivation and management in the Northwestern Himalayan region involve a complex interplay of challenges and opportunities. MPTs in the Indian Himalayan region present numerous opportunities that enhance both environmental sustainability and rural development. The key benefits include biodiversity conservation, provision of goods and ecosystem services, enhanced carbon sequestration within agroecosystems, improved soil fertility and land productivity, promotion of eco-tourism, strengthened food security, reduction in outmigration through local employment generation, improved social and economic well-being of rural communities (Pandey 2007). These multifaceted advantages position agroforestry as a promising land-use strategy for achieving sustainable development in the Himalayan region.

One of the key benefits of MPTs is the opportunity for income diversification. By cultivating MPTs, farmers can access multiple revenue streams through the sale of various products, such as fruits, nuts, timber, fodder, and medicinal herbs. For example, the Indian butter tree (*Diploknema butyracea*), which is native to the Himalayan foothills, produces chiuri ghee from its seeds, a product that has significant commercial value. Moreover, MPTs substantially improve fodder availability, which is particularly vital in mountainous regions where seasonal shortages often affect livestock productivity and rural livelihoods. Species such as *Grewia optiva*, *Celtis australis*, and *Bauhinia variegata* are positively valued in agroforestry systems due to their nutritional benefits as fodder, which supports livestock health and productivity (Navale et al. 2022). Their deep and extensive root systems stabilize slopes, reduce soil erosion, and improve water infiltration, functions that are particularly important in the fragile and erosion-prone landscapes of hilly regions. Certain MPTs, such as *Albizia chinensis*, enhance soil fertility through nitrogen fixation, which enhances nutrient availability for surrounding crops. MPT-based agroforestry systems improve soil fertility, capture atmospheric carbon, and regulate local microclimates, thereby significantly contributing to climate change mitigation. These systems help buffer temperature extremes, reduce evapotranspiration, and create more stable growing conditions. These benefits both crop productivity and livestock health, especially in the face of increasingly variable climatic conditions.

5.2 Challenges

The region is marked by significant variations in topography, soil conditions, climate, and land-use practices. Due to its predominantly hilly terrain, the ecosystems of the Northwestern Himalayas are inherently fragile and susceptible to a range of challenges, including geological hazards, land degradation, shifts in land use and land cover, and biodiversity loss. These vulnerabilities highlight the importance of adopting sustainable land management practices, such as agroforestry, which can help maintain ecological balance while improving the livelihoods of local communities (Saha and Kumar 2019). Human activities, including deforestation, overexploitation of natural resources, and unsustainable agricultural practices, have significantly impacted the Himalayan region. These actions, combined with the area's challenging and unstable terrain, have led to serious issues such as soil erosion, land degradation, and decreased agricultural productivity. Given the inherent vulnerability of this region, it is crucial to adopt sustainable land-use practices and implement restorative measures. These steps are essential for maintaining ecological balance and ensuring the long-term sustainability of the Himalayan ecosystem. Incorporating MPTs into agroforestry systems in the Northwestern Himalayas has significant potential to improve ecological sustainability and enhance rural livelihoods. Achieving optimal outcomes through MPT-based agroforestry in the Himalayan region requires a holistic approach encompassing targeted research, supportive policy frameworks, and active community participation. This section addresses key challenges hindering the widespread adoption of agroforestry, including limited awareness, the impacts of climate change, land-use conflicts and degradation, inadequate access to quality planting materials, and insufficient technical support.

5.2.1 Climate Change Impacts

Climate change significantly challenges the phenology and growth patterns of MPTs. Changes in temperature and precipitation regimes can disrupt flowering and fruiting cycles, which threatens the availability of essential resources like fodder, fruits, and fuelwood that local communities rely on (Panda et al. 2021). These phenological shifts may also trigger broader ecological consequences, affecting plant–pollinator interactions, altering community composition, and disrupting ecosystem functions such as nutrient cycling and carbon sequestration in agroforestry systems (Hughes 2000). Changing climatic conditions are expected to profoundly affect the survival, distribution, and productivity of MPTs in the Northwestern Himalayan region. As a highly sensitive mountain ecosystem, this area is already witnessing considerable shifts in climate patterns, including reduced frequency and intensity of snowfall, a widening gap between minimum and maximum temperatures, particularly in dry temperate zones, increased summer rainfall alongside lower summer temperatures, a general rise in mean annual temperatures, growing aridity, warmer winters, and more frequent extreme weather events such as cloudbursts, unseasonal frosts, and storms. These changing climatic trends pose significant risks not just to the growth and phenology of MPTs but also to the region's agriculture and horticulture sectors (Singh et al. 2010; Basannagari and Kala 2013). Based on an analysis of farmers' perceptions, Basannagari and Kala (2013) found that rising temperatures in Himachal Pradesh are likely to cause a noticeable shift in apple cultivation from lower to higher elevations. As temperatures increase, the climatic suitability of lower altitudes for apple farming is expected to decline, prompting farmers to seek higher, cooler zones for continued production. Additionally, the harvesting period for apples is projected to be delayed by 1–2 weeks. In regions where conditions become unsuitable for apple cultivation, farmers may increasingly switch to growing coarse grains and seasonal vegetables as alternative sources of income.

5.2.2 Land-Use Conflicts and Degradation

The rapid population growth and expanding developmental activities in the Western Himalayan region have significantly intensified land-use conflicts and environmental degradation. The increasing pressure on natural resources driven by overgrazing, deforestation, and unplanned agricultural expansion has resulted in a series of ecological consequences, including soil erosion, declining soil fertility, and substantial biodiversity loss. These environmental challenges directly threaten the long-term sustainability of AFS, which are closely tied to the integrity and health of surrounding ecosystems. Land degradation has emerged as a particularly critical issue in this fragile mountainous landscape. This problem in upland areas leads to shortages of food, fodder, fuelwood, and water for hill communities. It also contributes to flooding in the Indo-Gangetic plains and poses a serious threat to Himalayan biodiversity, which scientists recognize as one of the world's key biodiversity hotspots (Myers 1988). Several organizations have implemented land restoration initiatives, but they have achieved limited overall effectiveness due to unsuitable technologies, inadequate policies, and ineffective execution strategies (Maikhuri et al. 1997). They recommended establishing agroforestry systems on degraded community lands as a strategic solution to address both immediate and long-term concerns.

5.2.3 Tree Crop Interaction Studies

The adoption of new technologies is primarily driven by their economic benefits. In agroforestry, farmers often prioritize the income generated from tree components, while overlooking the broader socioecological advantages. In the Western Himalayan region, a distinctive feature of traditional rural farming systems is the continued presence of numerous unmanaged trees on agricultural land (Pant et al. 2022). Farmers often resist adopting integrated farming systems in the state because they perceive trees as negatively affecting crop growth and yield. Various studies highlight that tree canopies can have both positive and negative effects on the performance of crops grown beneath them (Pratap and Pant 2020; Sharma and Thakur 2020). Therefore, researchers and practitioners need to identify agroforestry systems according to the nature of their components to effectively address societal needs. They must also understand how these components interact to enhance profitability and improve adoption rates. By studying these interactions, they can appropriately select tree and crop species, design effective systems, and implement suitable management practices at the right time (Toppo and Toppo 2018).

5.2.4 Shortage of Quality Planting Material

The Government of India aims to double farmers' income. Since land and productivity are limited, integrated production systems offer a promising way to achieve this goal (Handa et al. 2019). To enhance the productivity of agroforestry systems, stakeholders must ensure the availability of high-quality planting material. As part of a long-standing tradition, farmers often preserve various MPTs along the boundaries of their fields (Sharma et al. 2017a, b, c). The rapidly increasing population and rising demand are putting significant pressure on resources, highlighting the need for high-quality planting material in agriculture. Currently, producers supply reliable planting material for only about 10% of the demand, leaving the majority of it unreliable (NAFP 2014). Experts estimate that the annual demand for quality planting materials is approximately ten million seedlings each for eucalyptus, poplar, and other multipurpose tree species. Additionally, there is an estimated requirement of about eight million seedlings for Casuarina (Handa et al. 2019). In total, the annual demand for seedlings across various tree species amounts to around 6.16 billion (Chavan et al. 2015). Stakeholders should establish certification systems for both nurseries and the planting stock they produce to guarantee the availability of quality planting materials. Collaborations between public and private sectors can help meet this demand while also creating job opportunities for local communities.

5.2.5 Topographical Challenges

The Western Himalayan region, which includes the Indian states of Himachal Pradesh, Jammu and Kashmir, and Uttarakhand, boasts rich biodiversity, with elevations ranging from 300 meters to over 6000 meters (Padma 2014). The steep and uneven mountainous terrain poses challenges in minimizing soil erosion and water loss from agro-ecosystems. Steep slopes accelerate soil erosion in catchment areas, leading to on-site degradation and off-site problems such as the sedimentation of rivers, streams, lakes, and most critically, dams, which ultimately reduces their operational lifespan. The state government is working to reduce silt accumulation in rivers and streams by restoring vegetation in the catchment areas of the cold desert through the plantation of trees, shrubs, and grasses. Traditional agroforestry offers a promising approach, as it can tackle off-site silting issues while providing both economic and ecological benefits at the source. Scientific agroforestry is challenging in the Indian Himalayan Region due to its rugged terrain. In contrast, traditional agroforestry remains the most prevalent and successful approach, helping maintain the ecological balance and preserve the soil characteristics of the region. Research indicates that adopting agroforestry practices, such as planting contour-aligned tree rows (hedgerows), can significantly contribute to soil and water conservation in the Western Himalayas. Pandey (2007) reported that these practices can reduce surface runoff by nearly 40% and lower soil erosion by around 48%.

5.2.6 Knowledge Gaps and Extension Services

Future research should aim to address four critical knowledge gaps to enhance the effectiveness and scalability of agroforestry in the Himalayan region: poor methodological approaches, limited focus on climate resilience, lack of robust policy and regulatory frameworks, and neglect of interdisciplinary integration. Filling these gaps will be instrumental in shaping robust, evidence-based interventions and policy frameworks that maximize agroforestry's potential for sustainable development. A major challenge in adopting MPT-based agroforestry systems is that many farmers are not aware of or lack the technical knowledge necessary for selecting appropriate species, propagation methods, and management practices. This gap underscores the need for strong extension services and targeted capacity-building programs (Bhatt et al. 2010). These programs should actively raise awareness about the benefits of integrating trees into agricultural systems, provide practical training in tree planting, and offer continuous technical support to help farmers address specific challenges (Chitakira and Torquebiau 2010). To ensure the long-term success and scalability of agroforestry initiatives, it is essential for policymakers and institutions to provide robust support. This support includes securing land tenure, clarifying the ownership rights of trees, and improving farmers' access to markets (FAO 2017).

5.2.7 Market Infrastructure and Value Chains

MPTs provide a range of valuable products, yet farmers often find it challenging to earn adequate income due to poorly structured markets and inefficient value chains. To enhance profitability, it is essential for stakeholders to improve infrastructure for processing, storage, and marketing. Moreover, the lack of an effective market information system limits farmers' awareness of market demand and supply trends, frequently leading to financial losses. Complex procedures for tree felling, harvesting, obtaining transit permits, and limited marketing rights for timber along with unclear regulatory frameworks, discourage farmers from cultivating trees. To address this, stakeholders must strengthen marketing infrastructure by encouraging active participation from the private sector, ensuring timely access to market data, and improving supply chain management. A competitive and accessible market, supported by adequate post-harvest and marketing infrastructure, can significantly enhance producer returns (Government of India 2021). However, in the absence of such systems, farmers often hesitate to adopt timber-based agroforestry, opting instead for fruit- and fodder-based methods.

6 Future Perspective

MPTs play a crucial role in AFS, especially in the fragile ecosystems of the Northwestern Himalayas. These species offer multiple benefits, such as soil conservation, carbon sequestration, and the provision of fodder, fuelwood, and timber, thereby supporting both environmental sustainability and rural livelihoods. However, the full potential of MPTs remains unrealized due to policy gaps, limited awareness, and inadequate research and extension services. MPTs in AFS offer significant potential for ecological sustainability and rural development in the Northwestern Himalayas. However, unlocking these potential demands targeted policy interventions, robust institutional support, and intensified research efforts. Future policies should prioritize climate resilience, biodiversity conservation, and sustainable livelihoods while ensuring farmers' economic viability. Successful integration of MPTs into Himalayan AFS requires a collaborative effort among government agencies, research institutions, and local communities. These are some key highlights of future MPTs in AFS in Northwestern Himalayas:

- MPTs offer promising solutions for climate change adaptation in the Himalayas by stabilizing ecosystems and reducing soil erosion. Future research should focus on selecting species with high carbon sequestration potential, carbon trading, and resilience to drought (Chauhan et al. 2019).
- Integrating MPTs with cash crops and medicinal plants can enhance farmers' economic benefits through market linkage strategies, value addition, and targeted incentives that promote sustainable agricultural practices. Simultaneously, developing fodder tree-based systems helps address fodder scarcity, increase land-use carbon stock potential, and support livestock rearing.
- In cold desert regions, the selection of tree species for incorporation into farming systems is limited primarily to *Salix* spp. and *Populus* spp. Therefore, research should focus on the genetic improvement of these species and the development of superior clones with enhanced productivity and adaptability for local communities.
- Tree breeding techniques for producing high-quality planting material for large-scale propagation, distributing it to farmers, and contributing to socioeconomic development.
- Tree-based farming systems should be promoted among farmers in fragile areas through well-tested models that demonstrate their feasibility in mitigating natural hazards and their effectiveness as sustainable livelihood sources.
- Enhancing research and extension activities focused on ecologically and economically promising land-use systems, such as horticulture-based and pasture-based systems, can significantly improve agricultural sustainability and livelihoods.

7 Conclusion

Multipurpose trees are essential to the sustainability of AFS in the Northwestern Himalayan region and play a vital role in providing food, financial stability, and vital ecosystem services, including regulatory, supporting, and cultural function to local communities. Agroforestry practices encourage improved livelihood opportunities for mountain communities, support sustainable land management, and strengthen livelihood security. In other land-use scenarios, integrating fruit trees or other MPT has demonstrated potential for carbon storage while also serving as a viable option for mitigating carbon dioxide and supporting the livelihoods of marginal farmers in the area. Overall, it can be concluded that tree-based systems should be promoted on agricultural lands to mitigate climate change, combat land degradation, improve soil health, and generate income through diverse products and carbon credits. However, widespread adoption of MPTs necessitates collaboration among policymakers, researchers, and local communities, with key strategies such as strengthened institutional frameworks, enhanced knowledge dissemination, and participatory approaches. These measures are crucial for addressing user needs and sociocultural factors, thereby maximizing the potential of MPTs in the region.

References

Ahmad S, Khan PA, Verma DK, Mir N, Singh JP, Dev I, Roshetko JM (2017) Scope and potential of hortipastoral systems for enhancing livestock productivity in Jammu and Kashmir. Indian J Agrofor 19(1):48–56

Assessment, ME (2005) Synthesis report. Island, Washington, DC, 748

Arunachalam A, Handa AK, Dev I, Kaushal R, Panwar P, Dhyani SK (2020) Alternate land use and agroforestry systems for resource conservation and enhanced productivity in the hills. Indian J Hill Farm 33(1):176–193

Atul A, Punam P, Khosla PK (1994) Himalayan agro ecosystems status- a case study. Biol Agric Hortic 104:271–286

Azim A, Khan AG, Ahmad J, Ayaz M, Mirza IH (2011) Nutritional evaluation of fodder tree legumes with goats. Asian Australas J Anim Sci 15:34–37

Baboo B, Sagar R, Bargali SS, Verma H (2017) Tree species composition, regeneration and diversity within the protected area of Indian dry tropical forest. Trop Ecol 58:409–423

Banyal R, Qaiser KN, Abidi RA (2016) Temperate agroforestry systems in Jammu and Kashmir. In: Agroforestry for increased production and livelihood security, pp 41–59

Basannagari B, Kala CP (2013) Climate change and apple farming in Indian Himalayas: a study of local perceptions and responses. PLoS One 8(10):e77976

Bhardwaj DR, Salve A, Kumar J, Kumar A, Sharma P, Kumar D (2024) Biomass production and carbon storage potential of agroforestry land use systems in high hills of north-western Himalaya: an approach towards natural based climatic solution. Biomass Convers Biorefinery 14(15):18079–18092

Bhatt GD, Parihaar RS (2020) Use of remote sensing and geographic information system on agroforestry ecosystem in Himalayan Region of Uttarakhand. In: Seema S, Singh RB, Kumar M (eds) Geoecology of landscape dynamics. Springer, Singapore, pp 167–185

Bhatt V, Purohit V, Negi V (2010) Multipurpose tree species of Western Himalaya with an agroforestry perspective for rural needs. J Am Sci 6(1):73–80

Bhatt BP, Rathore SS, Lemtur M, Sarkar B (2016) Fuelwood energy pattern and biomass resources in Eastern Himalaya. Renew Energy 94:410–417

Butola JS, Malik AR, Baba JA (2012) Livelihood diversification in cold arid desert of Indian Himalaya: urgent need of transforming traditional agroforestry system with emphasis on adoption of herbal farming. Int J Med Aromatic Plants 2(3):544–548

Chambial A (2016) Carbon sequestration potential of different multipurpose tree species. IOSR J Agric Vet Sci 9(5):11–14

Chauhan SK, Singh S, Sharma S, Sharma R, Saralch HS (2019) Tree biomass and carbon sequestration in four short rotation tree plantations. Range Manage Agrofor 40(1):77–82

Chavan SB, Keerthika A, Dhyani SK, Handa AK, Newaj R, Rajarajan K (2015) National Agroforestry Policy in India: a low hanging fruit. Curr Sci 25:1826–1834

Chavan SB, Newaj R, Rizvi RH, Ajit PR, Alam B, Handa AK, Dhyani SK, Jain A, Tripathi D (2021) Reduction of global warming potential vis-à-vis greenhouse gases through traditional agroforestry systems in Rajasthan, India. Environ Dev Sustain 23:4573–4593

Chavan SB, Dhillon RS, Sirohi C, Uthappa AR, Jinger D, Jatav HS, Chichaghare AR, Kakade V, Paramesh V, Kumari S, Yadav DK (2023) Carbon sequestration potential of commercial agroforestry systems in Indo-Gangetic Plains of India: poplar and eucalyptus-based agroforestry systems. Forests 14(3):559

Chisanga K, Bhardwaj DR, Sharma S (2013) Bio-economic appraisal of agroforestry systems in dry temperate Western Himalayas. J Tree Sci 32(1&2):1–7

Chisanga K, Bhardwaj DR, Pala NA, Thakur CL (2018) Biomass production and carbon stock inventory of high-altitude dry temperate land use systems in North-Western Himalaya. Ecol Process 7:22

Chitakira M, Torquebiau E (2010) Barriers and coping mechanisms relating to agroforestry adoption by smallholder farmers in Zimbabwe. J Agric Educ Ext 16(2):147–160

Choudhary P, Tandon V, Sharma R (2012) Traditional agroforestry practices for sustainable livelihood in Rajouri district of Jammu and Kashmir. J Tree Sci 31(1&2):102–107

Costanza R, D'Arge R, De Groot R, Farber S, Grasso M, Hannon B, Limburg K, Naeem S, O'Neill RV, Paruelo J, Raskin RG, Sutton P, Van den Belt M (1997) The value of the world's ecosystem services and natural capital. Nature 387:253–260

DAHD & F (2021) Livestock census, department of animal husbandry, dairying & fisheries, Ministry of Agriculture & Farmers' Welfare, Government of India, New Delhi, India [cited 2023 April 30]. Available from: https://www.nddb.coop/information/stats/pop

Dar M, Qaisar KN, Ahmad S, Wani AA (2018) Inventory and composition of prevalent agroforestry systems of Kashmir Himalayas. Adv Res 14(1):1–9

Datt C, Datt M, Singh NP (2008) Assessment of fodder quality of leaves of multipurpose trees in subtropical humid climate of India. J For Res 19:209–214

Devi NB, Lepcha NT, Bhutia PT, Rocky P, Sahoo UK, Pandey R, Nath AJ (2023) Biodiversity and ecosystems services of the agroforestry systems of the Himalayan region: an overview. Agrofor Sustain Intensific Agric Asia Africa 20:487–513

Dogra R, Eepsa HG, Singh H, Jatav LKJ, Khan MA (2024) Dynamics and vulnerability of complex agroforestry systems to farmers well-being in North-Western Himalayas. In: Soil health management for sustainable development goals

FAO (2009) How to feed the world in 2050. Food and Agricultural Organization of the United Nations. Available online: line: https://www.fao.org/fileadmin/templates/wsfs/docs/expert_paper/How_to_Feed_the_World_in_2050.pdf. Accessed on 13 Mar 2022

FAO (2017) Agroforestry for landscape restoration: exploring the potential of agroforestry to enhance the sustainability and resilience of degraded landscapes. Rome

FAO (2020) Global Forest Resources Assessment 2020. Food and Agriculture Organization of the United Nations

Furtak K, Gałązka A (2019) Edaphic factors and their influence on the microbiological biodiversity of the soil environment. Adv Microbiol 58(4):375–384

Ganai AM, Bakshi MPS, Ahmed MA, Matto FA (2009) Evaluation of some top fodder foliage in Kashmir valley. Indian J Anim Nutr 26:142–145

Gariya K, Dwivedi GK, Kumar V, Tewari SK (2020) Agrobiodiversity, composition and functional characteristics of home gardens in Bhimtal block of Nainital district, Uttarakhand, India. Int J Agric Environ Biotechnol 13(2):1–11

Ghale B, Mitra E, Sodhi HS, Verma AK, Kumar S (2022) Carbon sequestration potential of agroforestry systems and its potential in climate change mitigation. Water Air Soil Pollut 233(7):228

Goswami S, Verma KS, Kaushal R (2014) Biomass and carbon sequestration in different agroforestry systems of a Western Himalayan watershed. Biol Agric Hortic 30(2):88–96

Government of India (2016) Sub mission on agroforestry operational guidelines. Department of Agriculture, Cooperation and Farmer Welfare, Government of India, p 30

Government of India (2021) National Agroforestry Policy Review Report. Ministry of Agriculture & Farmers Welfare

Gupta RD, Arora S (2015) Agroforestry as alternate land use system for sustaining rural livelihoods in Himalayan ecosystem. In: Bhan S, Arora S (eds) Advances in soil and water resource management for food and livelihood security in changing climate. Soil Conservation Society of India, New Delhi, pp 656–669

Gupta B, Sarvade S, Singh M (2017) Species composition, biomass production, and carbon storage potential of agroforestry systems in Himachal Pradesh. In: Gupta SK, Panwar P, Kaushal R (eds) Agroforestry for increased production and livelihood security. New India Publishing Agency, New Delhi, pp 242–270

Handa AK, Dhyani SK, Rizvi J (2019) Guidelines to produce quality planting material of agroforestry species. ICAR (CAFRI) & WAC, p 42

Hepburn C, Adlen E, Beddington J, Carter EA, Fuss S, Mac Dowell N, Minx JC, Smith P, Williams CK (2019) The technological and economic prospects for CO_2 utilization and removal. Nature 575(7781):87–97

Himshikha H (2015) Existing agroforestry practices in Haridwar, India. Int J Sci Res 5(9):993–998

Himshikha H, Singh C (2017) Economics of poplar-based agroforestry models adopted by the farmers in Haridwar, Uttarakhand. Indian J Ecol 44(6):804–811

Himshikha NAK, Poonia PK, Kumar M, Daneva V (2020) Evaluation of existing agroforestry systems in plain region of Uttarakhand, India. Indian J Agrofor 22(2):44–50

Hoermann B, Choudhary D, Choudhury D, Kollmair M (2010) Integrated value chain development as a tool for poverty alleviation in rural mountain areas: an analytical and strategic framework. ICIMOD, Kathmandu

Hughes L (2000) Biological consequences of global warming: is the signal already apparent? Trends Ecol Evol 15:56–61

ISFR (2023). India State of Forest Report 2023. Forest Survey of India. Uttarakhand. https://fsi.nic.in/forest-report-2023-details

IPCC (2001) Climate change 2001: the scientific basis. In: Houghton JET, Ding Y, Griggs D, Noguer M, Van der Linden P, Dai X, Maskell M, Johnson C (eds) Contribution of working group I to the third assessment report of the Intergovernmental Panel on Climate Change (IPCC), p 881

Islam MA, Qaisar KN, Bhat GM (2017) Indigenous knowledge in traditional agroforestry systems of Kashmir valley: current challenges and future opportunities. Int J For Crop Improv 8(1):68–77

Jadhav SB, Pawar SN, Jadhav SB, Waghmare AA (2006) Evaluation of alley cropping system for soil and water conservation. Indian J Dryland Agric Res Dev 21(2):154–156

Jamnadass R, Place F, Torquebiau E, Malézieux E, Liyama M, Sileshi G, Kehlenbeck K, Masters E, McMullin S, Dawson I (2013) Agroforestry, food and nutritional security. World Agroforestry

Jana B, Biswas S, Majumder M, Roy P, Mazumdar A (2009) Carbon sequestration rate and aboveground biomass carbon potential of four young species. J Ecol Nat Environ 1:15–24

Jana P, Dasgupta S, Todaria NP (2021) Throughfall and stemflow nutrient flux in deodar and oak forests, Garhwal Himalaya, India. Water Supply 21(4):1649–1656

Jose S (2019) Environmental impacts and benefits of agroforestry. In: Oxford encyclopaedia of agriculture and environment. Oxford University Press, USA

Kaler NS, Gupta B, Negi V (2017) Traditional agroforestry systems in cold desert region of Himachal Pradesh, India. Popular Kheti 5(2):35–36

Kanime N, Kaushal R, Tewari S, Raverkar K, Chaturvedi S, Chaturvedi OP (2013) Biomass production and carbon sequestration in different tree-based systems of Central Himalayan Tarai region. For Trees Livelihoods 22:38–50

Kashyap SD, Dagar JC, Pant KS, Yewale AG (2014) Soil conservation and ecosystem stability: natural resource management through agroforestry in Northwestern Himalayan Region. In: Dagar JC et al (eds) Agroforestry systems in India: livelihood security & ecosystem services, advances in agroforestry. Springer, India, pp 21–55

Kaushal R, Kumar A, Mandal D, Tomar JM, Jinger D, Islam S, Panwar P, Jayaprakash J, Uthappa AR, Singhal V, Barh A (2024) Mulberry based agroforestry system and canopy management practices to combat soil erosion and enhancing carbon sequestration in degraded lands of Himalayan foothills. Environ Sustainability Indic 24:100467

Kehie M, Khamu S (2018) Indigenous alder-based farming practices in Nagaland, India: a sustainable agricultural model. J Tradit Folk Pract 5:25–29

Keprate A, Bhardwaj DR, Sharma P, Kumar D, Rana RK (2024) Biomass partitioning, carbon storage, and pea (*Pisum sativum* L.) crop production under a *Grewia optiva*-based agroforestry system in the Mid-Hills of the Northwestern Himalayas. Sustainability 16(17):7438

Khaki BA, Wani AA, Bhardwaj DR, Singh VRR (2016) Soil carbon sequestration under different agroforestry land use systems. Indian Forester 142(8):734–738

Khatta VK, Tewatia BS, Gupta PC (1999) Nutritional evaluation of tree leaves of humid subtropical region of Himachal Pradesh. Indian J Anim Nutr 16:220–223

Khosla PK, Toky OP (1996) Status of agroforestry in the western Himalaya. Curr Res Plant Sci 1(1):367–377

Khosla PK, Toky OP (1999) Pattern and processes in agroforestry systems of the Western Himalaya. Indian J Agron 1(1):73–80

Khosla PK, Toky OP, Bisht RP, Hamidullah S (1992) Leaf dynamics and protein content of six important fodder trees of the western Himalaya. Agrofor Syst 19:109–118

Khuman YSC, Pandey R, Rao KS (2011) Fuelwood consumption patterns in Fakot watershed, Garhwal Himalaya, Uttarakhand. Energy 36(8):4769–4776

Kisangau PD, Kimiti JM, Mburu MW, Snelder DJ (2021) Factors influencing use of multipurpose trees and shrubs in arid and semiarid lands of Kenya. Rangel Ecol Manag 74:125–134

Kumar M, Sharma CM (2009) Fuelwood consumption pattern at different altitudes in rural areas of Garhwal Himalaya. Biomass Bioenergy 33:1413–1418

Kumar A, Srivatava S (2018) Profitability of agroforestry system: the case of Udham Singh Nagar district of Uttarakhand state, India. Curr J Appl Sci Technol 25(4):1–15

Kumar GP, Murkute AA, Gupta S, Singh SB (2009) Carbon sequestration with special reference to agroforestry in cold deserts of Ladakh. Curr Sci 97:1063–1068

Kumar M, Anemsh K, Sheikh MA, Raj AJ (2012) Structure and carbon stock potential in traditional agro forestry system of Garhwal Himalaya. J Agric Technol 8(7):2187–2200

Kumar BM, Handa AK, Dhyani SK, Arunachalam A (2018a) Agroforestry in the India Himalayan region. In: Gordon AM, Newman SM, Coleman BRW (eds) Temperate agroforestry systems, 2nd edn. CAB International, Wallingford, pp 153–171

Kumar P, Thakur CL, Rai P, Attri K (2018b) Identification of existing agroforestry systems and socio-economic assessment in Kandaghat block of Solan district, HP, India. Int J Curr Microbiol App Sci 7(4):3815–3826

Kumar B, Singh K, Sharma J, Gairola S (2020) A comprehensive review of fuelwood resources and their use pattern in rural villages of western Himalaya, India. Plant Arch 20(2):1949–1958

Kumar A, Tewari S, Singh H, Kumar P, Kumar N, Bisht S, Devi S, Nidhi KR (2021) Biomass accumulation and carbon stock in different agroforestry systems prevalent in the Himalayan foothills, India. Curr Sci 120(6):1083–1088

Kumari A, Sehgal RN, Kumar S (2008) Traditional agroforestry systems practiced in Lahul (Lahul &Spiti) and Kinnaur districts of Himachal Pradesh. Indian Forester 8:1003–1010

Mace GM, Norris K, Fitter AH (2012) Biodiversity and ecosystem services: a multilayered relationship. Trends Ecol Evol 27:19–26

Madhu GS, Harish D, Mir JI, Nabi SU, Sharma OC (2023) Conservation and use of temperate fruit and nut genetic resources. In: Fruit and nut crops. Springer Nature Singapore, Singapore, pp 1–25

Mahato S, Dasgupta S, Todaria NP, Singh VP (2016) Agroforestry mapping and characterization in four districts of Garhwal Himalaya. Energy Ecol Environ 1(2):86–97

Maikhuri RK, Senwal RL, Rao KS, Saxena KG (1997) Rehabilitation of degraded community lands for sustainable development in Himalaya: a case study in Garhwal Himalaya, India. Int J Sustainable Dev World Ecol 4(3):192–203

Maikhuri RK, Semwal RL, Rao KS, Singh K, Saxena KG (2000) Growth and ecological impacts of traditional agroforestry tree species in Central Himalaya, India. Agrofor Syst 48:257–271

Maurya R, Kumar N, Gupta M, Sood KK, Singh C (2025) Agroforestry systems in the North-Western Himalayas: a nexus of biodiversity conservation, livelihood enhancement, and ecosystem services. Ind J Eco 52(5):1281–1287

Mbow C, Smith P, Skole D, Duguma L, Bustamante M (2014) Achieving mitigation and adaptation to climate change through sustainable agroforestry practices in Africa. Curr Opin Environ Sustain 6:8–14

Mey CB, Gore ML (2021) Biodiversity conservation and carbon sequestration in agroforestry systems of the Mbalmayo forest reserve. J For Environ Sci 37(2):91–103

Myers N (1988) Threatened biotas: 'hot spots' in tropical forests. Environmentalist 8:187–208

NAFP (2014) National Agroforestry Policy. Ministry of Agriculture, GoI, pp 14–16

Nair PKR (1993) State-of-the-art of agroforestry research and education. Agrofor Syst 23:95–119

Nair PR, Kumar BM, Nair VD, Nair PR, Kumar BM, Nair VD (2021) Multipurpose trees (MPTs) and other agroforestry species. In: An introduction to agroforestry: four decades of scientific developments, pp 281–351

Namgial J, Prabhakar M, Bishist R, Gautam KL, Sharma H, Sharma A (2020) Identification of different land use systems in Leh region of Himalayan Cold Desert. Int J Curr Microbiol App Sci 9(11):1–11

NAP (2014) National Agroforestry Policy, Department of Agriculture and Cooperation, Ministry of Agriculture, Government of India, pp 1–13

NAS (2020) National Accounts Statistics-2020, Central statistical organisation, Government of India, New Delhi, India, 2021 [cited 2021 April 30]. Available from: https://www.nddb.coop/information/stats/GDPcontrib

Nath PC, Thangjam U, Kalita SS, Sahoo UK, Giri K, Nath AJ (2022) Tree diversity and carbon important species vary with traditional agroforestry managers in the Indian eastern Himalayan region. Environ Sci Pollut Res 29(43):64732–64744

Navale MR, Bhardwaj DR, Bishist R, Thakur CL, Sharma S, Sharma P, Kumar D, Probo M (2022) Seasonal variations in the nutritive value of fifteen multipurpose fodder tree species: a case study of north-western Himalayan mid-hills. PLoS One 17(10):e0276689

Osei-Tutu G, Abunyewa AA, Dawoe EK, Agbenyega O, Barnes RV (2018) Effect of multipurpose trees and shrubs on degraded mined-out soil in a semi-deciduous forest zone of West Africa. Land Degrad Dev 29(10):3432–3439

Padma TV (2014) Himalayan plants seek cooler climes. Nature 512:359

Panda S, Bhardwaj DR, Sharma P, Handa AK, Kumar D (2021) Impact of climatic patterns on phenophase and growth of multi-purpose trees of north-western mid-Himalayan ecosystem. Trees For People 6:100143

Pandey DN (2007) Multifunctional agroforestry systems in India. Curr Sci 92:455–463

Pant KS, Bishist R, Sharma PP, Sharma H (2022) Agroforestry in Himachal Pradesh: status, challenges and future perspectives. Indian J Agrofor 24(3):25–29

Panwar P, Mahalingappa DG, Kaushal R, Bhardwaj DR, Chakravarty S, Shukla G, Thakur NS, Chavan SB, Pal S, Nayak BG, Srinivasaiah HT (2022) Biomass production and carbon sequestration potential of different agroforestry systems in India: a critical review. Forests 13(8):1274

Parikh J (2011) Hardships and health impacts on women due to traditional cooking fuels: a case study of Himachal Pradesh, India. Energy Policy 39:7587–7594

Parthiban KT, Fernandaz CC, Sudhagar RJ, Sekar I, Kanna SU, Rajendran P, Devanaand PS, Vennila S, Kumar NK (2021) Industrial agroforestry- a sustainable value chain innovation through a consortium approach. Sustainability 13:1–14

Pateria HM, Sharma OP, Bhatt BP (2003) Sustainability through agroforestry in Himalayas: an overview. Proceeding. In: Bhatt BP, Bujarbaruah KM, Sharma YP, Patiram (eds) Approaches for increasing agricultural productivity in hill and mountain ecosystem. ICAR Research Complex for NEH Region, Umiam, Meghalaya, pp 129–138

PCI (2011) Report of the sub group III on fodder and pasture management (No. id: 6836). Constituted under the Working Group on Forestry and Sustainable Natural Resource Management. Planning Commission of India. Govt. of India. New Delhi, India, 2011 [cited 2021 April 30]. Available from: https://landportal.org/node/63094

Phondani PC, Maikhuri RK, Rawat LS, Negi VS (2020) Assessing farmers' perception on criteria and indicators for sustainable management of indigenous agroforestry systems in Uttarakhand, India. Environ Sustainability Indic 5:1–8

Pratap UB, Pant KS (2020) Performance of okra as influenced by tree spacing and organic manures under Melia composita based Agrisilviculture systems. Pharma Innov J 9(10):277–281

Purohit VK (2010) Multipurpose tree species of western Himalaya with an agroforestry prospective for rural needs. J Am Sci 6(1):73–80

Rajput BS, Bhardwaj DR, Pala NA (2017) Factors influencing biomass and carbon storage potential of different land use systems along an altitudinal gradient in temperate northwestern Himalayas. Agrofor Syst 91:479–486

Rao MR, Palada M, Becker BN (2004) Medicinal and aromatic plants in agroforestry systems. Agrofor Syst 61–62:107–122

Rawat YS, Vishvakarma SCR, Todaria NP (2009) Fuelwood consumption pattern of tribal communities in cold desert of the Lahaul valley, North-Western Himalaya, India. Biomass Bioenergy 33:1547–1557

Rois-Díaz M, Lovric N, Lovric M, Ferreiro-Domínguez N, Mosquera-Losada MR, Den Herder M, Graves A, Palma JH, Paulo JA, Pisanelli A, Smith J (2017) Farmers' reasoning behind the uptake of agroforestry practices: evidence from multiple case-studies across Europe. Agrofor Syst 92:811–828

Roy ML, Chandra N, Kharbikar HL, Joshi P, Jethi R (2013) Socio-economic status of hill farmers: an exploration from Almora District in Uttarakhand. Int J Agric Food Sci Technol 4(4):353–358

Roy AK, Agrawal RK, Bhardwaj NR, Mishra AK, Mahanta SK (2019) Revisiting national forage demand and availability scenario. In: Roy AK, Agrawal RK, Bhardwaj NR (eds) Indian fodder scenario: redefining state wise status. ICAR-AICRP on Forage Crops and Utilization, Jhansi, pp 1–21

Saha SK, Kumar AS (2019) Northwest Himalayan ecosystems: issues, challenges and role of geospatial techniques. In: Navalgund RR et al (eds) remote sensing of Northwest Himalayan Ecosystems. Springer, Singapore, pp 3–14

Salve A, Bhardwaj DR (2020) Soil carbon stock and nutrient study in different agroforestry systems at Kinnaur district, Himachal Pradesh. Plant Arch 20(2):4251–4260

Salvini G, Ligtenberg A, Van Paassen A, Bregt AK, Avitabile V, Herold M (2016) REDD+ and climate smart agriculture in landscapes: a case study in Vietnam using companion modelling. J Environ Manag 172:58–70

Semwal RL, Nautiyal S, Maikhuri RK, Rao KS, Saxena KG (2013) Growth and carbon stocks of multipurpose tree species plantations in degraded lands in central Himalaya, India. For Ecol Manag 310:450–459

Shah G-M, Karki Nepal A, Rasul G, Ahmad F (2018) Value chain development of bay leaf in Nepal: an impact assessment. J Dev Eff 10:1–18

Shah G-M, Ahmad F, Panwar S, Khadka MS, Ali A, Bisht S (2019) *Ocimum sanctum* [Tulsi]—an alternative additional livelihood option for the poor and smallholder farmers. Sustainability 11(1):227

Sharma SK (2004) Hortipastoral based land use systems for enhancing productivity of degraded lands under rain fed and partially irrigated conditions. Uganda J Agric Sci 9:320–325

Sharma H, Thakur CL (2020) Productivity of *Capsicum annuum* under *Grewia optiva*-based traditional agroforestry system in mid hill zone of Himachal Pradesh (India). Appl Biol Res 22(1):83–87

Sharma G, Hunsdorfer B, Singh KK (2016) Comparative analysis on the socio-ecological and economic potentials of traditional agroforestry systems in the Sikkim Himalaya. Trop Ecol 57:751–764

Sharma H, Thakur CL, Bhardwaj DR, Prabhakar M (2017a) Growth and yield performance of maize (*Zea mays*) and tomato (*Solanum lycopersicum*) under *Grewia optiva* Drummond based agroforestry system. Indian J Ecol 44(6):771–776

Sharma P, Singh MK, Tiwari P, Verma K (2017b) Agroforestry systems: opportunities and challenges in India. J Pharmacogn Phytochem 7(1):953–957

Sharma P, Singh MK, Tiwari P (2017c) Agroforestry: a land degradation control and mitigation approach. Bull Environ Pharmacol Life Sci 6(5):312–317

Sharma P, Bhardwaj DR, Singh MK, Nigam R, Pala NA, Kumar A, Verma K, Kumar D, Thakur P (2022) Geospatial technology in agroforestry: status, prospects, and constraints. Environ Sci Pollut Res 30:116459–116487

Sharma H, Pant KS, Bishist R, Gautam KL, Dogra R, Kumar M, Kumar A (2023a) Estimation of biomass and carbon storage potential in agroforestry systems of north western Himalayas, India. Catena 225:107009

Sharma P, Bhardwaj DR, Singh MK, Nigam R, Pala NA, Kumar A, Verma K, Kumar D, Thakur P (2023b) Geospatial technology in agroforestry: status, prospects, and constraints. Environ Sci Pollut Res 30(55):116459–116487

Sharma H, Pant KS, Bishist R, Prakash P, Gautam KL (2024) An insight into prevalent agroforestry land use systems of North Western Himalayan Region, India: challenges and future prospects. In: Agroforestry to combat global challenges: current prospects and future challenges, pp 419–439

Singavarapu B, Du J, Beugnon R, Cesarz S, Eisenhauer N, Xue K, Wang Y, Bruelheide H, Wubet T (2023) Functional potential of soil microbial communities and their subcommunities varies with tree mycorrhizal type and tree diversity. Microbiol Spectr 11(2):e0457822

Singh G, Rawat GS (2012) Depletion of oak (*Quercus spp.*) forests in the Western Himalaya: grazing, fuelwood and fodder collection. Global perspectives on sustainable forest management 29:36–43

Singh C, Gussain P, Maurya R (2023) Integrating soil health in effective land use planning strategies. In: Fundamentals of soil science. Golden Leaf Publisher, Lucknow, pp 207–227

Singh N, Riyal MK, Singh B, Khanduri VP, Rawat D, Singh C, Kumar M (2024) Carbon sequestration potential of agroforestry versus adjoining forests at different altitudes in the Garhwal Himalayas. Atmosphere 15(3):313

Singh SP, Singh V, Skutsch M (2010) Rapid warming in the Himalayas: ecosystem responses and development options. Clim Dev 2(3):221–232

Singh B, Bishnoi M, Baloch MR, Singh G (2014) Tree biomass, resource use and crop productivity in agri-horti-silvicultural systems in the dry region of Rajasthan, India. Arch Agron Soil Sci 60:1031–1049

Singh M, Gupta B, Sarvade S, Awasthe RK (2015) Biomass and carbon sequestration potential in different agroforestry systems in Giri catchment of North Western Indian Himalayas. India J Agrofor 17(2):42–48

TEEB (2010) The economics of ecosystems and biodiversity: mainstreaming the economics of nature: a synthesis of the approach, conclusions and recommendations of TEEB. The Economics of Ecosystems and Biodiversity, Malta

TERI (2015) Green growth and forestry in Himachal Pradesh. New Delhi, The Energy and Research Institute, p 16

Tewari DN (1995) Orienting multipurpose tree species research in India. In: Hegde NG, Daniel JN (eds) Proceedings of the National Workshop “multipurpose tree species for agroforestry in India”. BAIF Development Research Foundation, Pune, India

Tewari JC, Pareek K, Raghuvanshi MS, Kumar P, Roy MM (2016) Fodder production system- a major challenge in cold arid region of Ladakh, India. MOJ Ecol Environ Sci 1(1):00005

Tewari VP, Verma RK, Gadow KV (2017) Climate change effects in the western Himalayan ecosystems of India: evidence and strategies. For Ecosyst 4(13):1–9

Thakur NS, Gupta NK, Gupta B (2004) Phytosociology analysis of woody and non woody components under some agroforestry systems in Western Himalaya: case study. Indian J Agrofor 6:65–71

Timmermann L, Smith-Hall C (2019) Commercial medicinal plant collection is transforming high-altitude livelihoods in the Himalayas. Mt Res Dev 39(3):13–21

Tiwari P, Pant KS, Singh R (2018) System units under prevalent agroforestry systems in North Western Himalayas and their constraints. J Pharmacogn Phytochem 7(6):2758–2764

Toky OP, Kumar P, Khosla PK (1989) Structure and function of traditional agroforestry systems in the western Himalaya. I Biomass and productivity. Agrofor Syst 9:47–70

Toppo P, Toppo S (2018) Tree crop interaction in agroforestry: a review. Int J Chem Stud 7(1):2359–2361

Verma A, Kaushal R, Alam NM, Mehta H, Chaturvedi OP, Mandal D, Tomar JM, Rathore AC, Singh C (2014) Predictive models for biomass and carbon stocks estimation in Grewia optiva on degraded lands in western Himalaya. Agrofor Syst 88:895–905

Verma K, Sharma P, Kumar D, Vishwakarma SP, Meena NK (2021) Strategies sustainable management of agroforestry in climate change mitigation and adaptation. Int J Curr Microbiol App Sci 10(1):2439–2449

Verma K, Sharma P, Bhardwaj DR, Kumar R, Kumar NM, Singh AK (2023) Land and environmental management through agriculture, forestry and other land use (AFOLU) system. In: Land and environmental management through forestry, vol 4, pp 247–271

Vikrant K, Chauhan DS, Rizvi R (2018) Biomass and carbon potential in agroforestry systems along altitudes in Tehri District of Uttarakhand. Agroforestry for climate resilience and rural livelihood 2:171–178

Vikrant K, Chauhan DS, Rizvi R (2020) Carbon stock assessment along altitudes in agrisilviculture systems of Tehri District, Uttarakhand, North-Western Himalaya—case study of Pratapnagar block. Indian For 146:202–207

Yadav RP, Bisht JK (2013) Agroforestry: a way to conserve MPTs in North Western Himalaya. Res J Agric For Sci 1(9):8–13

Yadav RP, Bisht JK (2015) *Celtis australis* Linn: a multipurpose tree species in North West Himalaya. Int J Life-sci Sci Res 1(2):66–70

Yadav RP, Bisht JK, Bhatt JC (2017) Biomass, carbon stock under different production systems in the mid hills of Indian Himalaya. Trop Ecol 58(1):15–21

Yadav RP, Gupta B, Bhutia PL, Bisht JK, Pattanayak A (2019) Sustainable agroforestry systems and their structural components as livelihood options along elevation gradient in central Himalaya. Biol Agric Hortic 35(2):73–95

Yadav N, Singh B, Khanduri VP (2023) Seasonal variation in nutrient composition in the leaves of two Bauhinia species. Folia Forestalia Polonica A For 65(3):12–17

Young A (1989) Agroforestry for soil conservation: CAB International, Wallingford, UK, in cooperation with ICRAF, p 276

Zahnoor S, Dutt V, Mughal AH, Pala NA, Rashid M, Khan PA (2022) Influence of intercropping on soil properties and nutrient status of fruit-based agroforestry systems in north western region of Indian Himalaya. Indian J Agrofor 24(1):85–92

Chapter 2
Agri-Horticultural Systems in Uttarakhand: Opportunities, Challenges, and Progress in the Western Himalaya

J. M. S. Rawat, Krishan Kumar Singh, Vipin Panwar, Etalesh Goutam, and Vidyawati Rawat

Abstract Agri-horticultural systems combine agricultural and horticultural crops and offer significant potential for long-term growth in hilly areas. These methods improve land-use efficiency, diversify farmers' revenue sources, and help to maintain ecological stability. Agri-horticultural systems combine the age-old rhythms of farming with the intentional production of horticultural riches such as vegetables, fruits, medicinal herbs, and more. From the arid plains to the misty highlands, the agri-horticultural system serves as a beacon of hope, combining tradition and innovation to meet current demands. Agri-horticultural system that drives India's economy and rural fabric, is the right complement for the country's agrarian character. Nestled in the Himalayan embrace, Uttarakhand elevates agri-horticultural system to the level of an art form, with its rugged beauty and fluctuating temperatures providing the perfect backdrop. Uttarakhand's diverse agro-climatic zones support a wide range of fruits, vegetables, and traditional crops, making it an ideal location for fostering agri-horticultural activities. This chapter investigates the extent, types, and regional adaptations of the state's agri-horticultural systems, as well as their role in improving livelihoods, conserving resources, and building climate resilience. Furthermore, it highlights the problems encountered during system deployment and provides strategic recommendations for future development and government assistance.

Keywords Agri-horticultural system · Climate resilience · Western Himalaya · Zero hunger · Food security

J. M. S. Rawat (✉) · V. Rawat
Department of Agriculture, Government Degree College, Rudraprayag, Uttarakhand, India

K. K. Singh · E. Goutam
Faculty of Agriculture, Guru Kashi University, Talwandi Sabo, Bathinda, Punjab, India

V. Panwar
Department of Botany, Bharat Singh Rawat, Govt. Degree College,
Pauri (Garhwal), Uttarakhand, India

B. Singh et al. (eds.), *Himalayan Agroforestry Systems*, Advances in Agroforestry 15,
https://doi.org/10.1007/978-3-032-25096-4_2

1 Introduction

Agri-horticultural systems integrate traditional agricultural practices with the intentional cultivation of horticultural crops—such as fruits, vegetables, and medicinal plants—creating a diverse and sustainable land-use model (Colmenares et al. 2020; Robinson 2024). This approach relies on the complementary growth cycles of short-duration crops and long-lived perennials to enhance productivity, restore soil health, and maintain ecological balance. Functioning as an advanced form of agroforestry, these systems combine fruit trees or medicinal shrubs with staple crops, resulting in a resilient and adaptable farming framework. The value of agri-horticulture in sustainable agriculture lies in its ability to promote biodiversity, reduce dependence on chemical inputs, and mitigate the risks associated with monoculture (Nair 2019). By strategically pairing species with varying life cycles and nutrient requirements, these systems maximize land-use efficiency and provide a buffer against environmental stresses such as drought and pest outbreaks. In India, where agriculture is a cornerstone of livelihoods, agri-horticultural systems offer significant benefits: increased food production, stable incomes, and environmental sustainability. More than a mere technique, this approach embodies a philosophy that equates diversity in cultivation with resilience in rural life. From the lowland plains to the Himalayan foothills, agri-horticulture represents a promising blend of tradition and innovation, poised to address contemporary agricultural challenges.

1.1 Importance of Agri-Horticultural System

The Indian Himalayan region is characterized by a rich diversity of both traditional and modern agroforestry systems, including agri-silvicultural, agri-horticultural, silvopastoral, and agri-silvihorticultural models (Rasul 2014; Bijalwan et al. 2015; Kumar et al. 2021). Uttarakhand is especially notable for its rich natural resources, lively folk traditions, and diverse ecological settings, all shaped by unique geographic, weather, geological, and ecological factors. Local communities in the state depend directly and indirectly on natural resources to sustain their livelihoods. Within these agroforestry systems, a wide array of fruit and forest trees coexist with agricultural crops, reflecting the region's ecological and cultural diversity (Gairola and Biswas 2008; Pretty 2008). However, agriculture in Uttarakhand faces numerous challenges, such as shrinking landholdings, limited irrigation infrastructure, erratic rainfall, and frequent human–wildlife conflicts (Tiwari and Joshi 2015). In this context, agri-horticultural systems emerge as a sustainable solution, offering livelihood security by reducing vulnerability to climate change, supporting soil conservation and fertility, promoting nutrient cycling, enhancing microclimatic conditions, and contributing to bio-drainage and biofuel production—all of which collectively increase overall biomass productivity (Bijalwan 2014; Kumar et al. 2020). Agricultural crops remain central to these systems, playing a vital role in optimizing land use. The long-term success and sustainability of agri-horticultural systems rely on maintaining a balance between positive and negative interactions: benefits such as increased productivity, sustainability, and profitability must outweigh potential drawbacks like resource competition. Effective agroforestry hinges on four key factors—complexity, profitability, competition, and sustainability—which must be carefully balanced to achieve optimal outcomes for both people and the environment (Fanish and Priya 2013; Toppo and Toppo 2019; Government of India 2023, 2024).

1.2 Components of Agri-Horticultural System

At its core, agri-horticultural system thrives on a dynamic combination of crops, each selected to match the unique agro-climatic and soil conditions of the region. The foundation of these systems is built upon staple agricultural crops such as rice, wheat, and sorghum, which are commonly cultivated across different zones in Uttarakhand and other parts of India (Singh and Singh 2010). These staples are integrated with a diverse range of horticultural crops—fruits like mango, banana, apple, and orange, as well as vegetables such as carrots, green beans, and other regionally suitable varieties. This strategic crop combination not only optimizes land use but also enhances productivity, soil health, and resilience against climatic stresses. By blending annual crops with perennial fruit trees and vegetables, agri-horticultural systems create a synergistic farming model that supports both food security and income diversification for farming communities (Pandey and Tiwari 2022; Goutam et al. 2025).

Farmers in Uttarakhand select crop combinations based on the unique characteristics of their local climate, soil texture, and prevailing market trends, aiming to achieve a balance between household food security and income generation. The integration of fruit trees is a defining feature of these agri-horticultural systems. For example, apple, mango, and citrus trees are commonly intercropped with pulses, potatoes, and other vegetables, creating multilayered farming systems that are both productive and resilient. These fruit trees offer multiple benefits beyond their harvests. Their deep root systems help anchor the soil, significantly reducing erosion on the region's terraced and sloped farmlands. By tapping into deeper water reserves, fruit trees leave more surface moisture available for annual crops, thus optimizing water use in rainfed and drought-prone conditions. The canopy provided by these trees also moderates the microclimate, protecting understory crops from temperature extremes and enhancing overall ecosystem health. While the presence of trees can sometimes reduce the yields of annual crops due to competition for light and nutrients, the additional income and nutritional value from fruit harvests often compensate for these reductions, supporting both the livelihoods and food security of local communities. Thus, agri-horticultural systems in Uttarakhand exemplify a harmonious blend of ecological stability and economic opportunity, making them a sustainable choice for the region's smallholder farmers (Singh and Singh 2010; Singh et al. 2021).

Given Uttarakhand's diverse terrain and varied climatic zones, the practice of integrating fruit orchards as windbreaks can be highly effective. Just as guava orchards serve as wind protectors for cotton fields in Gujarat, in Uttarakhand, fruit trees like apples, plums, or walnuts can be strategically planted around staple crops and vegetable fields. This not only shields crops from strong winds and soil erosion, especially in hilly areas, but also provides additional income through fruit production (Grow Billion Trees 2024; Uttarakhand Open University 2024; ICFRE 2024). The inclusion of medicinal and aromatic plants (MAPs) such as basil, aloe, and lemongrass further enriches these systems. When intercropped with staples or orchards, MAPs not only meet the rising market demand for natural remedies, cosmetics, and wellness products, but also provide significant ecological benefits. Their aromatic properties can naturally deter pests, while their organic matter enhances soil fertility as it decomposes. This approach is mirrored in Tamil Nadu, where turmeric is cultivated alongside paddy fields, exemplifying how MAPs can elevate both land productivity and farmer livelihoods. The integration of MAPs into agroforestry systems is increasingly recognized for its role in diversifying farm income, supporting environmental sustainability, and preserving traditional agricultural knowledge. These systems also contribute to improved biodiversity, soil health, and resilience against climatic and economic challenges, making them a promising model for sustainable rural development across India (Verma and Yadav 2023; Singh et al. 2025).

1.3 Scope of Agri-Horticultural System in India

India's agrarian identity is deeply intertwined with agri-horticulture, a sector that not only sustains rural livelihoods but also significantly powers the national economy. Horticultural crops are far from peripheral; they are central to agricultural growth, contributing about 33% to the agricultural gross value added (GVA) while occupying just 18% of the cultivated area. This sector has outpaced food grain production in recent years, with horticultural output reaching over 331 million tons in 2020–21, surpassing food grains and demonstrating much higher productivity per hectare. Beyond boosting agricultural gross domestic product (GDP), horticulture is a major driver of rural employment and value addition, creating jobs across the supply chain—from farm labor and processing to packaging, logistics, and exports (Subburathinam 2024). The sector's rapid growth has spurred the development of allied industries and expanded opportunities in rural areas, making horticulture a heavyweight in both economic and employment terms. As demand for fruits, vegetables, and value-added products rises domestically and internationally, horticulture continues to anchor India's rural economy and fuel its broader economic ambitions (Anonymous 2022, 2023a, b). Agri-horticulture forms the backbone of many value-added industries, fueling the production of fruit preserves, herbal teas, and a wide array of processed products found on store shelves. India's horticulture sector, which contributes about 33% to the agricultural gross value added (GVA), plays a pivotal role in the economy by providing raw materials for food processing, pharmaceuticals, cosmetics, and wellness industries. The sector's vast production—over 320 million tons of fruits and vegetables annually—supports a growing demand for natural and health-oriented products, creating extensive opportunities for farmers and entrepreneurs alike. Value addition in horticulture, through processing, packaging, and product diversification, not only enhances shelf life and reduces postharvest losses but also significantly increases farmers' incomes and export potential. For example, fruit preserves and herbal teas rely heavily on horticultural produce, weaving a complex supply chain that connects rural producers to urban and global markets. This interconnected web of production and processing creates employment, stimulates rural economies, and strengthens India's position in the global horticulture trade (National Horticulture Board 2022).

In 2022, the value and production of horticultural crops surpassed that of traditional grains in several Indian states, highlighting the sector's rapid ascent and transformative impact on rural livelihoods. For farmers, this shift is significant while grains like wheat, maize, etc., remain essential for food security, high-value horticultural crops—such as apple, mango, banana, kinnow, tomato, capsicum, leafy greens, and aromatic herbs—offer far greater income potential, especially in urban and export markets. These crops help farmers cushion the financial risks posed by erratic monsoons or volatile grain prices. Examples like Punjab's kinnoworchards intercropped with wheat demonstrate how diversification into horticulture can turn even small landholdings into profitable enterprises, underscoring the power of mixed farming systems. India's vast climatic diversity—from Kerala's humid tropics to Himachal Pradesh's cool hills—enables the cultivation of a wide range of horticultural crops, making agri-horticulture adaptable across regions. In the Deccan, bananas thrive alongside millets, while in Assam, pineapples flourish on the hills, illustrating the system's flexibility and scalability. With over 20 agro-climatic zones, India's landscape is ideally suited for agri-horticultural expansion, offering farmers across the country new opportunities for sustainable growth and resilience. This approach not only boosts farm incomes but also makes food and nutrition more secure, creates more jobs, and raises export opportunities, shaping a bright future for Indian agriculture. With these changes, the sector is creating pathways for growth, shaping a promising future for farmers across the country (Anonymous 2022, 2023a, b).

2 Agri-Horticultural System in Uttarakhand

Nestled within the Himalayan landscape, Uttarakhand exemplifies agri-horticulture as both tradition and innovation, shaped by its dramatic topography and diverse climates. The state's elevation gradient—from the humid Terai plains to the cold alpine heights—creates a mosaic of agro-climatic zones, each supporting distinct crops and farming systems. Warm, fertile valleys in the lower regions are ideal for citrus fruits, mangoes, litchis, and papayas, while the mid-hills nurture apples, plums, peaches, and walnuts, alongside off-season vegetables like peas and spinach. In the highest reaches, subalpine and alpine zones foster the growth of unique medicinal and aromatic plants such as bay leaves, Picrorhiza, and spikenard, reflecting a legacy of mixed farming adapted to both the land and local traditions. This intricate interplay between climate, elevation, and crop diversity has enabled Uttarakhand to develop a resilient and productive agri-horticultural system, where ancient practices blend seamlessly with modern techniques, sustaining livelihoods and preserving the region's ecological and cultural heritage (Rawat and Kumar 2018). Tables 2.1 and 2.2 typically summarize how different altitude levels influence key physical and chemical properties of aonla and malta fruits grown in this region. These crops are not merely sources of food—they are vital lifelines for Uttarakhand's villagers, sustaining local diets and generating premium income in major markets like Delhi and beyond. For example, a single apple harvest from Chamoli can rival or even surpass the annual yield of grain crops in terms of economic return, highlighting the significant financial impact of horticulture for hill farmers.

Historically, the people of Uttarakhand practiced mixed cropping, interweaving staples like barley and lentils with fruit crop such as pear, which exhibits a resilient farming tradition etched into the landscape for centuries. This approach not only ensured food security through dietary diversity but also provided a buffer against crop failures and market fluctuations. Today, the legacy of these integrated systems continues, with high-value horticultural produce offering both sustenance and a crucial economic safety net for mountain communities (Negi 2005; Yadav et al. 2019). Figures 2.1, 2.2, 2.3, 2.4, 2.5, 2.6, 2.7, and 2.8 depict locally available fruit tree varieties found across Uttarakhand. Farmers in Uttarakhand are blending traditional wisdom with modern agricultural innovations to enhance the productivity and sustainability of agri-horticultural systems. Terraced slopes, a hallmark of hill farming, are now optimized with advanced techniques: terraces efficiently capture rainwater, minimizing runoff and soil erosion while maximizing water infiltration and soil fertility. The integration of drip irrigation systems ensures that young plants receive precise amounts of water, supporting healthy growth even during dry spells. The adoption of high-yielding hybrid saplings further boosts crop output and resilience. In Rudraprayag, for example, farmers are intercropping kiwi with pulses, demonstrating a thoughtful fusion of ancestral mixed-cropping practices and contemporary agronomic tools. This approach not only diversifies farm income but also improves soil health and resource use efficiency. The result is a dynamic, living system where tradition and innovation coexist, sustaining both the land and the cultural heritage of the region. Table 2.3 presents a comprehensive statistical overview of the district-wise survey data pertaining to the area and production of vegetables, flowers, and fruits cultivated in Uttarakhand during the year 2023, as per the reports compiled and published by the Directorate of Horticulture and Food Processing, Chaubatia.

Table 2.1 Variation on physical and chemical characteristics of aonla [*Emblica officinalis* Gaerth.] in Western Himalaya

S. no.	Altitude (msl)	Fruit diameter (cm)	Fruit weight (g)	Number of fruits/kg	Fruit length (cm)	Pulp thickness (cm)	Pulp:stone ratio	Moisture percentage	Total soluble solids (%)	Citric acid (%)	Tartaric acid (%)	Vitamin C (mg/100 g)	Sugar (mg/g)
1	400–800	2.11	5.39	194.00	1.84	0.57	4.89	78.52	15.89	2.79	6.54	337.17	20.82
2	800–1200	2.24	6.95	185.14	1.99	0.62	4.88	79.26	14.04	2.75	6.44	347.50	28.49
3	1200–1600	2.04	5.02	209.89	1.82	0.53	4.32	79.01	14.73	2.80	6.56	333.82	20.96

Table 2.2 Physicochemical properties of local malta [*Citrus sinensis* (L.) Osbeck] in Western Himalaya

S. no.	Altitude (msl)	Fruit weight (g)	Fruit length (cm)	Fruit width (cm)	Juice sacks	Juice weight (gm)	Total seeds number	Total soluble solids (%)	Citric acid (%)	Ascorbic acid (mg/100 ml)	Soluble sugar (%)
1.	900–1300	164.96	6.72	6.86	191.83	58.878	19.61	9.32	4.07	49.17	6.16
2.	1300–1700	181.85	6.70	6.94	10.07	63.49	16.13	10.37	3.67	53.23	4.98
3.	1700–2100	155.61	6.24	6.58	10.53	56.28	19.00	10.19	4.86	57.94	5.98

Fig. 2.1 Locally available malta in Uttarakhand

Fig. 2.2 Ripened fruits of malta

2.1 Challenges in Agri-Horticultural Systems

Uttarakhand's agri-horticultural landscape faces formidable challenges that threaten both livelihoods and long-term sustainability. In the hills, land is fragmented into tiny, irregular plots, making large-scale cultivation nearly impossible and exacerbating the struggle for productive space (Sati et al. 2025). Soil health is under constant threat: steep slopes and intense rainfall accelerate runoff and erosion, stripping away fertile topsoil, while decades of chemical use and deforestation have further degraded the land's vitality. In the plains, relentless monoculture and overuse have left soils exhausted and vulnerable to further decline. Climate change compounds these difficulties (IPCC 2021). The region is now experiencing more erratic rainfall, prolonged droughts, and extreme weather events such as hailstorms and heat

Fig. 2.3 Aonla (NA-7) tree bearing fruits in Uttarakhand

Fig. 2.4 Freshly harvested aonla fruits of NA-7

waves—all of which directly impact crop yields and farmer incomes. For instance, in 2023, unexpected hail decimated peach orchards, demonstrating how unpredictable weather can erase a season's work overnight. Water scarcity is intensifying, with drying streams and shrinking glaciers leaving orchards and fields parched, while storms and landslides destroy both crops and infrastructure. Market access remains a persistent bottleneck. Poor rural roads, limited cold storage, and distant buyers mean that much of the perishable produce is lost before it can be sold, eroding already slim profit margins (Swain and Kalamkar 2016). Middlemen often capture much of the value, leaving farmers with little reward for their efforts.

These systemic gaps highlight the urgent need for investment in rural infrastructure—such as better roads, decentralized cold storage, and local processing facilities—to reduce postharvest losses and improve farmer incomes. Despite these obstacles, there are pathways forward. Recommendations include adopting

Fig. 2.5 Mango tree in Western Himalayan

Fig. 2.6 Local plum tree varieties

Fig. 2.7 Peach tree of Western Himalayan

Fig. 2.8 High-density planting system of guava trees (meadow orcharding)

Table 2.3 District-wise survey data of area and production covered under vegetable, flower, and fruit in Uttarakhand in 2023

S. no.	District	Vegetable			Flower		Fruit		
					Production				
		Area (ha)	Production (MT)	Productivity (MT/ha)	Cut flower/spike (no. in Lakh)	MT	Area (ha)	Production (MT)	Productivity (MT/ha)
1	Pauri	3422.15	22718.55	6.64	1.86	58.20	9047.49	21474.18	2.37
2	Chamoli	1925.70	11882.76	6.17	0.35	17.18	3390.00	7161.95	2.11
3	Haridwar	5009.10	84344.60	16.84	434.98	844.10	6548.88	34972.72	5.34
4	Dehradun	5323.04	46475.46	8.73	–	146.26	8528.90	32784.35	3.84
5	Tehri	5776.01	53213.16	9.21	2.22	336.57	5108.89	21476.04	4.20
6	Uttarkashi	10770.16	56551.09	5.25	–	3.60	8644.30	38837.64	4.49
7	Rudraprayag	1278.81	1822.72	1.43	–	8.83	2266.69	2689.81	1.19
8	Almora	3508.40	17224.69	4.91	1.37	52.22	8020.29	28451.03	3.55
9	Champawat	2374.94	15651.00	6.59	6.20	6.60	4118.33	7941.91	1.93
10	Pithoragarh	3134.98	27120.57	8.65	0.55	9.35	5888.77	15990.87	2.72
11	Nainital	5386.35	40943.05	7.60	73.49	195.64	9684.86	86768.42	8.96
12	Bageshwar	1264.44	6452.37	5.10	0.03	5.40	2459.90	7698.70	3.13
13	Udham Singh Nagar	9094.20	117386.62	12.91	85.92	663.41	7985.28	63199.68	7.91

Data published by Directorate of Horticulture and Food Processing, Chaubatia, Uttarakhand climate-resilient crop varieties, improving irrigation efficiency, reviving traditional water management systems, and fostering community-based marketing initiatives (Sati et al. 2025). Strengthening policy support and investing in sustainable practices are crucial if Uttarakhand's agri-horticultural sector is to adapt and thrive in the face of mounting environmental and economic pressures.

2.2 *Government Initiatives for Agri-Horticultural System in Uttarakhand*

India's policymakers and the Government of Uttarakhand have implemented a range of targeted initiatives to strengthen agri-horticulture, focusing on holistic sector development, sustainability, and farmer welfare. Horticulture mission for north east and Himalayan states (HMNEH) was launched in 2003–04 and now part of the mission for integrated development of horticulture (MIDH), HMNEH is specifically tailored for Himalayan states like Uttarakhand. It adopts a cluster approach to cover the entire value chain—production, postharvest management, processing, and marketing—with active stakeholder participation. The scheme provides substantial funding (90% central, 10% state share) and has led to the expansion of orchards, rejuvenation of old plantations, establishment of nurseries, and creation of infrastructure such as community tanks and cold storage. While National Horticulture Mission (NHM) covers all states, its interventions in Uttarakhand focus on supplying quality planting material, area expansion, technology dissemination, human resource development, and infrastructure for postharvest management and marketing (Ministry of Agriculture and Farmers Welfare 2020).

Rashtriya Krishi Vikas Yojana (RKVY) scheme supports state-led agricultural development, including agri-horticulture projects, by providing flexible funding for area expansion, technology adoption, and capacity building. It also promotes organic certification and training through the Uttarakhand Organic Products Council (Planning Commission 2018). National Mission on Micro Irrigation (NMMI) was launched in Uttarakhand in 2011–12, with the aim to increase the area under efficient irrigation methods such as drip and sprinkler systems, with substantial subsidies for small and marginal farmers, especially in horticulture. Uttarakhand is a national leader in organic agriculture, enacting the organic agriculture act in 2019 to regulate and promote organic farming. The state has established the Uttarakhand Organic Commodities Board (UOCB) and runs certification, training, and marketing programs for organic products, including medicinal and aromatic plants. Programs like Paramparagat Krishi Vikas Yojana (PKVY) and the Namami Gange clean campaign further support organic clusters, certification, and market access (Government of Uttarakhand 2022).

The state provides assistance for the construction of polyhouses, greenhouses, and water management systems to enable year-round cultivation and climate resilience. These interventions help farmers grow high-value crops such as off-season vegetables and floriculture products, even in challenging terrains. In Pithoragarh, as part of broader efforts to promote protected cultivation, farmers are increasingly adopting polyhouses and low tunnels—structures that create controlled environments for high-value crops like roses, enabling them to thrive even during harsh winter months. These advancements are supported by significant government investment in horticulture infrastructure, including funding for polyhouse farming and irrigation projects. Such initiatives not only extend the growing season for floriculture and vegetables but also provide farmers with new income opportunities, demonstrating how policy support and local determination can transform traditional agriculture into a resilient, year-round enterprise. Multiple schemes include training, exposure visits, and demonstration projects to equip farmers with modern techniques, improved planting material, and postharvest technologies, ensuring the adoption of best practices in agri-horticulture. These combined efforts have expanded the area under horticulture, improved productivity, and enhanced farmer incomes. Uttarakhand's organic and sustainable farming initiatives are positioning the state as a model for climate-resilient and high-value agriculture in the Himalayan region.

2.3 Technological Advancements in Agri-Horticultural System

Technological innovation is transforming agri-horticulture across India, including Uttarakhand, by making farming smarter, more efficient, and environmentally sustainable. Precision farming is at the forefront of this change, leveraging data analytics, remote sensing, drones, and automation to optimize crop yields, reduce input costs, and minimize environmental impact. Technologies such as GPS-guided tractors, IoT sensors, and satellite imagery allow farmers to monitor soil health, crop growth, and resource needs in real time, enabling targeted interventions for water, nutrients, and pest management. Drones are increasingly used for crop monitoring, disease detection, and targeted spraying of fertilizers and pesticides (Jat and Shivran 2020). These tools save labor, reduce water and chemical use, and can cover large areas quickly and accurately. For example, drone spraying can reduce water and fertilizer consumption by up to 70% and increase yields by 30–35%. Smart irrigation systems, such as drip and mist irrigation, further conserve water—crucial in drought-prone and hilly regions—by delivering moisture directly to plant roots and minimizing waste (Sharma and Singh 2017; Singh et al. 2018).

Protected cultivation using greenhouses, polyhouses, and low tunnels allows farmers to grow high-value crops year-round, shielded from extreme weather and pests. These structures are increasingly adopted in Uttarakhand's hill districts, enabling the production of crops like cucumbers and roses even during winter (Kumar and Gupta 2019). Alongside, organic farming practices—such as composting, vermiculture, and the use of microbial biofertilizers—are gaining ground, reducing reliance on chemical inputs and meeting rising consumer demand for safe, residue-free produce. Digital platforms and e-marketplaces are helping farmers connect directly with buyers, improving price realization and reducing postharvest losses. Government initiatives are also supporting the adoption of advanced technologies, with significant investments in precision farming, protected cultivation, and organic agriculture. These technological advancements are not just modernizing tradition—they are creating resilient, resource-efficient, and market-responsive agri-horticulture systems, positioning Indian agriculture for a sustainable and prosperous future.

3 Future Prospects and Development Strategies

Uttarakhand's agri-horticulture sector stands at a crossroads, facing both significant challenges and promising opportunities. While climate change and declining yields of traditional temperate fruits like apples, pears, and plums have cast uncertainty over the sector, there is a clear path forward through diversification, innovation, and infrastructure development. Shifting toward high-value and climate-resilient crops—such as kiwis, dragon fruit, strawberries, and medicinal and aromatic plants (MAPs)—can help buffer farmers against unpredictable weather and market fluctuations. As warming temperatures make some temperate fruits less viable, adopting varieties and species suited to new climatic realities is essential (FAO 2022). Investment in cold storage, improved rural roads, and decentralized food processing facilities is critical to reduce postharvest losses and improve market access, especially for perishable horticultural produce. Grower cooperatives and self-help groups can play a vital role in collective marketing and value addition, turning waste into new revenue streams and empowering rural communities (Singh and Trivedi 2018).

Leveraging digital platforms and mobile apps to connect farmers directly with buyers can maximize returns, especially for time-sensitive crops. Precision agriculture, improved planting material, and climate-smart irrigation systems should be prioritized to boost productivity and sustainability. Continued investment in agricultural research is needed to develop and disseminate climate-resilient crop varieties, innovative organic practices, and adaptive management techniques. Hands-on training, extension camps, and agromet advisory services will help farmers adopt new technologies and respond proactively to changing weather patterns. Proactive and robust policy interventions, including climate financing, are essential to safeguard the sector from future risks (ICAR 2021). Coordination between research institutions, policymakers, and local communities will be key to translating scientific breakthroughs into practical benefits for farmers.

4 Conclusion

Agri-horticultural system remains a cornerstone of Uttarakhand's rural economy and cultural identity, intricately integrating grains, fruits, medicinal plants, and spices into resilient, diverse systems adapted to the state's varied agroecological zones. These integrated systems are not only vital for sustaining livelihoods but also play a critical role in nutritional security, ecological balance, and cultural heritage preservation. Despite growing pressures from climate change, including erratic rainfall, rising temperatures, and increased pest incidences alongside challenges such as shrinking landholdings and rural-to-urban migration, Uttarakhand's agro-ecological diversity and deep-rooted tradition of mixed farming provide a robust foundation for renewal and innovation. The path forward necessitates bold, coordinated action encompassing diversification of crop portfolios to include climate-resilient and high-value horticultural crops, modernization of rural infrastructure such as cold storage, postharvest processing, and supply chains, adoption of smart agricultural technologies, including precision farming, drone monitoring, and organic inputs, and strengthening market linkages through cooperative models and e-commerce platforms. Enhanced investment in research and extension services tailored to the unique Himalayan context, combined with effective policy support focusing on subsidies, insurance, and skill development, will empower farmers to improve productivity and income sustainably. Moreover, nurturing community participation and integrating traditional knowledge with scientific innovations will ensure the social acceptability and long-term sustainability of these agri-horticultural systems. Thus, with continued commitment, Uttarakhand has the opportunity to transform its agri-horticultural landscape into a model of sustainable rural development, where every hill and valley not only survives but thrives showing a legacy of sustainability, prosperity, and resilience for generations to come, bolstered by improved yields, enhanced farmer incomes, and strengthened food and nutritional security.

References

Anonymous (2022) Horticulture overview. Published on November 18th, 2022, retrieved on May 3rd 2025, available via https://agriwelfare.gov.in/en/Hirticulture

Anonymous (2023a) Key role of horticulture in doubling farmers income and ensuring essential nutritional security. Ministry of Agriculture & Farmers Welfare, Government of India, available via https://pib.gov.in/PressReleasePage.aspx?PRID=1901482

Anonymous (2023b) India's horticulture sector holds untapped potential despite challenges. Last updated on April 25th, 2023, available via https://economictimes.indiatimes.com/news/economy/agriculture/indias-horticulture-sector-holds-untapped-potential-despite-challenges/articleshow/99763533.cms

Bijalwan A (2014) Alteration of tree species in traditional agri-silvi-horticulture systems along with altitude and aspects of Garhwal Himalaya, India. Int J Agrofor Silvic 1:37–51

Bijalwan A, Upadhyay AP, Dobriyal MJR (2015) Tree crop combinations, biomass and carbon sequestration in conventional agrisilviculture (agroforestry) system along altitude and aspects in the hills of Uttarakhand Himalaya, India. Int J Curr Res Biosci Plant Biol 2:214–217

Colmenares OM, Brindis RC, Verduzco CV, Grajales MP, Gómez MU (2020) Horticultural agroforestry systems recommended for climate change adaptation: a review. Agric Rev 41(1):14–24

Fanish SA, Priya RS (2013) Review on benefits of agro forestry system. Int J Educ Res 1:1–12

FAO (2022) The future of food and agriculture: trends and challenges. Food and Agriculture Organization, Rome

Gairola Y, Biswas S (2008) Bioprospecting in Garhwal Himalayas, Uttarakhand. Curr Sci 94:1139–1143

Goutam E, Rana DK, Shah KN, Singh V, Bharti, Choudhary R (2025) Impact of foliar application of plant bio-regulators on vegetative, f loral and yield traits in bottle gourd (*Lagenaria siceraria*) undervalley condition of Garhwal hills. Vegetos:1–8. https://doi.org/10.1007/s42535-025-01297-9

Government of India (2023) Economic survey 2022–23. Ministry of Finance, New Delhi

Government of India (2024) Agricultural roadmap 2030. Ministry of Agriculture and Farmers Welfare, New Delhi

Government of Uttarakhand (2022) Uttarakhand horticulture policy 2022. Department of Horticulture, Dehradun

Grow Billion Trees (2024) Agroforestry in Uttarakhand: Balancing ecology and economy in the Himalayas. Retrieved on September 5th,2025, available via: https://growbilliontrees.com/pages/agroforestry-in-uttarakhand-balancing-ecology-and-economy-in-the-himalayas

ICAR (2021) Vision 2050: Indian agriculture. Indian Council of Agricultural Research, New Delhi

Indian Council of Forestry Research and Education (ICFRE). Agroforestry models developed by ICFRE. 2024. Retrievedon September 5th, 2025, available via: https://hindi.icfre.gov.in/UserFiles/File/Books/Agroforestry%20Models%20Developed%20by%20ICFRE.pdf

IPCC (2021) Climate change 2021: the physical science basis. Cambridge University Press

Jat RA, Shivran AC (2020) Precision farming: concepts and applications. Agrotech Publishing Academy, Udaipur

Kumar V, Gupta RK (2019) Protected cultivation technologies. CRC Press

Kumar S, Bijalwan A, Singh B, Yewale AG, Rawat D, Rana K, Riyal MK (2020) Diversity parameters of *Quercus leucotrichophora* based agroforestry systems and pure forest in Garhwal Himalaya. Indian J Agrofor 22:62–70

Kumar S, Bijalwan A, Singh B, Rawat D, Yewale AG, Riyal MK, Thakur TK (2021) Comparison of carbon sequestration potential of *Quercus leucotrichophora*–based agroforestry systems and natural forest in central Himalaya, India. Water Air Soil Pollut 232:350

Ministry of Agriculture and Farmers Welfare (2020) National Horticulture Mission (NHM): annual report 2019–20. Government of India, New Delhi

Nair KP (2019) Sustainable agriculture: principles and practices. Springer Nature

National Horticulture Board (NHB) (2022) Indian horticulture database 2021–22. Ministry of Agriculture and Farmers Welfare, New Delhi

Negi SS (2005) Traditional farming systems of Uttarakhand. Bishen Singh Mahendra Pal Singh, Dehradun

Pandey A, Tiwari P (2022) Potential of horticulture crops and trees in agroforestry systems in India. Int J Agric Sci 13:106–113

Planning Commission (2018) Rashtriya Krishi Vikas Yojana (RKVY): Guidelines and impact. NITI Aayog, New Delhi

Pretty J (2008) Agricultural sustainability: concepts, principles, and evidence. Philos Trans R Soc B 363:447–465

Rasul G (2014) Food, water, and energy security in South Asia: a nexus perspective from the Hindu Kush Himalayan region. Environ Sci Pol 39:35–48

Rawat AJS, Kumar P (2018) Horticulture development in Uttarakhand: prospects and challenges. Uttarakhand State Council for Science and Technology (UCOST), Dehradun

Robinson GM (2024) Global sustainable agriculture and land management systems. Geogr Sustainability 5(4):637–646

Sati A, Negi RS, Butola JS, Singh S (2025) Hill farming in Uttarakhand: navigating challenges, embracing opportunities. Bio-Think 2:14–21

Sharma SK, Singh A (2017) Organic farming and bio-fertilizers in India. Scientific Publishers, Jodhpur

Singh HP, Singh G (2010) Horticulture for sustainable income and employment. Westville Publishing House, New Delhi

Singh B, Trivedi PC (2018) Agri-horticulture supply chain management in India. AgriBio Publishers, Jaipur

Singh KK, Singh SP, Uniyal A (2018) Organic farming. In: Sethi VK, Bharti PK (eds) Organic farming and biofertilizers. Discovery Publication House Delhi, India, pp 11–20

Singh KK, Meena RK, Singh SP, Chauhan JS (2021) Approaches to relieve a worldwide global warming through agriculture. In: Uniyal A, Sharma I (eds) Trends in agriculture: traditional and modern approaches. AgrobiosResearch: an imprint of Agrobios, Jodhpur (India), pp 279–289

Singh P, Nayak SL, Sharkar R, Sarkar A, Dutta S, Paul D (2025) Application of modern techniques in cultivation of medicinal plants: a review. Plant Arch 25(1):1947–1962

Subburathinam P (2024) Transformative trends and employment in agriculture in India. Team Lease Services Limited. Retrieved on May 3rd 2025, available via https://group.teamlease.com/employment-in-agriculture-in-india/

Swain M, Kalamkar SS (2016) Agricultural marketing and infrastructure in India. Allied Publishers, New Delhi

Tiwari PC, Joshi B (2015) Local and regional institutions and environmental governance in Hindu Kush Himalaya. Environ Sci Policy 49:66–74. Frames on the move: Regional Governance in Mountain Areas

Toppo P, Toppo S (2019) Tree crop interaction in agroforestry system: a review. Int J Chem Stud 27:2359–2361

Uttarakhand Open University (2024) Social forestry and agro-forestry. 2024. Retrieved on September 5th, 2025. Available at: https://uou.ac.in/sites/default/files/slm/FR-06.pdf

Verma N, Yadav RK (2023) Medicinal and aromatic plant farming: a sustainable model for agro-ecology, livelihoods and innovation. J Emerg Technol Innov Res 12(11S):a438

Yadav RP, Gupta B, Bhutia PL, Bisht JK, Pattanayak A (2019) Sustainable agroforestry systems and their structural components as livelihood options along elevation gradient in central Himalaya. Biol Agric Hortic 35(2):73–95. https://doi.org/10.1080/01448765.2018.1457982

Chapter 3
Home Gardens and Their Role in Sustainable Development of Indian Himalayas

Abhishek Pratap Singh, S. K. Verma, and Sumit Tewari

Abstract Home gardens (HGs) are considered human-managed microenvironments characterized by complex structural composition and diverse functions, hosting significant species and genetic diversity within a compact farm unit. They play a crucial role in promoting sustainable development in the Indian Himalayas Region (IHR), where delicate ecosystems, biodiversity, and traditional ways of living are under threat due to climate change and unsustainable land use. Studying the potential of HGs in the context of Sustainable Development Goals (SDGs) is crucial as they appear to be an important strategy for attaining the SDGs and can be accomplished with proper planning, in addition to taking into consideration how the traditional societies have sustained it for long. HGs have significant potential to contribute to the SDGs, as they offer a range of social, economic, and environmental benefits. To better understand their role, metrics like nutrition and yield per hectare should be improved to assess how much nourishment HGs can provide. As small-scale systems, HGs can empower women and marginalized groups by involving them in production processes. Their potential to enhance livelihood security is another important aspect. Additionally, the development of public policies to conserve traditional HG systems is essential.

Keywords Home garden · SDGs · Rural livelihoods · Agrobiodiversity and climate resilience

A. P. Singh (✉) · S. K. Verma
Department of Silviculture and Agroforestry, Acharya Narendra Deva University of Agriculture and Technology, Ayodhya, Uttar Pradesh, India

S. Tewari
Department of Silviculture and Agroforestry, College of Forestry, Veer Chandra Singh Garhwali Uttarakhand University of Horticulture and Forestry, Ranichauri, Tehri Garhwal, Uttarakhand, India

B. Singh et al. (eds.), *Himalayan Agroforestry Systems*, Advances in Agroforestry 15,
https://doi.org/10.1007/978-3-032-25096-4_3

1 Introduction

Among the world's mountain systems, the Himalayas stand out as the most intricate and diverse, acting as a natural divide between the northern Asian continent and South Asia. As a distinct geographical and ecological unit, this region plays a crucial role in the Earth's biophysical dynamics. Stretching over 2500 km in length, with a width ranging from 80 to 300 km and elevations exceeding 8000 m above sea level, the Himalayas have a unique climatic system that significantly influences weather patterns across much of Asia (Zobel and Singh 1997; Nyaupane and Timothy 2022). The region's varied topography results in diverse climatic conditions and habitat types. Differences in geological formation over time have led to notable variations in climate and physiography, ultimately shaping the distribution of biodiversity. The complex biogeographical patterns of the region are recognized for their spatial configuration and the heterogeneous distribution of flora and fauna (Huang et al. 2019). The Eastern Himalayas, including Northeast India, are home to approximately 8000 species of flowering plants, earning recognition as a biodiversity hotspot, whereas the Western Himalayas support over 5000 species (Singh 2006). Overall, the Indian Himalayan Region (IHR) contains nearly 50% of India's total flowering plant species, with 30% being endemic (Bargali et al. 2022). Additionally, the IHR is rich in tree diversity, with 18,440 plant species (Singh et al. 2019); 675 edible plants (Samant and Dhar 1997); 118 plant species known for yielding essential oils (Samant and Palni 2000); and 155 sacred plant species (Samant and Shreekar Pant 2003). Given the increasing threats to biodiversity, effective conservation and sustainable use of these resources could provide significant economic benefits to local communities while promoting sustainable development (Khoshoo 1992; Dhar 1997).

Referred to as the "water tower of the earth," the Himalayas have approximately 10–20% of their landmass covered by glaciers, with 30–40% experiencing seasonal snow cover (Bahadur 2004), which varies between 0.48 ± 0.43 and 2.20 ± 1.25 million km^2. Despite its vast water resources generating an annual river flow of 1,200,000 million m^3, the region is witnessing changes, including a decline in the regulatory functions of glaciers, streams, and rivers (Singh 2006). The forests in the Himalayas are a crucial source of biodiversity, supporting the livelihoods of local populations. The biomass value of oak forests in Central Himalaya, ranging between 545 and 782 t ha^{-1} year^{-1}, reflects the region's ecological significance (Singh 1992; DST 2010). Additionally, the Himalayas act as a carbon sink, storing approximately 5.4 billion tons of carbon and sequestering about 65 million tons per year. The carbon sequestration values for the Western, Central, and Eastern Himalayan regions are 3697.05, 3898.10, and 4235.05 tons per hectare, respectively, highlighting the crucial ecosystem services provided by Himalayan forests (Tolangay and Moktan 2020). Beyond their ecological significance, the Himalayas attract global visitors due to their breathtaking landscapes, cascading rivers and streams, diverse cultural and religious traditions, and vibrant festivals of indigenous communities. Whether for nature enthusiasts, tourists, or spiritual seekers, the region holds immense appeal (Goyal et al. 2024).

Recognizing the critical role of mountain ecosystems in supporting both upland and downstream communities, global efforts for sustainable development have increasingly emphasized the need for mountain-specific strategies. While addressing local aspirations and national development priorities is essential, development efforts must be carefully planned to avoid excessive resource exploitation and environmental degradation. Given the fragile and vulnerable nature of Himalayan ecosystems, a unique approach to development is necessary (Jodha 1992; Veith and Shaw 2011). Unregulated interventions that disregard the delicate balance of the region can lead to negative consequences, such as landslides, river obstructions, flash floods in the mountains, and recurrent flooding in downstream areas. These adverse effects of poorly planned development are already evident (Nagamani et al. 2024). In addition to local development challenges, climate change—exacerbated by unsustainable practices and waste generation—poses a severe threat to the Himalayan ecosystem. Global warming and glacial retreat have long-term and widespread implications (Kattel 2022). The increasing vulnerability of mountain environments to human activities and climate-induced changes is a growing concern. Consequently, forums such as the World Summit on Sustainable Development (WSSD) and the Sustainable Mountain Development Summit (SMDS-XI) have emphasized the need for specialized approaches and dedicated resources to sustain livelihoods and enhance the quality of life in mountain regions. An integrated approach that accounts for the interdependence of human sociocultural and economic systems with natural ecosystem processes is essential for sustainable development (Tse-ring et al. 2012).

The IHR holds a distinctive position within global mountain ecosystems. As a geodynamically young range, it is vital for climate regulation and serves as a crucial water source for much of the Indian subcontinent. Additionally, it harbors rich biodiversity, diverse human communities, and cultural heritage (Xu et al. 2019). Despite the abundance of natural resources, many inhabitants of the region remain marginalized, relying on subsistence living. Unsustainable exploitation of these resources is leading to environmental degradation and increasing the severity of natural hazards. To balance economic interests with ecological imperatives, it is imperative to adopt a new paradigm that integrates scientific conservation principles with sociocultural considerations (Singh 2006). Home gardens (HGs) play a crucial role in promoting sustainable development in the Indian Himalayas, where delicate ecosystems, biodiversity, and traditional ways of living are under threat due to climate change and unsustainable land use. These small yet multifunctional agroecosystems combine food production with ecological conservation, serving as a viable strategy for sustainability in this ecologically vulnerable region. HGs are considered human-managed microenvironments characterized by high-energy input, complex structural composition, and diverse functions, hosting significant species and genetic diversity within a compact farm unit (Das and Das 2005; Vibhuti et al. 2018; Pala et al. 2019). They are an integral part of the rural poor's livelihoods and economy (Reang et al. 2021; Roy et al. 2022a, b), ranking second only to shifting cultivation as one of the earliest food production systems (Kumar and Nair 2004).

Fernandes et al. (1984) described HGs as a land-use system encompassing a spectrum of practices ranging from vegetable cultivation to intricate multistoried setups surrounding a residence. According to FAO, an HG is a "small-scale, supplementary food production system maintained by household members that mirrors a natural, multilayered ecosystem" (Hoogerbrugge and Fresco 1993). One of the most widely accepted definitions describes HGs as "intimate, multi-story combinations of various trees and crops, sometimes associated with domestic animals, situated around homesteads" (Kumar and Nair 2006; Sharma et al. 2022). These systems are frequently recognized as traditional agroforestry systems with intricate structures and diverse functionalities, intentionally integrating the conservation of multipurpose vegetation alongside agricultural crops and livestock within individual households (Fig. 3.1). They also facilitate the preservation of forest species alongside semi-domesticated plants and are maintained by household members (Rahman et al. 2013). One of the defining characteristics of HGs is their rich species diversity, directly benefiting the household (Bargali 2015). The sustained cultivation of both domesticated and wild plant species makes them valuable for in situ conservation of plant genetic resources (Bargali et al. 2015a). The year-round cultivation of multiple crops enables households to meet both subsistence and commercial needs while ensuring resilience against crop failures and market price fluctuations (Sahoo and

Fig. 3.1 Components of home garden

Rocky 2015). Additionally, crop diversification supports the conservation of plant biodiversity through continuous utilization (Abdoellah et al. 2006). The close interconnection between herbaceous and woody plant species facilitates nutrient recycling and mitigates soil erosion risks (Bargali 2016).

HGs offer a consistent and varied supply of socioeconomic benefits to the households managing those (Amberber et al. 2013). Their key attributes include biodiversity conservation, livelihood enhancement, efficient nutrient cycling due to their multispecies and multistrata composition, minimal reliance on external inputs, improved household income, and nonmarket benefits such as cultural and social significance, including gender-equal management (Kumar and Nair 2004; Bargali 2015; Putri et al. 2016). From a genetic and species perspective, HGs serve as in situ conservation sites, providing breeding grounds for a diverse plant community (Idohou et al. 2014). Their high production diversity and staggered harvesting approach allow households to meet environmental, economic, and social needs throughout the year. Wiersum (2006) suggested that a diverse assortment of useful plant species within an HG enhances adaptability to changing socioeconomic conditions while ensuring resource sustainability for current and future generations (Bargali 2016). These systems create a vegetation mosaic that functions as biological corridors within landscapes, fostering favorable habitats for various species (Perfecto and Vandermeer 2008). Their role in enhancing the availability of plant resources while simultaneously preserving and rehabilitating forest vegetation is well-documented (Noble and Dirzo 1997; Moreno-Calles et al. 2010). HGs are often characterized by their high diversity, which distinguishes them from conventional agroecosystems (Peyre et al. 2006; Sunwar et al. 2006; Tiyakoat et al. 2010). Compared to cultivated agroecosystems, gardening practices are more diverse and can play a role in conserving nearby natural forests (Sampanpanish and Jamroenprucksa 1994; Gajaseni and Gajaseni 1999).

In India, home gardening is a critical subsistence strategy for marginal and tribal communities in both rural and urban areas, providing essential resources such as food, vegetables, medicine, and economic support, particularly in the Himalayan regions (Sahoo and Rocky 2015; Bargali 2016). Several wild plant and crop species have been domesticated in these regions (Saikia et al. 2012; Padalia et al. 2015; Barbhuiya et al. 2016). Research on HGs in India has largely focused on Kerala (Kumar et al. 1994), Assam (Saikia et al. 2012), Karnataka (Shastri et al. 2002), and Northeast India (Das and Das 2005). However, studies on home gardening in the Indian Himalayas remain limited. To address this research gap, the present study aims to examine the role of HGs in fostering sustainable development in the IHR.

2 Significance of Home Gardens in Sustainable Development

Sustainable development has gained prominence as a concept that emphasizes the need to balance socioeconomic progress with environmental health (Agbedahin 2019). The Sustainable Development Goals (SDGs) have expanded to encompass not only people, planet, and prosperity but also peace and partnership, integrating the economic, social, and environmental aspects of sustainability (Sachs et al. 2019). While implementing SDGs can be challenging due to potential conflicts and differing perspectives among stakeholders, they have become a cornerstone of policy frameworks. Governments, researchers, civil society, and the private sector, among others, are collaborating to discuss and implement these goals (Caiado et al. 2018). This collaborative approach fosters a global vision of human development, where diverse indicators are seen as complementary and mutually reinforcing (Hickel 2020). However, the effectiveness of global or regional SDGs frameworks depends on their adaptation to local contexts, considering the unique characteristics of socioecological systems (Jiménez-Aceituno et al. 2020). Resilience, a key attribute of sustainable systems, supports this implementation by enabling reorganization after disruptions (Carpenter et al. 2012) and by enhancing ecosystem services that contribute to human well-being and health (Nelson et al. 2009).

With the global demographic shift toward urbanization, cities are becoming critical arenas for SDGs implementation (Leote et al. 2022). Currently, 54% of the world's population resides in urban areas, a figure expected to rise to 59% by 2030. To achieve the SDGs, governments must address the direct and indirect impacts of urban growth (Nathaniel and Adeleye 2021). Urban and suburban green spaces—including public gardens, urban forests, private backyards, and rural HGs in monocultural or industrial agricultural landscapes—should be recognized as essential infrastructures for building sustainable societies (Gavrilidis et al. 2019). These spaces provide vital ecosystem services that are increasingly scarce (Vieira et al. 2018). For instance, HGs, which often occupy more land than commercial crops or natural habitats in many countries, play a significant role in land use (Dewaelheyns et al. 2014). Despite their spatial importance, rural HGs are often underrepresented in traditional mapping studies, though research highlights their prevalence across tropical and temperate regions (Caballero-Serrano et al. 2016; Calvet-Mir et al. 2016; Ferdous et al. 2016; Ivanova et al. 2021; Buechler 2016). However, these gardens are declining due to abandonment and the expansion of large-scale intensive agriculture (Mwavu et al. 2016; Zimmerer et al. 2015; Giller et al. 2021). Despite their small size and noncommercial nature, HGs are vital components of socioecological systems, offering a range of social, economic, and environmental benefits (Mellisse et al. 2018; Rammohan et al. 2019). Studies have documented their contributions to food production, income generation, physical activity, well-being, and fostering a connection with nature (Galluzzi et al. 2010; Galhena et al. 2013; Diekmann et al. 2020). Additionally, networks of HGs can enhance resilience during disruptions by promoting social equity and justice (Porter 2018; Sanyé-Mengual et al. 2018). They also support biodiversity and nature conservation, though the complex factors influencing species diversity and abundance require further exploration (Raymond et al. 2019; Filazzola et al. 2019; Diduck et al. 2020).

HGs have significant potential to contribute to the SDGs, particularly in areas such as food security, agrobiodiversity conservation, and sustainable food systems (SDG 2). To better understand their role, metrics like nutrition and yield per hectare should be improved to assess how much nourishment HGs can provide (Cassidy et al. 2013). As small-scale systems, HGs can empower women (SDG 5) and marginalized groups (SDG 1, SDG 8) by involving them in production processes. Their potential to enhance livelihood security (SDG 2, SDG 8) is another critical area requiring research. Additionally, the development of public policies (SDG 16) to conserve traditional HG systems is essential. While the ecosystem services provided by HGs are well-documented, policy frameworks for integrating these systems into broader food systems remain limited. A comprehensive analysis of the economic, social, and ecological impacts of HGs is necessary to fully appreciate their value. Although HGs are predominantly found in tropical regions, they are gaining popularity in temperate areas of North America and Europe. For example, the European Agroforestry Federation (EURAF 2021) recognizes HGs as one of five agroforestry practices, alongside silvoarable systems, silvopasture, riparian buffers, and forest farming, reflecting practices similar to those in North America (Sharma et al. 2022).

3 Role of Home Gardens in Sustainable Development of Indian Himalayas

HGs play a pivotal role in the sustainable development of the Indian Himalayas by promoting ecological balance, food security, and socioeconomic well-being. These small-scale, diverse agroforestry systems integrate trees, crops, and livestock, enhancing biodiversity and soil fertility while mitigating climate change impacts through carbon sequestration. By integrating traditional practices with modern agroecological techniques, HGs serve as a model for achieving SDGs in the IHR.

3.1 Carbon Sequestration Potential of Home Gardens

Carbon sequestration is the process of removing excess carbon from the ecosystem and storing it in a reservoir (UNFCCC 2007). This method captures and retains atmospheric carbon dioxide over long periods, helping to mitigate climate change. In plants, carbon sequestration occurs through photosynthesis, where CO_2 is converted into biomass and stored in "sinks" rather than being released into the atmosphere. Additionally, plants contribute to soil carbon storage by adding organic matter such as decomposing roots, root exudates, and other organic residues (Namitha et al. 2022). Carbon sequestration plays a crucial role in climate change mitigation by transforming atmospheric carbon into plant biomass and soil organic carbon. Trees, in particular, store more carbon in their biomass compared to other plant types, making perennial plantations essential for reducing climate-related adversities. Among agroforestry systems, HGs stand out for their high carbon sequestration potential. These systems not only capture carbon in both biomass and soil but also help reduce fossil fuel consumption by promoting firewood use and preserving agricultural biodiversity (Kumar and Nair 2004; Namitha et al. 2022). The effectiveness of HG agroforestry in carbon sequestration depends on biologically driven CO_2 absorption and its conversion into long-lasting carbon-containing compounds, a process known as bio-sequestration (USDOE 2008; Vibhuti et al. 2020). This process helps conserve natural forests by reducing human pressure on these ecosystems (Kumar 2006).

Climate change mitigation and carbon sequestration potential of HG with reference to IHR is presented in Table 3.1. HGs maintain a relatively stable biomass pool since biomass removal is minimal, making them a sustainable system. These gardens serve as significant carbon sinks, capable of storing up to 20.28 t C ha^{-1}. With effective management, they can enhance local, national, and global efforts in carbon sequestration (Dey et al. 2014). Research by Roshetko et al. (2002) on 19 HGs (12–17 years old) found that, in terms of aboveground biomass, HGs stored approximately 34.7 Mg C ha^{-1} more (58 times greater) than Imperata-cassava systems and 21 Mg C ha^{-1} more (1.5 times greater) than young rubber agroforests. Their projections suggested that, depending on management practices, the time-averaged aboveground carbon stocks in HGs could range from 30 to 123 Mg C ha^{-1}. In the

Table 3.1 Carbon sequestration potential of home gardens

Region	Carbon sequestration (Mg C ha^{-1}/ *Mg C ha^{-1} year^{-1})		References
	Tree/crop biomass carbon	SOC	
Eastern Himalayan Region, India	55.18	52.15	Panwar et al. (2022)
	46.49 (old home gardens) 35.62 (young home gardens)	59.07 (old home gardens) 42.63 (young home gardens)	Singh et al. (2018)
	–	108.83–258.43	Singh and Sahoo (2021)
	–	17.72 (0–10 cm) 15.16 (10–20 cm) 13.94 (20–30 cm)	Manpoong et al. (2021)
	6.51–8.95 1.20–1.46*	–	Nath and Das (2011)
	–	10.45	Koul et al. (2011)
	0.112	–	Devi et al. (2019)
Central Himalaya Region, India	–	22.56–81.51	Vibhuti et al. (2020)
	32.60	–	Karki et al. (2024)
Western Himalayan Region, India	29.17	–	Dar et al. (2019)
	22.56	–	Saleem et al. (2023)
	44.53–104.86	–	Sarkar et al. (2023)

agroforestry systems of the Kumaun Himalayan region in Uttarakhand, Karki et al. (2024) reported the highest soil carbon sequestration (SCS) in HG systems (32.60 ± 4.24 Mg C ha^{-1}), followed by the agrihorticulture system (28.47 ± 2.46 Mg C ha^{-1}) and the agrisilviculture system (28.03 ± 3.35 Mg C ha^{-1}), with the lowest soil carbon sequestration observed in the agrihortisilviculture system (25.16 ± 2.29 Mg C ha^{-1}). Similar findings were also reported by Nair et al. (2009). Panwar et al. (2022) investigated biomass production and carbon sequestration potential across various agroforestry systems in different regions of India. Their findings indicated that HGs had the highest total biomass carbon (55.18 Mg C ha^{-1}) compared to other agroforestry systems in the eastern and northeastern regions. Additionally, HGs exhibited the greatest belowground biomass (34.68 Mg ha^{-1}) among all agroforestry systems, irrespective of location. The total biomass followed the order: HG>agrihorticulture>silvipasture>agrisilviculture> boundary plantations. The mean total biomass carbon (TBC) across different agroforestry systems ranged from 25.24 to 52.98 Mg C ha^{-1}, with the highest values recorded in HGs and the lowest in boundary plantations. Notably, HGs in northeastern India demonstrated a superior capacity for carbon sequestration in both biomass and soil. Among various land uses in agricultural landscapes, HGs and block plantation agroforestry systems exhibited the highest carbon content, contributing significantly to net carbon stock gains. These systems, along with plantation crop-based agroforestry, hold substantial potential for biomass production, biomass carbon accumulation, and soil organic carbon (SOC) sequestration (Montagnini and Nair 2004; Chauhan et al. 2019; Panwar et al. 2017; Das et al. 2019; Koul et al. 2011).

Singh et al. (2018) examined the impact of land-use changes on carbon stock dynamics, estimating tree biomass carbon and soil organic carbon for both old (46.49 and 59.07 Mg C ha^{-1}, respectively) and young HGs (35.62 and 42.63 Mg C ha^{-1}, respectively). Their study also found a positive rate of carbon stock change when current Jhum was converted into young HGs. However, the transition from old HGs to oil palm plantations resulted in a negative carbon stock change rate of −6.67 Mg C ha^{-1} year^{-1}. Similar findings were reported by Guo and Gifford (2002). Kumar (2011) reported that aboveground carbon stocks in tropical HGs of Kerala ranged from 16 to 36 Mg ha^{-1}, with smaller HGs generally storing more carbon per unit area than medium- and large-sized ones. The soil carbon content within a 1-meter profile ranged from 101.5 to 127.4 Mg ha^{-1}. Saha et al. (2009) also found that smaller HGs (less than 0.4 ha) with higher tree density and greater plant species richness have more soil carbon per unit area (119.3 Mg C ha^{-1}) compared to larger HGs, which have a lower soil carbon content of 108.2 Mg C ha^{-1}. The small HGs and rubber plantation exhibited higher soil organic carbon (SOC) levels compared to the coconut plantation. SOC in small HGs was 30% and 114% higher than in coconut plantations and rice paddies, respectively, while in large HGs, SOC was 18% and 94% higher than in these land uses. Although HGs had 10% more SOC than large HGs, this difference was not statistically significant. Overall, SOC content within a 1 m soil depth followed the order: Forest > Small HG = Rubber ≥ Large HG ≥ Coconut > Rice Paddy (Saha et al. 2010). Similar trends were also observed by Singh and Sahoo (2021), Mattsson et al. (2018), and Tilinti et al. (2025).

Vibhuti et al. (2020) investigated how size and altitude influence soil organic carbon stock (SOCS) in home garden agroforestry systems in the Central Himalayas. Their findings revealed that SCS varied between 22.56 and 81.51 Mg C ha^{-1}, with an increasing trend at higher altitudes for small home gardens. In contrast, SCS in medium and large home gardens increased up to mid-altitude before declining at higher elevations. Across all altitudes, medium-sized home gardens generally exhibited the highest SCS (38.56–66.16 Mg C ha^{-1}), except at high altitudes, where small home gardens recorded the maximum value (66.01 Mg C ha^{-1}).

3.2 *Home Gardens for Biodiversity Conservation*

HGs are a key component of urban green infrastructure worldwide, with their potential to support biodiversity gaining increasing recognition (Loram et al. 2007; Goddard et al. 2010; Delahay et al. 2023). Traditionally, HGs have played a crucial role in conserving diverse crop varieties, contributing significantly to agricultural biodiversity. Households cultivate a wide range of plant species primarily to meet their daily needs, ensure food security, and generate income. These gardens are typically arranged in a multilayered vertical structure with three to four strata. While most species are deliberately cultivated, wild plants are also preserved for specific purposes, such as fuel and animal fodder (Sarkar et al. 2020). The practice of cultivating fruits, vegetables, and ornamental plants in HGs has been deeply rooted in the traditions of northeast India, particularly in the hilly regions of Uttarakhand, Assam, Manipur, Nagaland, and Meghalaya (Vibhuti et al. 2019). As urbanization expands, the role of HGs in biodiversity conservation is expected to become even more significant. However, in densely populated areas, the pressure on space may lead to the reduction or loss of HGs. Recognizing their value in enhancing biodiversity and delivering essential ecosystem services, global policy frameworks—such as the United Nations Sustainable Development Goals—have increasingly integrated HGs into urban planning and design (Beumer 2018). While the contribution of HGs in rural areas remains less well-documented, evidence suggests they also provide important ecosystem services, benefiting adjacent agricultural landscapes (Samnegard et al. 2011). In recent decades, growing media attention and public interest—driven by conservation organizations and gardening groups—have promoted "wildlife-friendly" gardening. This has led to increased recognition of garden biodiversity in both local and national government policies (Goddard et al. 2010; Loram et al. 2011). Consequently, urban ecologists, planners, and policymakers must engage with the current body of research to develop effective strategies for safeguarding and enhancing the biodiversity benefits of residential gardens (Delahay et al. 2023).

The plant diversity found in home gardens of the Indian Himalayan Region is detailed in Table 3.2. The HGs in the Eastern Himalayan Region of West Bengal host a rich diversity of plant species, comprising 262 species across 88 families and 211 genera. Herbs are the most dominant group, with 93 species spanning 34

Table 3.2 Plant diversity in home gardens of the Indian Himalayan Region

Location	Plant types	Species	References
West Bengal	Trees	*Areca catechu*, *Litsea monopetala*, *Ailanthus integrifolia*, *Alnus nepalensis*, *Ficus auriculata*, *Persea americana*, and *Ficus neriifolia*	Sarkar et al. (2020)
	Shrubs	*Citrus aurantium*, *Nyctanthes arbor-tristis*, *Abroma augusta*, *Codiaeum variegatum*, *Euphorbia pulcherrima*, *Leucosceptrum canum*, *Musa* spp., etc.	
	Herbs	*Brassica juncea* var. *rugosa*, *Colocasia esculenta*, *Raphanus raphanistrum*, *Chrysanthemum morifolium*, *Tagetes erecta*, *Cucurbita maxima*, *Lycopersicon esculentum*, etc.	
	Others	*Sechium edule*, *Cucurbita maxima*, *Momordica subangulata*, *Phaseolus vulgaris*, *Vigna umbellate*, etc.	
Kumaun Himalaya	Trees	*Mangifera indica*, *Artocarpus hetrophyllum*, *Bombex ceiba*, *Litchi chinensis*, *Syzygium cumini*, *Mallotus philippensis*, *Carica papaya*, *Ficus hispida*, *Cinnamomum tamala*, *Citrus limetta*, *Malus domestica*, *Prunus dulis*, *Sarcaasoca*, *Achras sapota*, *Morus alba*, *Elaeocarpus ganitrus*, *Punica granatum*, *Psidium guajava*, *Prunus persica*, etc.	Vibhuti et al. (2019)
	Shrubs	*Murraya koenigii*	
	Herbs	*Ageratum conyzoides*, *Cynodon dactylon*, *Curcuma longa*, *Galinsoga parviflora*, *Stellaria media*, *Oxalis corniculata*, *Poa* spp., *Zingiber officinale*, etc.	

(continued)

Table 3.2 (continued)

Location	Plant types	Species	References
Northeast India	Trees	*Aegle marmelos*, *Albizia odoratissa*, *Albizia procera*, *Anacardium occidentale*, *Anogeissus acuminate*, *Aralia foliosa*, *Artocarpus chama*, *Artocarpus heterophyllus*, *Averrhoa carambola*, *Azadirachta indica*, *Baccaurea ramniflora*, *Bauhinia variegate*, *Bruinsmia polysperma*, *Callistemon lanceolatus*, *Carallia brachiata*, *Cassia fistula*, *Cassia tora*, *Castanopsis tribuloides*, *Celtis australis*, *Cinnamomum tamala*, *Citrus grandis*, *Citrus macroptera var anamensis*, *Clerodendrum colebrookianum*, *Clerodendron serratum*, *Delonix regia*, *Dillenia pentagyna*, *Elaeocarpus floribundus*, *Emblica officinalis*, *Erythrina indica*, *Eurya simplocina*, *Ficus elastic*, *Ficus hirta*, *Ficus hispida*, *Ficus racemosa*, *Garcinia lanceaefolia*, *Gmelina arborea*, *Grevillea robusta*, *Haldina cordifolia*, *Holarrhena antidysentrica*. *Ilex umbellulata*, *Jacaranda mimosaefolia*, *Lagerstroemia speciosa*, *Leucaena leucocephala*, *Litchi chinensis*, *Litsea monopetala*, *Malus domestica*, *Mangifera indica*, *Mangifera sylvatica*, *Melia azaderach*, *Mesua ferrea*, *Michelia champaca*, *Moringa oleifera*, *Morus alba*, *Oroxylum indicum*, *Parkia timoriana*, *Persea Americana*, *Phyllanthus acidus*, *Pithecolobium montanum*, *Polyalthia pendula*, *Prunus domestica*, *Prunus persica*, *Psidium guajav*, *Pyrus communis*, *Rhus semialata*, *Samanea saman*, *Saracaasoca*, *Schima wallichii*, *Semecarpus anacardium*, *Spondias pinnata*, *Sterculia alata*, *Sterculia villosa*, *Syzygium cumini*, *Tamarindus indica*, *Tectona grandis*, *Terminalia bellirica*, *Thuja compacta*, *Thuja orientalis*, *Toona ciliate*, *Trema orientalis*, *Trevesia palmate*, *Ulmus lancifolia*, *Vitex penducularis*, *Zanthoxylum budrunga*, *Zizyphus jujuba*, *Areca cathechu*, *Arenga pinnata*, *Calamus tenius*, *Caryota urens*, *Cocus nucifera*, and *Daemonorops jenkinsian*	Sahoo and Jeeceelee (2015)
	Shrubs	*Acacia pinnata*, *Adhatoda vasica*, *Allamanda cathartica*, *Antidesma acidum*, *Bougainvillea spectabilis*, *Caesalpinia pulcherrima*, *Cajanus cajan*, *Camellia sinensis*, *Carica papaya*, *Cassia alata*, *Citrus aurantifolia*, *Citrus limon*, *Citrus reticulate*, *Coffea khasiana*, *Coffea Arabica*, *Crotolaria juncea*, *Duranta repens*, *Eleagnus latifolia*, *Euphorbia pulcherrima*, *Jatropha curcus*, *Kadsura heteroclite*, *Murraya koenigii*, *Nerium indicum*, *Opuntia stricta*, *Psychotria calocarpa*, *Punica granatum*, *Ricinus communis*, *Sida acuta*, *Solanum anguivi*, *Solanum melongena*, and *Solanum nigrum*	
	Herbs	*Abelmoschus esculentus*, *Acalypha indica*, *Agave Americana*, *Allium cepa*, *Allium hookerii*, *Allium sativum*, *Amomum dealbatum*, *Amorphophallus paeonifolius*, *Anannas comosus*, *Anoectochilus luteus*, *Arachis hypogaea*, *Brassica juncea*, *Bambusa tulda*, *Brassica oleracea*, *Brassica oleracea*, *Canavalia ensiformis*, *Capsicum annum*, *Capsicum frutescens*, *Catharanthus roseus*, *Celosia argentea*, *Centella asiatica*, *Chrysanthemum indicum*, *Colocasia affinis*, *Colocasia esculenta*, *Costus speciosa*, *Cucumis melo var saccharinus*, *Cucumis sativus*, *Cucurbita maxima*, *Curcuma longa*, *Daucus carota*, *Dendrocalamus longispathus*, *Dioscorea alata*, *Dioscorea glabra*, *Diplazium maxima*, *Elettaria cardamomum*, *Elsholtzia communis*, *Ensete superbum*, *Entada pursaetha*, *Eryngium foetidum*, *Fagopyrum dibotrys*, *Glycine max*, *Hedychium coronarium*, *Helianthus annuus*, *Hibiscus rosa-sinensis*, *Hibiscus sabdariffa*, *Hodgsonia macrocarpa*, *Homalomena aromatic*, *Houttuynia cordata*, *Ipomea batatas*, *Kaempferia rotunda*, *Lablab purpureus*, *Lagenaria siceraria*, *Latuca indica*, *Luffa aegyptiaca*, *Lycianthes laevis*, *Lycopersicon esculentum*, *Melocanna baccifera*, *Mirabilis jalapa*, *Momordica charantia*, *Momordica mixta*, *Mentha viridis*, *Musa paradisiacal*, *Passiflora edulis*, *Phaseolus vulgaris*, *Plantago major*, *Psophocarpus tetragonolobus*, *Pueraria montana* var. *chinensis*, *Saccharum officinarum*, *Sesamum indicum*, *Raphanus sativus*, *Sansevieria zeylanica*, *Sechium edule*, *Spilanthes acmella*, *Tagetes patula*, *Thysanolaena maxima*, *Vigna unguiculata*, *Vitis vinifera*, *Zea mays*, and *Zingiber officinalis*	

Mizoram	Trees	*Mangifera indica*, *Mangifera sylvatica*, *Rhus semialata*, *Semecarpus anacardium*, *Spondias pinnata*, *Alstonia scholaris*, *Anodendron paniculatum*, *Holarrhena antidysenterica*, *Wrightia angustifolia*, *Borassus flabellifer*, *Oroxylum indicum*, *Bombax ceiba*, *Bombax insigne*, *Cordia dichotoma*, *Bursera serrata*, *Canarium bengalense*, *Tamarindus indica*, *Delonix regia*, *Chukrasia velutina*, *Trema orientalis*, *Anogeissus acuminate*, *Terminalia bellirica*, *Terminalia chebula*, *Alangiumb begoniifolium*, *Cryptomeria japonica*, *Dillenia indica*, *Dillenia pentagyna*, *Elaeagnus caudate*, *Elaeocarpus aristatus*, *Elaeocarpus floribundus*, *Croton wallichii*, *Emblica officinalis*, *Phyllanthus acidus*, *Vernicia fordii*, *Acacia nilotica*, *Albizia myriophylla*, *Albizia saman*, *Albizia lebbeck*, *Bauhinia variegate*, *Butea monosperma*, *Cassia alata*, *Cassia fistula*, *Cassia nodosa*, *Cassia tora*, *Dalbergia spinosa*, *Erythrina arborescens*, *Erythrina indica*, *Erythrina stricta*, *Parkia timoriana*, *Castanopsis indica*, *Quercus griffithii*, *Mesua ferrea*, *Garcinia cowa*, *Garcinia lanceifolia*, *Itea macrophylla*, *Itea chinensis*, *Leucosceptrum canum*, *Premna racemosa*, *Cinnamomum verum*, *Cinnamomum tamala*, *Cinnamomum glanduliferum*, *Persea americana*, *Phoebe attenuate*, *Duabanga grandiflora*, *Lagerstroemia speciosa*, *Michelia champaca*, *Pterygota alata*, *Azadirachta indica*, *Melia azadirachta*, *Toona ciliate*, *Albizia procera*, *Artocarpus chama*, *Artocarpus nitidus*, *Artocarpus heterophyllus*, *Artocarpus lakoocha*, *Ficus cunia*, *Ficus elastica*, *Ficus geniculata*, *Ficus racemosa*, *Morus australis*, *Myrica esculenta*, *Callistemon lanceolatus*, *Eucalyptus globules*, *Psidium guajava*, *Syzygium jambos*, *Syzygium cumini*, *Ligustrum robustum*, *Olea dioica*, *Averrhoa carambola*, *Areca cathechu*, *Baccaurea ramiflora*, *Grevillea robusta*, *Punica granatum*, *Ziziphus jujuba*, *Carallia brachiata*, *Malus pumila*, *Prunus cerasoides*, *Prunus domestica*, *Pyrus communis*, *Hymenodictyon excelsum*, *Psychotria calocarpa*, *Sapros maternatum*, *Wendlandia grandis*, *Wendlandia tinctoria*, *Aegle marmelos*, *Atalantia monophylla*, *Citrus aurantifolia*, *Citrus grandis*, *Citrus jambhiri*, *Citrus karna*, *Citrus limon*, *Citrus limonia*, *Citrus macroptera*, *Citrus medica*, *Citrus nobilis*, *Citrus paradise*, *Citrus reshni*, *Citrus reticulate*, *Citrus rugulosa*, *Citrus sinensis*, *Zanthoxylum budrunga*, *Pyrularia edulis*, *Lepisanthes senegalensis*, *Litchi chinensis*, *Mimusops elengi*, *Sterculia villosa*, *Symplocos laurina*, *Tetrameles nudiflora*, *Schima wallichii*, *Celtis timorensis*, *Ulmus lanceifolia*, *Tectona grandis*, and *Vitex peduncularis*	Barbhuiya et al. (2016)

(continued)

Table 3.2 (continued)

Location	Plant types	Species	References
	Shrubs	*Justicia adhatoda*, *Strobilanthes flaccidifolia*, *Viburnum mullaha*, *Amaranthus caudatus*, *Amaranthus viridis*, *Catharanthus roseus*, *Rauvolfia serpentina*, *Trevesia palmate*, *Calamus acanthospathus*, *Calamus erectus*, *Calamus guruba*, *Licuala peltata*, *Livistona chinensis*, *Artemisia vulgaris*, *Helianthus annuus*, *Tithonia diversifolia*, *Cannabis sativa*, *Carica papaya*, *Rhododendron arboreum*, *Rhododendron formosum*, *Rhododendron veitchianum*, *Vaccinium sprengelii*, *Codiaeum variegatum*, *Euphorbia royleana*, *Jatropha curcas*, *Ricinus communis*, *Securinega virosa*, *Bauhinia scandens*, *Cajanus cajan*, *Crotalaria juncea*, *Desmodiumgyroides*, *Zea mays*, *Mentha viridis*, *Ocimum americanum*, *Ocimum sanctum*, *Vitex negundo*, *Lawsonia inermis*, *Abelmoschus esculentus*, *Gossypium arboreum*, *Hibiscus macrophyllus*, *Hibiscus rosa-sinensis*, *Hibiscus sabdariffa*, *Hibiscus surattensis*, *Melastomanepalense*, *Acacia pennata*, *Ensete glaucum*, *Musa acuminate*, *Musa balbisiana*, *Musa glauca*, *Musa nagensium*, *Musa paradisiacal*, *Musa sanguine*, *Musa velutina*, *Pandanus pseudofoetidus*, *Arundo donax*, *Saccharum officinarum*, *Thysanolaena maxima*, *Polygonum barbatum*, *Polygonum convolvulus*, *Polygonum orientale*, *Polygonum plebeium*, *Rubus lasiocarpus*, *Rubus niveus*, *Coffea arabica*, *Psychotriacalocarpa*, *Murrayakoenigii*, *Capsicum annuum*, *Capsicum frutescens*, *Nicotiana tabacum*, *Solanum aethiopicum*, *Solanum anguivi*, *Solanum esculentum*, *Solanum khasianum*, *Solanum melongena*, *Solanum nigrum*, *Solanum spinosum*, *Solanum torvum*, *Solanum villosum*, *Solanum violaceum*, *Cammellia sinensis*, *Boehmeria penduliflora*, *Boehmeria rugulosa*, *Gmelina arborea*, *Clerodendrum colebrookianum*, *Clerodendrum infortunatum*, and *Lantana camara*	
	Herbs	*Thunbergia grandiflora*, *Amaranthus spinosus*, *Celosia argentea*, *Allium sativum*, *Centella asiatica*, *Coriandrum sativum*, *Daucus carota*, *Eryngium foetidum*, *Trachyspermum roxburghiaum*, *Colocasia esculenta*, *Aralia racemosa*, *Colocasia gigantea*, *Colocasia lihengiae*, *Colocasia macrorrhizos*, *Colocasia marnii*, *Colocasia obtusiloba*, *Ageratum conyzoides*, *Bidens biternata*, *Blumea alata*, *Chromolaena odorata*, *Spilanthes acmella*, *Stevia rebaudiana*, *Impatiens balsamina*, *Ananas comosus*, *Canna orientalis*, *Chenopodium album*, *Ipomoea batatas*, *Ipomoea aquatic*, *Costus speciosus*, *Costusviridis*, *Brassica juncea*, *Brassica botrytis*, *Brassica oleracea*, *Brassica campestris*, *Brassica oleracea*, *Brassica rapa*, *Raphanus sativus*, *Cyperus rotundus*, *Microlepia strigosa*, *Manihot esculenta*, *Phyllanthus niruri*, *Canavalia ensiformis*, *Curculigo crassifolia*, *Elsholtzia communis*, *Mentha spicata*, *Allium cepa*, *Allium hookeri*, *Urena lobata*, *Phrynium capitatum*, *Phrynium placentarium*, *Mimosa pudica*, *Arundina graminifolia*, *Dendrobium chrysotoxum*, *Vanda coerulea*, *Oxalis corniculata*, *Plantago major*, *Cynodon dactylon*, *Imperata cylindrica*, *Polygonum nepalense*, *Portulaca oleracea*, *Adiantum caudatum*, *Adiantum philippense*, *Pteris amoena*, *Paederia foetida*, *Paederia scandens*, *Lindernia ruellioides*, *Torenia peduncularis*, *Alpinia nigra*, *Amomum dealbatum*, *Curcuma caesia*, *Curcuma grandiflora*, *Curcuma latiflora*, *Curcuma longa*, *Curcuma longispica*, *Curcuma amada*, *Curcuma trichosantha*, *Curcumorpha longiflora*, *Hedychium spicatum*, *Kaempferia rotunda*, *Zingiber officinale*, and *Zingiber zerumbet*	
	Others	*Mikania micrantha*, *Benincasa hispida*, *Cucumis sativus*, *Cucurbita maxima*, *Cucurbita siceraria*, *Cucumis melo*, *Luffa cylindrical*, *Momordica charantia*, *Sechium edule*, *Thladiantha cordifolia*, *Trichosanthes anguina*, *Dioscorea alata*, *Dioscorea glabra*, *Asparagus gonoclados*, *Asparagus racemosus*, *Bougainvillea spectabilis*, *Passiflora edulis*, *Passiflora nepalensis*, *Piper diffusum*, *Piper betle*, *Piper boehmerrifolium*, *Piper thomsonii*, *Smilax glabra*, and *Vitis vinifera*	

West Bengal	Trees	*Mangifera indica*, *Polyalthia longifolia*, *Thespesia populnea*, *Araucaria araucana*, *Areca catechu*, *Borassus flabellifer*, *Caryota urens*, *Phoenix sylvestris*, *Bombax ceiba*, *Ceiba pentandra*, *Ehretia acuminate*, *Calophyllum inophyllum*, *Dillenia indica*, *Shorea robusta*, *Diospyros malabarica*, *Elaeocarpus floribundus*, *Elaeocarpus serratus*, *Mallotus tetracoccus*, *Acacia auriculiformis*, *Albizia odoratissima*, *Albizia procera*, *Albizia saman*, *Butea monosperma*, *Cassia fistula*, *Pongamia pinnata*, *Delonix regia*, *Erythrina variegate*, *Leucaena leucocephala*, *Saracaasoca*, *Sesbania grandiflora*, *Tamarindus indica*, *Cinnamomum verum*, *Tectona grandis*, *Machilus fasciculate*, *Lagerstroemia speciosa*, *Magnolia champaca*, *Amoora rohituka*, *Amoora wallichii*, *Aphanamixis polystachya*, *Chukrasia tabularis*, *Melia azedarach*, *Swietenia macrophylla*, *Toona australis*, *Artocarpus heterophyllus*, *Artocarpus lacucha*, *Ficus benghalensis*, *Ficus carica*, *Ficus elastic*, *Ficus hispida*, *Ficus lacor*, *Ficus racemosa*, *Ficus religiosa*, *Morus alba*, *Psidium guajava*, *Syzygium aromaticum*, *Syzygium cumini*, *Syzygium jambos*, *Syzygium samarangense*, *Nyctanthes arbor-tristis*, *Dalbergia sissoo*, *Baccaurea ramiflora*, *Bischofia javanica*, *Grevillea robusta*, *Neolamarckia cadamba*, *Neolamarckia cadamba*, *Citrus × aurantium*, *Citrus reticulate*, *Citrus sinensis*, *Zanthoxylum rhetsa*, *Populus deltoids*, *Salix tetrasperma*, *Mimusopselengi*, *Ailanthus integrifolia*, *Aquilaria malaccensis*, *Trema orientalis*, *Aegle marmelos*, *Alstonia scholaris*, *Averrhoa carambola*, *Azadirachta indica*, *Carica papaya*, *Cinnamomum tamala*, *Cinnamomum*, *Verum*, *Cocus nucifera*, *Lannea coromandelica*, *Moringa oleifera*, *Oroxylum indicum*, *Phyllanthus emblica*, *Punica granatum*, *Streblus asper*, *Terminalia arjuna*, *Terminalia bellirica*, *Terminalia chebula*, and *Ziziphus mauritiana*	Pala et al. (2019)
	Shrubs	*Barleria cristata*, *Phlogacanthus thyrsiflorus*, *Allamanda cathartica*, *Asclepias curassavica*, *Carissa carandas*, *Tabernaemontana divaricata*, *Wrightia antidysenterica*, *Calamus erectus*, *Cajanus cajan*, *Senna tora*, *Crotalaria pallid*, *Desmodium laxiflorum*, *Bixa orellana*, *Codiaeum variegatum*, *Euphorbia pulcherrima*, *Lawsonia inermis*, *Abelmoschus angulosus*, *Malvaviscus arboreus*, *Sida acuta*, *Sida rhombifolia*, *Triumfetta rhomboidea*, *Melastoma malabathricum*, *Callistemon lanceolatus*, *Maesa indica*, *Rosa* spp., *Ixora coccinea*, *Mussaenda philippica*, *Mussaenda roxburghii*, *Citrus aurantiifolia*, *Murraya paniculata*, *Cestrum nocturnum*, *Solanum abancayense*, *Dendrocnide sinuate*, *Leea aequata*, *Abroma augusta*, *Hibiscus rosa-sinensis*, *Jatropha curcas*, *Murraya koenigii*, and *Vitex negundo*	

(continued)

Table 3.2 (continude)

Location	Plant types	Species	References
	Herbs	*Achyranthes aspera*, *Alternanthera brasiliana*, *Amaranthus spinosus*, *Amaranthus tricolor*, *Amaranthus viridis*, *Gomphrena globosa*, *Coriandrum sativum*, *Daucus carota*, *Catharanthus roseus*, *Alocasia macrorrhiza*, *Amorphophallus paeoniifolius*, *Anthurium andraeanum*, *Typhonium trilobatum*, *Caladium* spp., *Ageratum houstonianum*, *Artemisia vulgaris*, *Centratherum punctatum*, *Chrysanthemum indicum*, *Crassocephalum crepidioides*, *Dahlia apiculata*, *Eclipta prostrate*, *Emilia sonchifolia*, *Gerbera ambigua*, *Spilanthes calva*, *Parthenium hysterophorus*, *Tagetes erecta*, *Diplazium esculentum*, *Diplazium molokaiense*, *Heliotropium indicum*, *Brassica juncea*, *Brassica oleracea*, *Brassica rugosa*, *Raphanus caudatus*, *Cannabis sativa*, *Chenopodium album*, *Commelina benghalensis*, *Ipomoea aquatic*, *Cheilocostus speciosus*, *Bryophyllum pinnatum*, *Cyperus rotundus*, *Hippochaete debilis*, *Eryngium foetidum*, *Pisum sativum*, *Heliconia abalos*, *Curculigo annamitica*, *Clerodendrum infortunatum*, *Abelmoschus esculentus*, *Malva verticillata*, *Mentha sylvestris*, *Musa* spp., *Nelumbo nucifera*, *Oxalis corniculata*, *Oxalis debilis*, *Plantago ovate*, *Scoparia dulcis*, *Axonopus compressus*, *Brachiaria ramose*, *Brachiaria reptans*, *Dactyloctenium aegyptium*, *Polypogon monspeliensis*, *Setaria viridis*, *Thysanolaena latifolia*, *Persicaria chinensis*, *Persicaria hydropiper*, *Persicaria maculosa*, *Eichhornia crassipes*, *Houttuynia cordata*, *Capsicum annuum*, *Lycopersicon esculentum*, *Physalis angulata*, *Corchorus olitorius*, *Boehmeria allophylla*, *Pouzolziahirta*, *Acorus calamus*, *Aloe vera*, *Ananas comosus*, *Andrographis paniculata*, *Asparagus racemosus*, *Bacopa monnieri*, *Bryophyllum pinnatum*, *Centella asiatica*, *Clerodendrum*, *Infortunatum*, *Curcuma aeruginosa*, *Curcuma longa*, *Cynodon dactylon*, *Datura metel*, *Equisetum arvense*, *Euphorbia hirta*, *Hygrophila schulli*, *Justicia Adhatoda*, *Justicia gendarussa*, *Leucas aspera*, *Mentha arvensis*, *Mimosa pudica*, *Ocimum gratissimum*, *Ocimum sanctum*, *Rauvolfia serpentina* and *Zingiber officinale*	
	Others	*Epipremnum aureum*, *Mikania micrantha*, *Basella alba*, *Benincasa hispida*, *Cucumis sativus*, *Cucurbita maxima*, *Lagenaria siceraria*, *Luffa aegyptiaca*, *Luffa cylindrical*, *Momordica charantia*, *Momordica dioica*, *Trichosanthes cucumerina*, *Trichosanthes dioica*, *Dioscorea belophylla*, *Clitoriaternatea*, *Lablab purpureus*, *Lygodium japonicum*, *Bougainvillea spectabilis*, *Digitaria sanguinalis*, *Eleusine indica*, *Cymbopogon flexuosus*, *Saccharum officinarum*, *Bambusa balcooa*, *Bambusa nutans*, *Paederiafoetida*, *Ageratum conyzoides*, *Cuscuta europaea*, *Piper betle*, *Piper nigrum*, and *Tinospora sinensis*	

families and 78 genera, followed by trees, which include 81 species from 37 families and 64 genera. The most prevalent plant families in these HGs are Lamiaceae (15 species), Leguminosae (12 species), and Solanaceae (12 species). Among genera, *Ficus* is the most dominant, followed by *Brassica* and *Solanum*. The biodiversity indices calculated indicate that HGs support a highly diverse plant assemblage. The cultivated and wild plant species found in the Darjeeling Himalayas provide significant, cost-effective nutritional security to rural communities. Therefore, conserving HG ecosystems is essential for ensuring the sustainable use of these species for future generations (Sarkar et al. 2020). These home gardens serve as vital refuges for both cultivated and wild plant species, playing a significant role in conserving regional biodiversity while maintaining ecological stability and safeguarding genetic resources (Panwar and Chakravarty 2010; Gao et al. 2012; Bajpai et al. 2013; Subba et al. 2018; Lepcha et al. 2019; Pala et al. 2019). Similar patterns of species diversity have been reported by Kumari et al. (2009), Tesfaye et al. (2010), Larios et al. (2013), and Shukla et al. (2017). The presence of diverse cultivated and wild species in these home gardens highlights their significance for sustainable resource management. Policymakers should prioritize their conservation and sustainable utilization to enhance nutritional security and improve the livelihoods of hill communities (Thrupp 2000; Eyzaguirre and Linares 2004; Perrault-Archambault and Coomes 2008; Senanayake et al. 2010). These traditional agroecosystems also foster crucial biocultural linkages, integrating local communities into sustainable development efforts while preserving invaluable traditional knowledge (Sthapit et al. 2004). Vibhuti et al. (2019) investigated the floristic composition and species diversity of HGs in the Kumaun Himalaya, documenting a total of 70 plant species across 35 families from 120 surveyed HGs. These included 22 tree species, 1 shrub, and 45 herbs. The dominant plant families were Brassicaceae, Cucurbitaceae, Fabaceae, Rosaceae, and Poaceae, each of which was represented by five species. Solanaceae contributed four species, while Moraceae, Malvaceae, and Rutaceae each had three species. Additionally, four families were represented by two species each, and 21 families contained only a single species. Vegetable crops and fruit plants exhibited the highest density, and a notable similarity in species composition was observed across different HGs. *Mangifera indica* was the most commonly found tree species across all HGs. However, significant variation in species density was noted among different HG categories. Herb density was highest in large HGs (479 individuals/m^2), followed by medium (316 individuals/m^2) and small (249 individuals/m^2) HGs. Tree density followed a similar trend, being highest in large (825 individuals/ha), medium (750 individuals/ha), and small (317 individuals/ha) HGs. Additionally, species frequency also increased with garden size. The basal area of tree species was also highest in large gardens (24.33 m^2/ha), followed by medium (24.09 m^2/ha) and small (6.32 m^2/ha) gardens. Species richness was positively correlated with HG size, with large gardens housing 27 species, medium-sized ones 22 species, and small gardens 13 species. These findings align with those of Kabir and Webb (2009), who observed a similar positive relationship between HG size and species richness in Bangladesh.

Sahoo and Jeeceelee (2015) analyzed plant species diversity in 40 HGs across Mizoram, recording 198 species from 69 families and 169 genera. The species composition included 82 tree species, 31 shrubs, 79 herbs, and 6 palms. The family Papilionaceae was the most diverse, with 11 species, followed by Cucurbitaceae (9), and Caesalpiniaceae, Euphorbiaceae, and Rutaceae (8 each). Thirty families were represented by a single species, while 14 families had more than five species. The most dominant families in Aizawl district's HGs included Caesalpiniaceae, Cucurbitaceae, Euphorbiaceae, Mimosaceae, Moraceae, Musaceae, Papilionaceae, Rutaceae, Solanaceae, Verbenaceae, and Zingiberaceae. Among the plant species recorded, 78% were perennials, 16% annuals, and 6% biennials. Regarding plant habit, 79 species (40%) were herbs, 31 species (15%) were shrubs, 82 species (41%) were trees, and 6 species (4%) were palms. The most frequently observed species included *Parkia timoriana*, *Psidium guajava*, *Mangifera indica*, *Trevesiapalata*, and *Artocarpus heterophyllus* among trees; *Acacia pinata*, *Carica papaya*, and *Murrayakoenigii* among shrubs; and *Cucurbita maxima*, *Colocasia esculenta*, *Brassica juncea*, *Phaseolus vulgaris*, and *Zea mays* among herbs. *Parkiatimoriana*, a protein-rich species, was particularly significant for its green pods, while the latter two species provided marketable fruits. Dominant plant families such as Cucurbitaceae, Caesalpiniaceae, Solanaceae, Poaceae, Papilionaceae, and Euphorbiaceae played a crucial role in HG biodiversity. A defining characteristic of these HGs, regardless of size, was their layered canopy structure and integration of compatible species. Barbhuiya et al. (2016) assessed plant diversity in 90 HGs in the Eastern Himalayan Region of Mizoram through direct observation and farmer discussions. A total of 333 plant species, including 133 trees, 92 shrubs, and 108 herbs, were identified across 128 plant families, with an average of 78 species per HG. The study recorded 96 tree genera across 52 families, 59 shrub genera across 36 families, and 59 herbaceous genera across 52 families. HG sizes ranged from 0.10 to 0.60 ha and showed a significant positive correlation between garden size and plant species diversity. The species diversity index was calculated as 4.76 for trees, 4.39 for shrubs, and 4.58 for herbs, with species similarity recorded at 50% for trees, 38% for shrubs, and 49% for herbs. The most commonly found plant families included Fabaceae, Rutaceae, Zingiberaceae, Solanaceae, Asteraceae, Euphorbiaceae, and Cucurbitaceae. Fabaceae contributed the highest number of food plants, while Rutaceae dominated in fruit and medicinal plant diversity. The most abundant tree species were *Areca catechu*, *Artocarpus heterophyllus*, *Mangifera indica*, *Parkia timoriana*, and various *Citrus* species. Dominant shrub species included *Amaranthus viridis*, *Cajanus cajan*, *Calamus erectus*, *Capsicum annuum*, *Carica papaya*, *Clerodendrum colebrookianum*, *Hibiscus macrophyllus*, *Murrayakoenigii*, and several *Musa* and *Solanum* species. Common herbaceous species comprised *Ageratum conyzoides* and several *Allium*, *Brassica*, and *Cucurbita* species. The number of plant species per HG ranged from 36 to 167, with an average of 78 species, highlighting significant intra-garden variation in species composition and richness. Pala et al. (2019) conducted a study on floristic composition in Eastern Himalayan HGs, identifying 260 plant species across 45 genera. Fabaceae was the most dominant family, with 21 species, followed by Asteraceae and Poaceae (14 each) and Rubiaceae (13). HGs were recognized for their role in conserving agro-biodiversity in compact spaces and as socially significant elements of biodiversity preservation. Studies from Tehuacan Valley, Mexico (Larios et al. 2013), which reported 281 HG plant species, and the Thimbirigaskatuwa suburban area in Sri Lanka (Kumari et al. 2009), which documented 289 species from 106 suburban HGs, further corroborate these findings. Tiwari et al. (2016) examined plant biodiversity in the HGs of Uttarakhand's Garhwal Himalayan Region, compiling a comprehensive checklist of 105 species from 50 families.

Among them, 35 species were ornamental plants, 19 were fruit-bearing, 16 were vegetables, 11 were spices, 6 were medicinal plants, 5 were leafy vegetables, 5 were fodders, 4 were cereals, 3 were aesthetic plants, and 1 was a pulse crop.

3.3 Home Gardens for Livelihood Security

The impacts of climate change, driven by complex weather-related phenomena, have threatened the livelihoods of agricultural and local communities (Gifawesen et al. 2020). Homesteads, which are typically small-scale enterprises designed for both subsistence production and income generation, play a crucial role in these communities (Dar et al. 2018). These homesteads have a complex structure, both vertically and horizontally, with the entire crop-tree-animal system being managed by household members, particularly women (Islam et al. 2017a). On a global scale, forests provide essential socioeconomic benefits, livelihood security, and ecological sustainability, while homesteads on household land serve similar functions (Haque et al. 2018). Homesteads contribute to a wide range of ecosystem services, including provisioning, regulating, supporting, and cultural services, which are vital not only locally but also globally (Yasin et al. 2018). As such, homesteads are increasingly recognized for their role in addressing resource crises, energy shortages, biodiversity conservation, deforestation, and climate change mitigation (George and Christopher 2020). Homesteads provide a diverse array of resources that support livelihoods, including revenue, employment, shelter, housing materials, fuel, fodder, timber, food, vegetables, medicinal plants, fertilizer, fiber, floss, oilseeds, cottage industries, handicrafts, and other nontimber forest products (NTFPs) (Devi and Das 2013; Islam et al. 2017b). Additionally, homesteads serve as a safety net, offering alternative livelihoods during emergencies, including natural hazards (Darcha et al. 2015). The reliance of rural households on homestead resources and their varied uses has become a key issue in developing economies (Foysal et al. 2013). Studies (Foysal et al. 2013; Haque et al. 2018; Bhat et al. 2019; Musa et al. 2019) conducted globally have highlighted the significance of resource production and their contribution to livelihoods. Given the potential of these land-use systems for further development, there is room for introducing new livelihood opportunities. However, despite their importance, homestead resources have often been undervalued by scientists and planners, who have failed to prioritize them in research, extension services, credit facilities, and market access. Understanding the production of resources, subsistence consumption, economic functions, and the role of homestead resources in reducing income inequality is crucial for effective policy reform and sustainable growth (Islam et al. 2021).

Sahoo (2009) explored the livelihood security of 45 traditional HGs in three villages in northeast India during 2006–2007. The results showed that HGs are dynamic, with high crop diversity, including both annuals and perennials, and serve various economic and ecological functions. Eight plant species—*A. hookeri*, *A. catechu*, *C. reticulate*, *C. maxima*, *P. edulis*, *M. paradisiacal*, *S. eduli*, and *Z. officinales*—were found to contribute over Rs. 10,000 annually to household income. The sale of these plant products in local markets ranged from 51% (*S. eduli*) to 90% (*A. catechu*), generating substantial income for farmers (Sahoo 2009). The economic returns from these HGs surpassed those from other land uses (Sahoo 2007). A four-year-old *P. roxburghii* stand intercropped with *Zea mays* yielded Rs. 33,280 per hectare annually (Kharshi-ing et al. 2004). Another study found that *P. roxburghii* produced between 9400 and 28,300 pods in HGs, with the price of the pods ranging from Rs. 150–180 per 100 unripe pods and Rs. 400–500 per 100 ripe pods. If 75% of the pods sold were unripe, households could earn between Rs. 3854 and Rs. 11,453 per month for 6 months (Rocky et al. 2004). Fuelwood is another valuable product from HGs, and many households sell broom grass and bamboo from HG boundaries, with over 80% of households engaged in such sales. However, medicinal plants are typically consumed at the household level during emergencies (Lalremruata et al. 2007). These findings underscore the importance of various products in supporting farmers' livelihoods. In a study conducted by Islam et al. (2021) in the Budgam district of Kashmir, tree resources in homesteads were found to contribute significantly to both subsistence and cash income. The average annual production and consumption of homestead tree resources included fuel wood (production = 11.86 quintals, consumption = 9.13 quintals), tree browse (production = 10.50 quintals, consumption = 7.16 quintals), timber (production = 3.15 m^3, consumption = 1.80 m^3), wicker (production = 0.22 quintals, consumption = 0), fruits (production = 7.91 quintals, consumption = 4.50 quintals), and leaf litter (production = 5.02 quintals, consumption = 5.02 quintals). Homestead tree resources generated a total income of Rs. 3,680,437 per year, with Rs. 2,159,414 from subsistence consumption and Rs. 1,521,023 from cash income. Fruits were the largest contributor, accounting for 47.49% of total tree income, followed by wicker (18.74%), tree browse (13.97%), fuelwood (11.44%), and timber (8.36%).

3.4 Role of Home Gardens in Women Empowerment

In traditional rural Indian society, women's roles are often confined to household tasks, childbearing, and child-rearing. Globally, approximately 43% of agricultural workers are women (Akter et al. 2017). Research by Meinzen-Dick et al. (2012) has shown that empowering women positively influences the health and nutrition of their households. HGs play a significant role in enhancing the status of rural women by increasing their knowledge of farm management and improving their economic and social standing (Bhandari et al. 2021). These gardens are frequently a shared household activity (Reyes-García et al. 2010; Bargali et al. 2015b), with women in India taking an active role in managing them (Pandey et al. 2011). In fact, 78% of all regularly working women are involved in agriculture (Singh and Tewari 2013). Their contributions include food crop production, resource management, and domestic responsibilities within HGs (Bargali et al. 2009). Because HGs are typically located closer to homes than agricultural fields, it is common for rural women to manage these gardens alongside their household duties. Women in farming households who do not receive a salaried income may find themselves cultivating off-season crops in these gardens, thus not only earning income but also meeting their households' food needs. Providing rural women with training and technology can have substantial socioeconomic benefits, potentially eliminating food insecurity and alleviating poverty in their communities (Bargali et al. 2015a).

Women play a central role in HG management, which is crucial for livelihoods. While most of the work is carried out by women, men also contribute (Keatinge et al. 2012; Khan et al. 2020). HGs hold promise for increasing gender participation, developing marketable products, and improving food availability and community protection (Khan et al. 2021). In the Kumaun Himalaya, 68% of women independently manage HG agroforestry activities, while 26% work jointly with men, and 6% do not participate. Additionally, majority of women make decisions about what to grow in HGs. Around 55% of women independently engage in activities such as seed selection, nursery preparation, transplanting, weeding, watering, and applying fertilizers, while 40% share responsibilities with men, and 5% have partial responsibilities. On average, 76% of women independently handle postharvest activities, such as cutting, drying, threshing, winnowing, surplus product management, marketing, and cash management, with 24% sharing these duties with men (Bargali 2015). Studies have shown that women in hilly regions possess extensive knowledge about seed collection, storage, and utilization. Women are also known for establishing and managing HGs, making decisions about crops, and determining how to use the garden products (Finerman and Sackett 2003). In Himachal Pradesh, women independently carry out crop production tasks such as transplanting, weeding, and transporting manure (Kishtwaria et al. 2009). HGs are often regarded as women's enterprises, relying on household knowledge, labor, plant material exchange, and the marketability of produce (Petterson 2000). In hill region, women play a more active role than men in cultivating and managing HGs, significantly contributing to agricultural growth. Introducing crop production technologies in HGs could enhance rural women's incomes while promoting agricultural development in these regions (Bargali 2015). Bargali et al. (2015b) analyzed rural women's contributions to vegetable cultivation in HGs in Nainital district, Kumaun Himalaya, India. Women were found to be the primary contributors to vegetable production, managing activities such as land preparation, nursery preparation, planting, watering, harvesting, and postharvest handling. Men typically assist with tasks like plowing, applying fertilizers and pesticides, and marketing the produce. Eighty percent of respondents reported that women have the largest share in growing and managing vegetables in HGs. In rural areas where malnutrition and insufficient vegetable supply are persistent challenges, women's contributions are particularly valuable. Research by Koopman (2009) and Sanyang et al. (2009) further supports this, highlighting that women predominantly tend to HG vegetables, which have proven to be vital for both the nutritional and economic well-being of households (Parihaar et al. 2015).

4 Conclusion

Sustainable development has emerged as a key framework for ensuring a balance between economic growth, social well-being, and environmental preservation. The Indian Himalayan Region (IHR) holds a unique and vital place among global mountain ecosystems. As a geologically young and dynamic range, the Himalayas play a crucial role in climate regulation and act as a lifeline by supplying water to vast areas of the Indian subcontinent. However, the unsustainable exploitation of the region's natural resources is accelerating environmental degradation and amplifying the risk of natural disasters. In this context, home gardens (HGs) offer promising solutions for promoting sustainable development in the IHR. These small-scale agroecosystems contribute to climate change mitigation by sequestering carbon in both biomass and soil. Moreover, they enhance climate resilience by fostering microclimates. HGs also play a critical role in conserving biodiversity and supporting the livelihoods of local communities. Notably, they empower women, who are more actively involved in managing home gardens due to their proximity to the household and manageable scale. There is a pressing need to assess the population-supporting capacity of HGs relative to the land area of HG. To effectively advance the Sustainable Development Goals (SDGs), policymakers must take into account both the direct and indirect impacts of urbanization. Recognizing and promoting rural HGs as integral elements of sustainable infrastructure is crucial for building resilient and climate-adaptive societies. It is also imperative to formulate well-informed public policies focused on preserving traditional home garden systems.

References

Abdoellah OS, Hadikusumah HY, Takeuchi K, Okubo S, Parikesit P (2006) Commercialization of homegardens in Indonesian villages: vegetables composition and functional changes. Agrofor Syst 68(1):1–13. https://doi.org/10.1007/s10457-005-7475-x

Agbedahin A (2019) Sustainable development, education for sustainable development, and the 2030 agenda for sustainable development: emergence, efficacy, eminence, and future. Sustain Dev 27:669–680. https://doi.org/10.1002/sd.1931

Akter S, Rutsaert P, Luis J, Htwe NM, San SS, Raharjo B, Pustika A (2017) Women's empowerment and gender equity in agriculture: a different perspective from Southeast Asia. Food Policy 69:270–279. https://doi.org/10.1016/j.foodpol.2017.05.003

Amberber M, Argau M, Asfaw Z (2013) The role of homegardens for in situ conservation of plant diversity in Holeta town, Oromia National State, Ethiopia. Int J Biodivers Conserv 6(1):8–16

Bahadur J (2004) Himalayan snow and glaciers: associated environmental problems, progress, and prospects. Concept Publishing Co, New Delhi

Bajpai S, Sharma AK, Kanungo VK (2013) Traditional home gardens: a preserve of medicinal plants. Int J Herb Med 1(2):152–161

Barbhuiya AR, Sahoo UK, Upadhyaya K (2016) Plant diversity in the indigenous home gardens in the eastern Himalayan region of Mizoram, Northeast India. Econ Bot 70:115–131. https://doi.org/10.1007/s12231-016-9349-8

Bargali K (2015) Comparative participation of rural women in agroforestry home gardens in Kumaun Himalaya, Uttarakhand, India. Asian J Agric Ext Econ Sociol 6(1):16–22. https://doi.org/10.9734/AJAEES/2015/16115

Bargali K (2016) Traditional homegardens as a sustainable ecosystem for maintenance of biodiversity: a case study from Kumaun Himalaya, India. J Biodivers 7(2):88–100

Bargali SS, Pandey K, Singh L, Shrivastava SK (2009) Participation of rural women in rice-based agroecosystems. Int Rice Res Notes 33(1):1–2

Bargali K, Vibhuti V, Shahi C (2015a) Contribution of rural women in vegetable cultivation in homegardens of Nainital District, Kumaun Himalaya, India. Curr Agric Res J 3(2):90–100

Bargali K, Parihaar RS, Bargali SS (2015b) Traditional agroforestry systems practiced in Kumaun Himalaya, India. In: abstracts of the climate change: social-economic and environment issues-problem and challenges, 21–22 April 2013

Bargali H, Kumar A, Singh P (2022) Plant studies in Uttarakhand, Western Himalaya–a comprehensive review. Trees For People 8:100203. https://doi.org/10.1016/j.tfp.2022.100203

Beumer C (2018) Show me your garden and I will tell you how sustainable you are: dutch citizens' perspectives on conserving biodiversity and promoting a sustainable urban living environment through domestic gardening. Urban For Urban Green 30:260–279. https://doi.org/10.1016/j.ufug.2017.09.010

Bhandari S, Yadav PK, Rijal S (2021) Home garden; an approach for household food security and uplifting the status of rural women: a case study of Saptari, Nepal. Turk J Agri-Food Sci Tech 9(10):1792–1798. https://doi.org/10.24925/turjaf.v9i10.1792-1798.4283

Bhat GM, Islam MA, Malik AR, Rather TA, Khan FA, Mir AH (2019) Productivity and economic evaluation of Willow (*Salix alba* L.) based silvopastoral agroforestry system in Kashmir valley. J Appl Nat Sci 11(3):743–751. https://doi.org/10.31018/jans.v11i3.2104

Buechler S (2016) Gendered vulnerabilities and grassroots adaptation initiatives in home gardens and small orchards in Northwest Mexico. Ambio 45:322–334. https://doi.org/10.1007/s13280-016-0832-3

Caballero-Serrano V, Onaindia M, Alday JG, Caballero D, Carrasco JC, McLaren B, Amigo J (2016) Plant diversity and ecosystem services in Amazonian homegardens of Ecuador. Agric Ecosyst Environ 225:116–125. https://doi.org/10.1016/j.agee.2016.04.005

Caiado RGG, Filho WL, Quelhas OL, de Mattos Nascimento DL, Ávila LV (2018) A literature-based review on potentials and constraints in the implementation of the sustainable development goals. J Clean Prod 198:1276–1288. https://doi.org/10.1016/j.jclepro.2018.07.102

Calvet-Mir L, Riu-Bosoms C, González-Puente M, Ruiz-Mallén I, Reyes-García V, Molina JL (2016) The transmission of home garden knowledge: safeguarding biocultural diversity and enhancing social–ecological resilience. Soc Nat Resour 29(5):556–571. https://doi.org/10.1080/08941920.2015.1094711

Carpenter SR, Arrow KJ, Barrett S, Biggs R, Brock WA, Crépin AS, Engström G, Folke C, Hughes TP, Kautsky N, Li CZ (2012) General resilience to cope with extreme events. Sustainability 4(12):3248–3259. https://doi.org/10.3390/su4123248

Cassidy ES, West PC, Gerber JS, Foley JA (2013) Redefining agricultural yields: from tonnes to people nourished per hectare. Environ Res Lett 8(3):034015. https://doi.org/10.1088/1748-9326/8/3/034015

Chauhan SK, Singh S, Sharma S, Sharma R, Saralch HS (2019) Tree biomass and carbon sequestration in four short rotation tree plantations. Range Manage Agrofor 40(1):77–82

Dar M, Qaisar KN, Ahmad S, Wani AA (2018) Inventory and composition of prevalent agroforestry systems of Kashmir Himalaya. Adv Res 14(1):1–9

Dar MU, Qaisar KN, Masoodi TH, Mughal AH, Khan PA (2019) Biomass production and carbon stock potential under home gardens of Kashmir Himalaya. J Plant Dev Sci 11:256–271

Darcha G, Birhane E, Abadi N (2015) Woody species diversity in Oxytenantheraabyssinica based homestead agroforestry systems of serako, northern Ethiopia. J Nat Sci Res 5(9):18–27

Das T, Das AK (2005) Inventorying plant biodiversity in homegardens: a case study in Barak Valley, Assam, North East India. Curr Sci 10:155–163. https://www.jstor.org/stable/24110441

Das M, Nath PC, Reang D, Nath AJ, Das AK (2019) Tree diversity and the improved estimate of carbon storage for traditional agroforestry systems in north East India. Appl Ecol Environ Sci 8(4):154–159

Delahay RJ, Sherman D, Soyalan B, Gaston KJ (2023) Biodiversity in residential gardens: a review of the evidence base. Biodivers Conserv 32(13):4155–4179. https://doi.org/10.1007/s10531-023-02694-9

Devi NL, Das AK (2013) Diversity and utilization of tree species in Meitei homegardens of Barak Valley. Assam J Environ Biol 34(2):211–217

Devi NL, Singha D, Tripathi SK (2019) Phenology, population structure and carbon sequestration potential of Parkia timoriana: a heirloom tree in traditional Meitei homegarden of northeast India. Vegetos 33(2):222–228. https://doi.org/10.1007/s42535-019-00096-3

Dewaelheyns V, Rogge E, Gulinck H (2014) Putting domestic gardens on the agenda using empirical spatial data: the case of Flanders. Appl Geogr 50:132–143. https://doi.org/10.1016/j.apgeog.2014.02.011

Dey A, Islam M, Masum KM (2014) Above ground carbon stock through palm tree in the homegarden of Sylhet City in Bangladesh. J For Environ Sci 30(3):293–300

Dhar U (ed) (1997) Himalayan biodiversity: action plan. Gyanodaya Prakashan, Nainital

Diduck AP, Raymond CM, Rodela R, Moquin R, Boerchers M (2020) Pathways of learning about biodiversity and sustainability in private urban gardens. J Environ Plan Manag 63(6):1056–1076. https://doi.org/10.1080/09640568.2019.1633288

Diekmann LO, Gray LC, Baker GA (2020) Growing 'good food': urban gardens, culturally acceptable produce and food security. Renewable Agric Food Syst 35(2):169–181. https://doi.org/10.1017/S1742170518000388

DST (2010) National mission for sustaining the Himalayan eco-system. Department of Science and Technology, GOI, New Delhi

Eyzaguirre PB, Linares OF (2004) Introduction to homegardens. In: Eyzaguirre PB, Linares OF (eds) Home gardens and agro-biodiversity. Smithsonian Books, Washington, pp 1–28

Ferdous Z, Datta A, Anal AK, Anwar M, Khan AM (2016) Development of home garden model for year round production and consumption for improving resource-poor household food security in Bangladesh. NJAS-Wageningen J Life Sci 78:103–110. https://doi.org/10.1016/j.njas.2016.05.006

Fernandes EC, Oktingati A, Maghembe J (1984) The Chaggahomegardens: a multistoried agroforestry cropping system on Mt. Kilimanjaro (Northern Tanzania). Agrofor Syst 2(2):73–86. https://doi.org/10.1007/BF00131267

Filazzola A, Shrestha N, MacIvor JS (2019) The contribution of constructed green infrastructure to urban biodiversity: a synthesis and meta-analysis. J Appl Ecol 56(9):2131–2143. https://doi.org/10.1111/1365-2664.13475

Finerman R, Sackett R (2003) Using home gardens to decipher health and healing in the Andes. Med Anthropol Q 17(4):459–482. https://doi.org/10.1525/maq.2003.17.4.459

Foysal MA, Hossain ML, Rubaiyat A, Hasan MB (2013) Economics of homestead forestry and their management activities at Fatickchari Upazila of Chittagong district, Bangladesh. Agric For Fish 2(4):161–176

Gajaseni J, Gajaseni N (1999) Ecological rationalities of the traditional homegarden system in the Chao Phraya Basin, Thailand. Agrofor Syst 46:3–23. https://doi.org/10.1023/A:1006188504677
Galhena DH, Freed R, Maredia KM (2013) Home gardens: a promising approach to enhance household food security and wellbeing. Agric Food Secur 2:1–3. https://doi.org/10.1186/2048-7010-2-8
Galluzzi G, Eyzaguirre P, Negri V (2010) Home gardens: neglected hotspots of agro-biodiversity and cultural diversity. Biodivers Conserv 19:3635–3654. https://doi.org/10.1007/s10531-010-9919-5
Gao J, He T, Li QM (2012) Traditional home-garden conserving genetic diversity: a case study of Acacia pennata in Southwest China. Conserv Genet 13:891–898. https://doi.org/10.1007/s10592-012-0338-x
Gavrilidis AA, Niţă MR, Onose DA, Badiu DL, Năstase II (2019) Methodological framework for urban sprawl control through sustainable planning of urban green infrastructure. Ecol Indic 96:67–78. https://doi.org/10.1016/j.ecolind.2017.10.054
George MV, Christopher G (2020) Structure, diversity and utilization of plant species in tribal homegardens of Kerala, India. Agrof Syst 94(1):297–307. https://doi.org/10.1007/s10457-019-00393-5
Gifawesen ST, Tola FK, Duguma MS (2020) Review on role of home garden agroforestry practices to improve livelihood of small scale farmers and climate change adaptation and mitigation. J Plant Sci 8(5):134–145
Giller KE, Delaune T, Silva JV, Descheemaeker K, Van De Ven G, Schut AG, Van Wijk M, Hammond J, Hochman Z, Taulya G, Chikowo R (2021) The future of farming: who will produce our food? Food Secur 13(5):1073–1099. https://doi.org/10.1007/s12571-021-01184-6
Goddard MA, Dougill AJ, Benton TG (2010) Scaling up from gardens: biodiversity conservation in urban environments. Trends Ecol Evol 25(2):90–98. https://doi.org/10.1016/j.tree.2009.07.016
Goyal A, Bahuguna R, Junaid KC, Divya NV, Dhodi RK (2024) Sustainability in the Indian Himalayan region: a thematic analysis of tourism and hospitality research over four decades. International J of Commu Netw and Info Secu 16(1):342–362
Guo LB, Gifford RM (2002) Soil carbon stocks and land use change: a meta analysis. Glob Change Biol 8(4):345–360. https://doi.org/10.1046/j.1354-1013.2002.00486.x
Haque MA, Naher N, Sultana T (2018) Homestead tree species diversity and its impact on the livelihood of the farmers in Bangladesh. Int J Agric Environ Food Sci 2(4):148–154. https://doi.org/10.31015/jaefs.18025
Hickel J (2020) The sustainable development index: measuring the ecological efficiency of human development in the anthropocene. Ecol Econ 167:106331. https://doi.org/10.1016/j.ecolecon.2019.05.011
Hoogerbrugge I, Fresco LO (1993) Homegarden systems: agricultural characteristics and challenges. International Institute for Environment and Development, London. https://edepot.wur.nl/73745
Huang S, Meijers MJ, Eyres A, Mulch A, Fritz SA (2019) Unravelling the history of biodiversity in mountain ranges through integrating geology and biogeography. J Biogeogr 46(8):1777–1791
Idohou R, Fandohan B, Salako VK, Kassa B, Gbedomon C (2014) Biodiversity conservation in homegardens: traditional knowledge, use patterns and implications for management. Int J Biodivers Sci Ecosyst Serv Manag 10(2):89–100
Islam MA, Qaisar KN, Bhat GM (2017a) Indigenous knowledge in traditional agroforestry systems of Kashmir valley: current challenges and future opportunities. Int J For Crop Improv 8(1):68–77
Islam MA, Sofi PA, Bhat GM, Wani AA, Gatoo AA, Singh A, Malik AR (2017b) Public assessment for socioeconomic and environmental services of agroforestry networks in Kashmir Himalaya, India. Int J Curr Microbiol Appl Sci 6(10):410–420

Islam MA, Wani AA, Bhat GM, Gatoo AA, Murtaza S, Anjum K, Atta U (2021) Homestead tree resource production for rural livelihood security in Kashmir Himalaya. Range Manag Agrofor 42(2):334–340

Ivanova T, Bosseva Y, Chervenkov M, Dimitrova D (2021) Enough to feed ourselves!—food plants in Bulgarian rural home gardens. Plants 10(11):2520. https://doi.org/10.3390/plants10112520

Jiménez-Aceituno A, Peterson GD, Norström AV, Wong GY, Downing AS (2020) Local lens for SDG implementation: lessons from bottom-up approaches in Africa. Sustain Sci 15:729–743. https://doi.org/10.1007/s11625-019-00746-0

Jodha NS (1992) Sustainability issues in the mountain context: emerging scenario. Paper presented in the workshop on Approaches to Sustainable Development of the Indian Himalaya, Manali, Himachal Pradesh, 1–4 August 1992

Kabir ME, Webb EL (2009) Household and homegarden characteristics in southwestern Bangladesh. Agrofor Syst 75(2):129–145. https://doi.org/10.1007/s10457-008-9142-5

Karki H, Bargali K, Bargali SS (2024) Carbon sequestration potential of soils under different agroforestry systems of Kumaun Himalaya India and its implication for climate change mitigation. https://doi.org/10.2139/ssrn.4819865

Kattel GR (2022) Climate warming in the Himalayas threatens biodiversity, ecosystem functioning and ecosystem services in the 21st century: is there a better solution? Biodivers Conserv 31(8):2017–2044. https://doi.org/10.1007/s10531-022-02417-6

Keatinge JD, Chadha ML, Hughes JD, Easdown WJ, Holmer RJ, Tenkouano A, Yang RY, Mavlyanova R, Neave S, Afari-Sefa V, Luther G (2012) Vegetable gardens and their impact on the attainment of the Millennium Development Goals. Biol Agric Hortic 28(2):71–85. https://doi.org/10.1080/01448765.2012.681344

Khan AU, Choudhury MA, Tarapder SA, Maukeeb AR, Ema IJ (2020) Status of mango fruit infestation at home garden in Mymensingh, Bangladesh. Curr Res Agric Farm 1(4):35–42. https://doi.org/10.18782/2582-7146.119

Khan AU, Ema IJ, Afsana AS, Khan AU (2021) Home garden and women participation: a mini review. Curr Res Agric Farm 2(4):46–52. https://doi.org/10.18782/2582-7146.152

Kharshi-ing A, Rocky P, Khatri PK (2004) Tree bean (*Parkia roxburghii* G. Don): a commercially lesser-known tree from Northeast. In: Abstracts of the national workshop on sustainable utilization of lesser known tree species, FRI, Dehradun, 8–10 March 2004

Khoshoo TN (1992) Plant diversity in the Himalaya: conservation and utilization. G.B. Pant Institute of Himalayan Environment and Development, Kosi-Katarmal, Almora

Kishtwaria J, Rana A, Sood S (2009) Work pattern of hill farm women–a study of Himachal Pradesh. Stud Home Community Sci 3(1):67–70

Koopman JE (2009) Globalization, gender, and poverty in the Senegal River valley. Fem Econ 15(3):253–285. https://doi.org/10.1080/13545700902920370

Koul DN, Shukla G, Panwar P, Chakravarty S (2011) Status of soil carbon sequestration under different land use systems in Terai Zone of West Bengal. Environ We Int J Sci Technol 6:95–100

Kumar BM (2006) Carbon sequestration potential of tropical homegardens. In: Kumar BM, Nair PKR (eds) Tropical homegardens: a time-tested example of sustainable agroforestry. Springer, Dordrecht, pp 185–204. https://doi.org/10.1007/978-1-4020-4948-4_11

Kumar BM (2011) Species richness and aboveground carbon stocks in the homegardens of Central Kerala, India. Agric Ecosyst Environ 140(3–4):430–440. https://doi.org/10.1016/j.agee.2011.01.006

Kumar BM, Nair PR (2004) The enigma of tropical homegardens. Agrofor Syst 61(1):135–152. https://doi.org/10.1023/B:AGFO.0000028995.13227.ca

Kumar BM, Nair PKR (2006) Introduction. In: Kumar BM, Nair PKR Tropical homegardens: a time-tested example of sustainable agroforestry. Springer, Dordrecht, p 1–10. https://doi.org/10.1007/978-1-4020-4948-4_1

Kumar BM, George SJ, Chinnamani S (1994) Diversity, structure and standing stock of wood in the homegardens of Kerala in peninsular India. Agrofor Syst 25:243–262. https://doi.org/10.1007/BF00707463

Kumari MAS, Kansuntisukmongkol K, Brockelman WY (2009) Plant diversity in home gardens and its contribution to household economy in suburban areas in Sri Lanka. Environ Nat Resour J 7:12–30

Lalremruata J, Sahoo UK, Lalramnghinglova H (2007) Inventory on non-timber forest products of Mizoram in North—East India. J Non-timber For Prod 14(3):173–180. https://doi.org/10.54207/bsmps2000-2007-XIKSDX

Larios C, Casas A, Vallejo M, Moreno Calles AI, Blancas J (2013) Plant management and biodiversity conservation in Náhuatlhomegardens of the Tehuacán Valley, Mexico. J Ethnobiol Ethnomed 9:74. https://doi.org/10.1186/1746-4269-9-74

Leote P, Cajaiba RL, Moreira H, Gabriel R, Santos M (2022) The importance of invertebrates in assessing the ecological impacts of hiking trails: a review of its role as indicators and recommendations for future research. Ecol Indic 137:108741. https://doi.org/10.1016/j.ecolind.2022.108741

Lepcha LD, Shukla G, Sarkar BC, Tamang M, Chettri R, Abha MK, Pala NA, Chakravarty S (2019) Ethno-botanical plant diversity in home gardens based agroforestry system in kanchendzonga biosphere reserve, Sikkim, India. J Tree Sci 38(2):59–69

Loram A, Tratalos J, Warren PH, Gaston KJ (2007) Urban domestic gardens (X): the extent & structure of the resource in five major cities. Landsc Ecol 22(4):601–615. https://doi.org/10.1007/s10980-006-9051-9

Loram A, Warren P, Thompson K, Gaston K (2011) Urban domestic gardens: the effects of human interventions on garden composition. Environ Manag 48(4):808–824

Manpoong C, Wapongnungsang, Tripathi SK (2021) Soil carbon stock in different land-use systems in the hilly terrain of Mizoram, Northeast India. J App Nat Sci 13(2):723–728. https://doi.org/10.31018/jans.v13i2.2615

Mattsson E, Ostwald M, Nissanka SP, Pushpakumara DKNG (2018) Quantification of carbon stock and tree diversity of homegardens in a dry zone area of Moneragala district, Sri Lanka. Agrofor Syst 89(3):435–445. https://doi.org/10.1007/s10457-014-9780-8

Meinzen-Dick R, Behrman J, Menon P, Quisumbing A (2012) Gender a key dimension linking agricultural programs to improved nutrition and health. In: Fan S, Pandya-Lorch R (eds) Reshaping agriculture for nutrition and health. International Food Policy Research Institute, Washington, DC. https://hdl.handle.net/10568/153988

Mellisse BT, Descheemaeker K, Giller KE, Abebe T, van de Ven GW (2018) Are traditional home gardens in Southern Ethiopia heading for extinction? Implications for productivity, plant species richness and food security. Agric Ecosyst Environ 252:1–13. https://doi.org/10.1016/j.agee.2017.09.026

Montagnini F, Nair PKR (2004) Carbon sequestration: an underexploited environmental benefit of agroforestry systems. Agrofor Syst 61:281–295. https://doi.org/10.1023/B:AGFO.0000029005.92691.79

Moreno-Calles A, Casas A, Blancas J, Torres I, Masera O, Caballero J, García-Barrios L, Pérez-Negrón E, Rangel-Landa S (2010) Agroforestry systems and biodiversity conservation in arid zones: the case of the Tehuacán Valley, Central México. Agrofor Syst 80:315–331. https://doi.org/10.1007/s10457-010-9349-0

Musa F, Lile NA, Hamdan DDM (2019) Agroforestry practices contribution towards socioeconomics: a case study of Tawau communities in Malaysia. Agric For 65(1):65–72

Mwavu EN, Ariango E, Ssegawa P, Kalema VN, Bateganya F, Waiswa D, Byakagaba P (2016) Agrobiodiversity of homegardens in a commercial sugarcane cultivation land matrix in Uganda. Int J Biodivers Sci Ecosyst Serv Manag 12(3):191–201. https://doi.org/10.1080/21513732.2016.1177595

Nagamani K, Mishra AK, Meer MS, Das J (2024) Understanding flash flooding in the Himalayan region: a case study. Sci Rep 14:7060. https://doi.org/10.1038/s41598-024-53535-w

Nair PKR, Kumar BM, Nair VD (2009) Agroforestry as a strategy for carbon sequestration. J Soil Sci Plant Nutr 172:10–23. https://doi.org/10.1002/jpln.200800030

Namitha VV, Raj SK, Prathapan K (2022) Carbon sequestration potential in coconut based cropping system: a review. Agric Rev 46(1):143–146

Nath AJ, Das AK (2011) Carbon storage and sequestration in bamboo-based smallholder homegardens of Barak Valley, Assam. Curr Sci 25:229–233

Nathaniel SP, Adeleye N (2021) Environmental preservation amidst carbon emissions, energy consumption, and urbanization in selected African countries: implication for sustainability. J Clean Prod 285:125409. https://doi.org/10.1016/j.jclepro.2020.125409

Nelson E, Mendoza G, Regetz J, Polasky S, Tallis H, Cameron D, Chan KM, Daily GC, Goldstein J, Kareiva PM, Lonsdorf E (2009) Modeling multiple ecosystem services, biodiversity conservation, commodity production, and tradeoffs at landscape scales. Front Ecol Environ 7(1):4–11. https://doi.org/10.1890/080023

Noble IR, Dirzo R (1997) Forests as human-dominated ecosystems. Science 277(5325):522–525. https://doi.org/10.1126/science.277.5325.522

Nyaupane GP, Timothy DJ (2022) Introduction: tourism and development in the Himalaya. In: Nyaupane GP, Timothy DJ (eds) Tourism and development in the Himalaya: social, environmental, and economic forces, 1st edn. Routledge, London

Padalia K, Bargali K, Bargali SS (2015) How does traditional home-gardens support ethnomedicinal values in Kumaun Himalayan Bhabhar belt, India? Afr J Tradit Complement Altern Med 12(6):100–112. https://doi.org/10.4314/ajtcam.v12i6.10

Pala NA, Sarkar BC, Shukla G, Chettri N, Deb S, Bhat JA, Chakravarty S (2019) Floristic composition and utilization of ethnomedicinal plant species in home gardens of the Eastern Himalaya. J Ethnobiol Ethnomed 15:14. https://doi.org/10.1186/s13002-019-0293-4

Pandey K, Bargali SS, Kolhe SS (2011) Adoption of technology by rural women in rice based agroecosystem. Int Rice Res Notes 36:1–4

Panwar P, Chakravarty S (2010) Floristic structure and ecological function of homegardens in humid tropics of West Bengal, India. Indian J Agrofor 12:69–78

Panwar P, Chauhan S, Kaushal R, Das DK, Ajit, Arora G, Chaturvedi OP, Jain AK, Chaturvedi S, Tewari S (2017) Carbon sequestration potential of poplar-based agroforestry using the CO2FIX model in the Indo-Gangetic Region of India. Trop Ecol 58(2):439–447

Panwar P, Mahalingappa DG, Kaushal R, Bhardwaj DR, Chakravarty S, Shukla G, Thakur NS, Chavan SB, Pal S, Nayak BG, Srinivasaiah HT (2022) Biomass production and carbon sequestration potential of different agroforestry systems in India: a critical review. Forests 13(8):1274. https://doi.org/10.3390/f13081274

Parihaar RS, Bargali K, Bargali SS (2015) Status of an indigenous agroforestry system: a case study in Kumaun Himalaya, India. Indian J of Agri Sci 85(3):442–447

Perfecto I, Vandermeer J (2008) Biodiversity conservation in tropical agroecosystems: a new conservation paradigm. Ann N Y Acad Sci 1134(1):173–200. https://doi.org/10.1196/annals.1439.011

Perrault-Archambault M, Coomes OT (2008) Distribution of agrobiodiversity in home gardens along the Corrientes River, Peruvian Amazon. Eco Bot 62:109–126. https://doi.org/10.1007/s12231-008-9010-2

Petterson ML (2000) Agroforestry in Blege: Maya homegardens in San Lucas. Master's Thesis, Department of Renewable resources, University of Albeta

Peyre A, Guidal A, Wiersum KF, Bongers FJ (2006) Dynamics of homegarden structure and function in Kerala, India. Agrofor Syst 66:101–115. https://doi.org/10.1007/s10457-005-2919-x

Porter C (2018) What gardens grow: outcomes from home and community gardens supported by community-based food justice organizations. J Agric Food Sys Community Dev 8:187–205. https://doi.org/10.5304/jafscd.2018.08A.002

Putri WK, Hakim L, Batoro J (2016) Ethnobotanical survey of homegardens in Pandansari and Sumberejo to support ecotourism program in Bromo Tengger Semerru National Park, Indonesia. Int J Res Stud in Agric Sci 2(1):6–12

Rahman SA, Baldauf C, Mollee EM, Abdullah-Al-Pavel M, Abdullah-Al-Mamun M, Toy MM, Sunderland T (2013) Cultivated plants in the diversified homegardens of local communities in Ganges Valley, Bangladesh. Sci J Agric Res Manag 2013:1–6

Rammohan A, Pritchard B, Dibley M (2019) Home gardens as a predictor of enhanced dietary diversity and food security in rural Myanmar. BMC Public Health 19:1145. https://doi.org/10.1186/s12889-019-7440-7

Raymond CM, Diduck AP, Buijs A, Boerchers M, Moquin R (2019) Exploring the co-benefits (and costs) of home gardening for biodiversity conservation. Local Environ 24(3):258–273. https://doi.org/10.1080/13549839.2018.1561657

Reang D, Sahoo UK, Giri K, Hazarika A, Nath AJ (2021) Ethnic homestead forests of north-East India revealed as diverse land-use systems. Agrofor Syst 96(3):465–478. https://doi.org/10.1007/s10457-021-00700-z

Reyes-García V, Vila S, Aceituno-Mata L, Calvet-Mir L, Garnatje T, Jesch A, Lastra JJ, Parada M, Rigat M, Vallès J, Pardo-de-Santayana M (2010) Gendered homegardens: a study in three mountain areas of the Iberian Peninsula. Econ Bot 64:235–247. https://doi.org/10.1007/s12231-010-9124-1

Rocky P, Sahoo UK, Thapa HS (2004) Livelihood generation through tree bean (Parkia roxburghii g. Don) in Imphal. J Non-Timber Forest Prod 11(2):135–139

Roshetko JM, Delaney M, Hairiah K, Purnomosidhi P (2002) Carbon stocks in Indonesian homegarden systems: can smallholder systems be targeted for increased carbon storage? Am J Altern Agric 17(3):138–148. https://doi.org/10.1079/AJAA200116

Roy M, Sarkar BC, Manohar KA, Shukla G, Vineeta, Nath AJ, Bhat JA, Chakravarty S (2022a) Fuelwood species diversity and consumption pattern in the homegardens from foothills of Indian Eastern Himalayas. Agrofor Syst 96(3):453–464. https://doi.org/10.1007/s10457-021-00728-1

Roy M, Sarkar BC, Shukla G, Debnath MK, Nath AJ, Bhat JA, Chakravarty S (2022b) Traditional homegardens and ethnomedicinal plants: insights from the Indian Sub-Himalayan region. Trees For People 8:100236. https://doi.org/10.1016/j.tfp.2022.100236

Sachs JD, Schmidt-Traub G, Mazzucato M, Messner D, Nakicenovic N, Rockström J (2019) Six transformations to achieve the sustainable development goals. Nat Sustainability 9:805–814. https://doi.org/10.1038/s41893-019-0352-9

Saha SK, Nair PKR, Nair VD, Kumar BM (2009) Soil carbon stock in relation to plant diversity of homegardens in Kerala, India. Agrofor Syst 76:53–65. https://doi.org/10.1007/s10457-009-9228-8

Saha SK, Nair PKR, Nair VD, Kumar BM (2010) Carbon storage in relation to soil size-fractions under tropical tree-based land-use systems. Plant Soil 328:433–446. https://doi.org/10.1007/s11104-009-0123-x

Sahoo UK (2007) Agroforestry systems and practices prevailing in Mizoram. In: Puri S, Panwar P (eds) Agroforestry systems and practices. New India Publishing Agency, New Delhi

Sahoo UK (2009) Traditional home gardens and livelihood security in North-East India. J Food Agric Environ 7(2):665–670

Sahoo UK, Jeeceelee L (2015) Plant species diversity and its utilization pattern in homegardens of Mizo community of North-East India. In: Tripathi SK (ed) Biodiversity in tropical ecosystems. Today and Tomorrow's Printers and Publishers, New Delhi, pp 203–224

Sahoo UK, Rocky P (2015) Species composition and plant diversity as influenced by altitude and size of homegardens in Mizoram, north-East India. Int J Ecol Environ Sci 41(3–4):195–215

Saikia P, Choudhury BI, Khan ML (2012) Floristic composition and plant utilization pattern in homegardens of Upper Assam, India. Trop Ecol 53(1):105–118

Saleem I, Mugloo JA, Pala NA, Bhat GM, Masoodi TH, Mughal AH, Baba AA, Mehraj B (2023) Biomass production, carbon stock and sequestration potential of prominent agroforestry systems in north-western Himalaya, India. Front For Glob Change 14(6):1192382. https://doi.org/10.3389/ffgc.2023.1192382

Samant SS, Dhar U (1997) Diversity, endemism and economic potential of wild edible plants of Indian Himalaya. Int J Sustain Dev World Ecol 4(3):179–191. https://doi.org/10.1080/13504509709469953
Samant SS, Palni LM (2000) Diversity, distribution and indigenous uses of essential oil-yielding medicinal plants of the Indian Himalayan region. J Med Aromat Plant Sci 22(1B):671–684
Samant SS, Shreekar Pant SP (2003) Diversity, distribution pattern and traditional knowledge of sacred plants in Indian Himalayan Region. Indian J For 26(3):222–234
Samnegard U, Persson AS, Smith HG (2011) Gardens benefit bees and enhance pollination in intensively managed farmland. Biol Conserv 144(11):2602–2606. https://doi.org/10.1016/j.biocon.2011.07.008
Sampanpanish P, Jamroenprucksa M (1994) Ecological characteristics of homegarden agroforestry system in Amphoe Muang, Changwat Nonthaburi. Thai J For 13:114–124
Sanyang SE, Kao TC, Huang WC (2009) The impact of agricultural technology transfer to women vegetable production and marketing groups in The Gambia. World J Agric Sci 5(2):169–179
Sanyé-Mengual E, Gasperi D, Michelon N, Orsini F, Ponchia G, Gianquinto G (2018) Eco-efficiency assessment and food security potential of home gardening: a case study in Padua, Italy. Sustainability 10(7):2124. https://doi.org/10.3390/su10072124
Sarkar BC, Shukla G, Debnath MK, Pala NA, Chakravarty S (2020) Diversity and utilization pattern of homegardens: a case study from eastern Himalayan region of West Bengal. Indian J Agrofor 22(2):31–36
Sarkar BC, Shukla G, Suresh CP, Chakravarty S (2023) A review on structure, floristic diversity and functions of homegardens. In: Uppaluri RVS, Rangan, L (eds) Conservation of biodiversity in the north eastern states of India. Springer, Singapore, pp 291–308. https://doi.org/10.1007/978-981-99-0945-2_16
Senanayake RL, Sangakkara UR, Pushpakumara DK, Stamp P (2010) Vegetation composition and ecological benefits of home gardens in the Meegahakiula region of Sri Lanka. Trop Agric Res 1:1–9. https://doi.org/10.4038/tar.v21i1.2581
Sharma R, Mina U, Kumar BM (2022) Homegarden agroforestry systems in achievement of sustainable development goals. A review. Agron Sustain Dev 42:44. https://doi.org/10.1007/s13593-022-00781-9
Shastri CM, Bhat DM, Nagaraja BC, Murali KS, Ravindranath NH (2002) Tree species diversity in a village ecosystem in Uttara Kannada district in Western Ghats, Karnataka. Curr Sci 82:1080–1084
Shukla G, Vineeta AK, Chakravarty S (2017) Plant diversity, structure and uses of the plants in home garden of Jharkhand, India. Indian J Trop Biodivers 25(1):40–50
Singh JS (1992) Forests of Himalaya: structure, functioning and impact of man. Gyanodaya Prakashan, Nainital
Singh JS (2006) Sustainable development of the Indian Himalayan region: linking ecological and economic concerns. Curr Sci 25:784–788. https://www.jstor.org/stable/24089189
Singh P, Tewari P (2013) Intra-zonal differences in roles among gender in farming activities. Int J Social Work Human Serv Pract 1:15–20
Singh SL, Sahoo UK, Gogoi A, Kenye A (2018) Effect of land use changes on carbon stock dynamics in major land use sectors of Mizoram, Northeast India. J Environ Prot 9(12):1262–1285. https://doi.org/10.4236/jep.2018.912079
Singh P, Sinha BK, Dash SS (2019) Plants of Indian Himalayan region: an annotated checklist & pictorial guide. Botanical Survey of India, Kolkata
Singh SL, Sahoo UK (2021) Tree species composition, diversity and soil organic carbon stock in homegardens and shifting cultivation fallows of Mizoram, Northeast India. Vegetos 34(1):220–228. https://doi.org/10.1007/s42535-021-00194-1
Subba M, Pala NA, Shukla G, Chakravarty S (2018) Study of the variability of home gardens influencing carbon stock under sub-humid tropical zone of West Bengal, India. Indian Forester 144(1):60–66

Sthapit B, Gautam R, Eyzaguirre PB (2004) The value of home gardens to small farmers. In: Gautam R, Sthapit B, Shrestha P (eds) Proceeding of National Workshop on "Enhancing the contribution of home garden to on-farm management of plant genetic resources and to improve the livelihoods of Nepalese farmers: lessons learned and policy implications", Pokhara, Nepal, 6–7 August 2004

Sunwar S, Thornström CG, Subedi A, Bystrom M (2006) Home gardens in western Nepal: opportunities and challenges for on-farm management of agrobiodiversity. Biodivers Conserv 15:4211–4238. https://doi.org/10.1007/s10531-005-3576-0

Tesfaye A, Wiersum KF, Bongers F (2010) Spatial and temporal variation in crop diversity in agroforestry homegardens of Southern Ethiopia. Agrofor Syst 78:309–322. https://doi.org/10.1007/s10457-009-9246-6

Thrupp LA (2000) Linking agricultural biodiversity and food security: the valuable role of agrobiodiversity for sustainable agriculture. Int Aff 76:265–281

Tilinti B, Negash M, Asfaw Z, Woldeamanuel T (2025) Variations in carbon stocks across traditional and improved agroforestry in reference to agroforestry and households' characteristics in Southeastern Ethiopia. Heliyon 11(2):e42127. https://doi.org/10.1016/j.heliyon.2025.e42127

Tiwari V, Rawat R, Negi KS, Chandra S (2016) Inventorying plant biodiversity in home gardens: a case study of Alaknanda and Mandakini Valleys, Garhwal Region of Uttarakhand, India. Asian Agri-History 20(4):275–290

Tiyakoat W, Chllachakkawat C, Dhompongsa G, Sarobol S (2010) The migration of Tai Yai from Shan State-Myanmar, into Thailand. In: Abstracts of the 1st International Conference on Culture, Toruism& Economy in Salween River Basin Reion, Mae Hong Son, Thailand, 10–11 August 2010

Tolangay D, Moktan S (2020) Trend of studies on carbon sequestration dynamics in the Himalaya hotspot region: a review. J Appl Nat Sci 12(4):647–660. https://doi.org/10.31018/jans.v12i4.2426

Tse-Ring K, Sharma E, Chettri N, Shrestha A (2012) Climate change vulnerability of mountain ecosystems in the Eastern Himalayas. Working Papers id:5000, eSocialSciences

UNFCCC (2007) Report of the Conference of the Parties on its thirteenth session held in Bali. https://unfccc.int/resource/docs/2007/cop13/eng/06a01.pdf

USDOE (2008) Carbon cycling and biosequestration: integrating biology and climate through systems science; report from the march 2008 workshop, DOE/SC-108, U.S. Department of Energy Office of Science. https://www.genomicscience.energy.gov/wp-content/uploads/2021/10/CarbonCycle012609HR.pdf

Veith C, Shaw J (2011) Why invest in sustainable mountain development? Food and Agriculture Organization, Rome

Vibhuti, Bargali K, Bargali SS (2018) Effects of homegarden size on floristic composition and diversity along an altitudinal gradient in central Himalaya, India. Curr Sci 25:2494–2503. https://www.jstor.org/stable/26495762

Vibhuti, Bargali K, Bargali SS (2019) Species composition, diversity and traditional uses of homegarden in Kumaun Himalaya, India. Indian J Agric Sci 89(9):1415–1418

Vibhuti, Bargali K, Bargali SS (2020) Effect of size and altitude on soil organic carbon stock in homegarden agroforestry system in central Himalaya, India. Acta Ecologica Sinica 40(6):483–491. https://doi.org/10.1016/j.chnaes.2020.10.002

Vieira J, Matos P, Mexia T, Silva P, Lopes N, Freitas C, Correia O, Santos-Reis M, Branquinho C, Pinho P (2018) Green spaces are not all the same for the provision of air purification and climate regulation services: the case of urban parks. Environ Res 160:306–313. https://doi.org/10.1016/j.envres.2017.10.006

Wiersum KF (2006) Diversity and change in homegarden cultivation in Indonesia. In: BM Kumar, Nair PKR Tropical homegardens: a time-tested example of sustainable agroforestry. Springer, Dordrecht, p 13–24. https://doi.org/10.1007/978-1-4020-4948-4_2

Xu J, Badola R, Chettri N, Chaudhary RP, Zomer R, Pokhrel B, Hussain SA, Pradhan S, Pradhan R (2019) Sustaining biodiversity and ecosystem services in the Hindu Kush Himalaya. In: Wester P, Mishra A, Mukherji A, Shrestha A (eds) The Hindu Kush Himalaya assessment: mountains, climate change, sustainability and people. Springer Cham, pp 127–165. https://doi.org/10.1007/978-3-319-92288-1_5

Yasin H, Kebebew Z, Hundera K (2018) Woody species diversity, regeneration and socioeconomic benefits under natural forest and adjacent coffee agroforests at Belete forest, Southwest Ethiopia. Ekológia (Bratislava) 37(4):380–391

Zimmerer KS, Carney JA, Vanek SJ (2015) Sustainable smallholder intensification in global change? Pivotal spatial interactions, gendered livelihoods, and agrobiodiversity. Curr Opin Environ Sustain 14:49–60. https://doi.org/10.1016/j.cosust.2015.03.004

Zobel DB, Singh SP (1997) Himalayan forests and ecological generalizations. Bioscience 11:735–745. https://doi.org/10.2307/1313096

Chapter 4
Perceptions of Climate Change and Mitigation Among Traditional Large Cardamom Agroforestry Growers in the Darjeeling Himalaya

Bisleshna Tamang, Vineeta, Prabhat Kumar Pal, Gopal Shukla, and Sumit Chakravarty

Abstract Understanding the perceptions of large cardamom growers in the Darjeeling Himalayas regarding climate change and its socioeconomic impacts is essential for enhancing their adaptive capacity. This study is based on responses from 125 randomly selected growers and reveals that younger farmers have smaller landholdings, are more educated, and economically better off. Growers widely perceived rising temperatures, erratic and reduced rainfall, stronger winds, and increased humidity—factors associate with declining crop productivity and threats to their livelihoods. Most respondents favored mitigation strategies such as mulching, improved planting materials, and adoption of better technology to enhance productivity. The findings highlight the need for climate-resilient practices to support sustainable livelihoods in large cardamom farming under changing climatic conditions.

Keywords Darjeeling Himalayas · Climate change perception · Large cardamom · Socioeconomic impacts · Adaptive strategies

B. Tamang · Vineeta (✉) · S. Chakravarty
Department of Forestry, Uttar Banga Krishi Viswavidyalaya, Cooch Behar, West Bengal, India

P. K. Pal
Department of Agricultural Extension, Uttar Banga Krishi Viswavidyalaya, Cooch Behar, West Bengal, India

G. Shukla
Department of Forestry, North Eastern Hill University, Tura, Meghalaya, India

B. Singh et al. (eds.), *Himalayan Agroforestry Systems*, Advances in Agroforestry 15,
https://doi.org/10.1007/978-3-032-25096-4_4

1 Introduction

The concern about global climate change is abundantly reported with a fear that it can potentially restrict the achievement of Millennium Development Goals (Jagermeyr 2020; Tamang et al. 2021; Pradhan et al. 2022). Climate change is progressively impacting critical sectors such as water resources, biodiversity, agriculture, human health, and food security, with these effects being felt across various nations (Wangchuk and Wangdi 2018; Rasul et al. 2019; Singh et al. 2020). These environmental shifts have far-reaching consequences on the livelihoods of local communities, particularly marginalized and economically disadvantaged groups who heavily depend on natural resources for their subsistence (Dey et al. 2017a, b; Meena et al. 2019). In India, more than 400 million rural poor, including indigenous populations, farming communities, and forest-dependent households, are particularly vulnerable to the unpredictable nature of climate variability (Shukla et al. 2016; Dey et al. 2017a, b). Mountainous regions, especially, are experiencing heightened susceptibility to the effects of climate change, rendering Himalayan communities increasingly vulnerable to its adverse consequences (Gioli et al. 2019; Chhogyel et al. 2020).

The Eastern Himalayan regions, too, face mounting climate-related challenges, amplifying both environmental and socioeconomic vulnerabilities (Mukherjee 2013; Vineeta et al. 2021, 2023). In Sikkim and the Darjeeling Himalayan regions, farmers are witnessing the escalating negative impacts of climate change, which is threatening traditional agroforestry systems, particularly the cultivation of large cardamom (*Amomum subulatum* Roxb.). As a vital cash crop native to the moist, semi-evergreen forests of the Eastern Himalayas are highly sensitive to changing climatic conditions (Sharma et al. 2016; Vineeta et al. 2021, 2023). Traditionally, this crop has been cultivated under the canopy of Himalayan alder (*Alnus nepalensis* D. Don) in mixed agroforestry systems. The cultivation of large cardamom significantly contributes to the regional economy, representing about 30% of household income in these areas (Vineeta et al. 2021, 2023).

Large cardamom cultivation not only holds significant economic value by influencing local trade, employment, and industrial growth, but also supports an extensive network of farmers, laborers, traders, exporters, and processing units (Sharma et al. 2016; Vineeta et al. 2021, 2023). Beyond its economic value, large cardamom-based traditional agroforestry systems (TAF) offer substantial ecological benefits. These systems contribute to biodiversity conservation, enhance carbon sequestration, improve soil fertility, regulate microclimates, and mitigate the broader impacts of climate change (Vineeta et al. 2021; Rai et al. 2022). Its demand for high-quality, organic large cardamom has risen, particularly in South Asian and Middle Eastern markets, where the spice is integral to traditional cuisines and medicinal practices (Vineeta et al. 2021, 2023). However, over the past decade, large cardamom cultivation has experienced a sharp decline in both the area under cultivation and its productivity, driven by a multitude of stressors. These include the direct effects of climate change, pest infestations, the spread of fungal and viral diseases (such as

Colletotrichum blight, “chirkey,” and “foorkey”), price volatility, and rising anthropogenic pressures (Sharma et al. 2016; Negi et al. 2018; Rai et al. 2022).

In order to cope with these challenges, farmers have adapted by shifting plantations across altitudes to find more favorable conditions. Studies highlight their resilience and innovation through strategies such as crop diversification, adoption of climate-resilient varieties, adjusted planting schedules, use of organic fertilizers, soil and water conservation, contour farming, and financial tools like agricultural insurance (Diallo et al. 2020; Li et al. 2021; Bagheri and Teymouri 2022). Despite their critical role, the adaptation strategies of grassroots communities remain under-documented and under-recognized (Shukla et al. 2016; Dey et al. 2017a, b). Climate studies have largely focused on macro-landscapes, overlooking local-level insights (Gillies et al. 2013; Karmacharya et al. 2015). A bottom-up approach is essential, as local and indigenous knowledge offers valuable insights into site-specific climate impacts and responses (Makondo and Thomas 2018; Islam et al. 2020; Islam 2022). Understanding the perceptions and socioeconomic realities of large cardamom growers is crucial for designing effective, community-based adaptation and mitigation strategies, particularly in the context of the Darjeeling Himalayas (Brody et al. 2008).

2 Materials and Methods

The study was carried out in the Darjeeling Himalayan region of West Bengal, India, from April 2019 to April 2022 (Fig. 4.1). The extent of the study site was 3149 km^2 between 26° 27′ 05″-27° 13′ 10″ N latitude and 87° 59′ 30″-88° 53′ E longitude within the altitudinal range of 132–3660 meter above sea level. The region is humid and subtropical to sub-alpine but has a dry winter with a normal temperature of the coldest month between −3 °C and −18 °C while above 22 °C during the

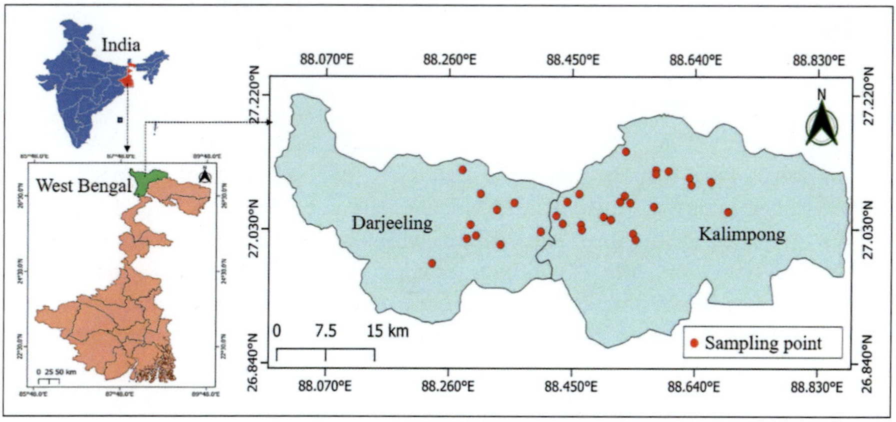

Fig. 4.1 Outline map of study area

warmest month (Saxena 2005). Rainfall of the region is through the southwest monsoon, with an average annual rainfall of 270–310 cm, of which about 80% is received between June and August (Moktan and Das 2013). The mountain and glacial soil, brown hill soil, forest soil, brown forest soil, tea soil, cinchona soil, and Terai soil types of the region are acidic, yellow to red-brown in color and silty loam to sandy loam textured but poor in calcium, magnesium, nitrogen, potassium, phosphate, and organic matter (De and Bera 1990; Froehlich and Sarkar 2000).

The population of the Darjeeling Himalayas was reported to be 1,842,034 (Census, 2011), with a sex ratio of 971 females per 1000 males, density of 585 people km^{-2}, and average literacy rate of 79.92%. The major land uses of the Darjeeling Himalayas are forests and agriculture, including tea gardens (Cajee 2018). The diverse population of Nepalis, Limbus, Lepchas, Tamang, Bhuita, and tea plantation laborers are marginally agrarian, cultivating paddy, maize, millet, chilies, ginger, orange, large cardamom, flowers, orchids, and other vegetables (www.darjeeling.gov.in).

A reconnaissance survey was conducted in the study area to explore the large cardamom-based TAF. The study employed both purposive and simple random sampling techniques, wherein large cardamom growers were selected as respondents using the simple random sampling method.

The survey revealed that of the eight community blocks, the large cardamom plantations were concentrated in six blocks, so all these blocks were selected. The plantations were mostly concentrated in particular locations. The survey found 15 such locations in the study area. One-tenth of the owners in each location were randomly selected as respondents of the study from the exhaustive list of the large cardamom owners aged 30 years or above. Thus, a total of 125 respondents were finally selected for the study (Dey et al. 2017a, b; Lepcha et al. 2018). Pre-tested, well-structured questionnaire was prepared for personal in-depth interviews with the selected respondents (Frechtling et al. 1997).

3 Results and Discussion

3.1 *Profile of the Respondents*

The majority of the respondents were male (75.22%), while the rest were female (24.68%). The majority (56%) of respondents were aged between 30 and 50 years, followed by 36% in the 50–70 age group, while the remaining participants were aged 70 years or older. Only 4% of the respondents did not attend any formal education, while the rest of them were formally educated, with a few up to master's level. More than three-fourths of the respondents were farmers. The respondents were prominently represented by scheduled tribe (44.8%), followed by backward classes (26.4%), scheduled caste (16%), and least respondents of general caste (12.8%). The majority of the respondents (60%) had landholding between one and three acres, while about 26% held less than an acre and about 13% owned more than three acres of land. Based on the classification of Indian households by annual income (Rathore 2022), the respondents of the present study were classified into only two groups as low, with 12% of respondents, and the remaining were in the moderate-income group. The total annual household income of 12% respondents have less than 10% share from large cardamom (low contribution) while large cardamom contributed more than 30% of the total annual household income (high contribution) of about 44% of respondents and the rest with contribution between 10% and 30% (medium). It was observed that more educated respondents, despite having smaller landholdings, had greater access to diverse income sources and demonstrated higher awareness of improved farming technologies and markets. This enabled them to adopt better cultivation practices for large cardamom, resulting in higher earnings. A direct and significant correlation was found between income from large cardamom and total annual household income.

3.2 Perception of Climate Change

Entire respondents of the study were aware of climate change manifested through indicators like temperature and precipitation, irrespective of their age, gender, formal education, occupation, landholding, income, and inheritance except caste. Overall, in the study area entire respondent growers were well aware about the phenomena related to climate change though with different levels of perception having a mean perception score related to climatic variability 0.98, which is high (Table 4.1). Of all the 29 indicator statements designed for this purpose, the majority of the respondents (92–100%) agreed all the statements, except for the statement IC4 where the majority of the respondents disagreed (65.6%) on the usage of ceiling or table fans. The perception of climate change of the Darjeeling Himalayan large cardamom growers is well validated with past meteorological trend analysis from recorded climatic data of the last few decades in the Himalayan region. However, analyzing trend of climatic data for Darjeeling Himalayas in the past four decades (1981–2022) indicates a decreasing trend of daily maximum and average day temperature but increase in daily minimum temperature with increasing trend of average annual relative humidity and annual rainfall.

The trend of the past historical climatic data of the Darjeeling Himalaya partially matches the climate change perception of the large cardamom growers in terms of the quantitative value of the data. The growers perceived climate change manifested through an increase in temperature and change in pattern and increase in the intensity of rainfall. With increasing rainfall, the relative humidity of the region also increased, which actually led to the perception of increased temperature by the large cardamom growers instead of a gradual decrease in temperature at Darjeeling Himalayas.

The spider diagram also clearly indicates low to high perception level on climate change (depending on six main indicators of climate change, i.e., G1–G6, i.e.,

Table 4.1 Distribution of respondents based on climate change perception score

IC	Indicators of climate change perception	Distribution (% of N)			MS
		Y	N	DK	
G1 General perception of climate change					
IC1	Climate is changing over the years	100.0	0	0	1.0
IC2	Seasons are changing	100.0	0	0	1.0
Average perception score on climate change					**1.0**
G2 Perception of temperature variability					
IC3	Temperature has increased	100.0	0	0	1.0
IC4	Ceiling fan or other fans are required now	34.4	65.6	0	0.67
IC5	Winter temperature has increased	98.4	1.6	0	0.99
IC6	Duration of winter has decreased	92.0	8.0	0	0.96
IC7	Difference between day and night temperatures has narrowed down	100.0	0	0	1.0
Average perception score on temperature variability					**0.92**
G3 Perception of wind variability					
IC8	Winds are getting warmer	99.2	0.8	0	0.99
IC9	Wind storm getting stronger	97.6	2.4	0	0.99
Average perception score on wind variability					**0.99**
G4 Perception of precipitation variability					
IC10	Change in intensity and pattern of rainfall	100.0	0	0	1.0
IC11	Rainfall intensity increased	97.6	1.6	0.8	0.98
IC12	Rainfall/snowfall is more unpredictable	99.2	0	0.8	0.99
IC13	Uneven distribution of rainfall	99.2	0.8	0	0.99
IC14	Early withdrawal of monsoon	99.2	0	0.8	0.99
IC15	Late onset of monsoon	100.0	0	0	1.0
IC16	Duration of rainy days decreased	98.4	1.6	0	0.99
IC17	Duration of cloudy days decreased	97.6	2.4	0	0.99
IC18	Snowfall has decreased on the mountains	98.4	0.8	0.8	0.99
IC19	Less duration of snow on the mountain	100.0	0	0	1.0
Average perception score on precipitation variability					**0.99**
G5 Perception of extreme weather events					
IC20	Heavy rainfall and storms/cyclone increased	100.0	0	0	1.0
IC21	Sudden occurrence of storm/cyclone	95.2	4.8	0	0.97
IC22	Hail storm occur beyond autumn and spring season	100.0	0	0	1.0
IC23	Incidents of drought and floods are increasing	100.0	0	0	1.0
IC24	Long dry spell	100.0	0	0	1.0
IC25	Extreme weather events increased	100.0	0	0	1.0

(continued)

Table 4.1 (continued)

IC	Indicators of climate change perception	Distribution (% of N)			MS
		Y	N	DK	
Average perception score on extreme weather events					**0.99**
G6 Perception of relation of climate change and weather variability					
IC26	Climate change affects the season	100.0	0	0	1.0
IC27	Climate change affects late onset of monsoon	100.0	0	0	1.0
IC28	Increase in temperature affects the late onset of monsoon	100.0	0	0	1.0
IC29	Climate change is responsible for the decrease in cloudy days, duration, intensity, and pattern of rainfall	100.0	0	0	1.0
Average perception score on the relation of climate change and weather variability					**1.0**
Overall perception score					**0.98**

N = 125; *IC* indicator code, *Y* yes, *N* no, *DK* don't know, *MS* mean score

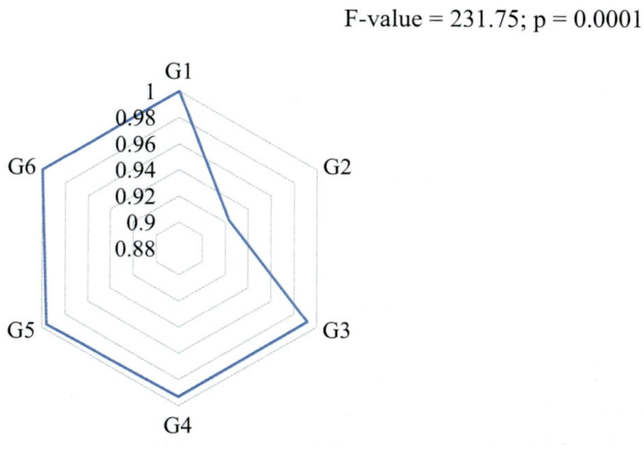

Fig. 4.2 Spider diagram depicting large cardamom grower's perception of different indicators of climate change. G1, general perception of climate change; G2, perception of temperature variability; G3, perception of wind variability; G4, perception of precipitation variability; G5, perception of extreme weather events; G6, perception of relation of climate change and weather variability

0.92–1.00) of large cardamom growers of the Darjeeling Himalayas but perception level varying significantly among these six indicators of climate change (Fig. 4.2).

Many studies across the globe also reported similar awareness of climate variability and perception of climate change in terms of increasing temperature and decreasing rainfall (Roy et al. 2021; Rai et al. 2022; Dahal et al. 2023). Studies from different parts of the Himalayas reported temperature rise; shift in seasons; reduced severity of cold during winter and changes in the number of cloudy days; and changes in timing, magnitude, and intensity of precipitation (Paudel et al. 2020; Rai et al. 2022; Dahal et al. 2023).

3.3 Perception of the Impact of Climate Change on Livelihood

The large cardamom growers who inherited the farming from their parents/forefathers have a higher level of perception regarding the impacts of climate change on their livelihood than those growers who were first-generation large cardamom growers. Given the perception of the impact and risk of climate change at the local level on biological patterns and activity, natural hazards and drastic climatic events, agriculture and large cardamom farming that were influenced by personal values, beliefs, and experiences, it is obvious for the respondents to also perceive their livelihood impacts through such indicators locally (Adger et al. 2007; Dey et al. 2017b; Vineeta et al. 2023).

Based on the average perception score of 0.927 after assigning scores to 18 different indicator statements related to the impact/risk of climate change on livelihood, it was found that the overall level of perception of the large cardamom growers on the impact of climate change on livelihood is medium (Table 4.2). Similarly, based on grouping these 18 indicator statements to five indicator groups, that is, overall livelihood (G1), income (G2), health-related expenditure (G3), occupation and migration (G4), and poverty and starvation (G5), the average perception score estimated was 0.985 with a high perception level, 0.902 (low), 0.868 (low), 0.917 (medium), and 0.961 (high), respectively. Almost all large cardamom growers perceived the decline in water bodies, reduced water availability, and rapid drying after the rainy season as significant impacts and risks associated with climate change.

All these growers also believed the increased incidences of natural calamities like landslides and flash floods over the past decades in their region were due to climate change and impending threats for them (Shukla et al. 2016; Dey et al. 2017a; Vineeta et al. 2023). The spider diagram (Fig. 4.3) also clearly indicates low to medium perception level on different indicators, that is, G1–G5 related to climate change impact and risk on livelihood but the perception level of the respondents varied significantly among these five indicators.

Entire respondents of the study believed that livelihood will be affected due to climate change with a reduced standard of living perceived by 92% of the respondents. This is because entire respondents believed that due to climate change their income will decrease particularly from agriculture that too from their traditional farming of large cardamom, Darjeeling Mandarin, ginger, and orchids, while about 83% respondents still believe they can rely on their NTFP collection from forest as they believe that income from NTFP has not decreased yet. However, about 74% of the respondents believed that their dependency on forest has decreased as they believe that climate change has also decreased the productivity of the forest.

Having belief that income from their primary occupation, that is, farming will not be profitable due to climate change impacts, 80–96% of the respondents are thinking of alternate occupation by leaving agriculture or abandoning their traditional farming of large cardamom or migrating elsewhere. Additionally, almost entire respondents (89.6–99.2%) believe that climate change will increase drudgery and miseries in life due to poverty, food shortage, and starvation and increased

Table 4.2 Distribution of the respondents based on the perception score of climate change impact and risk

IC	Indicators of climate change impact on livelihood	Distribution (% of N)			MS
		Y	N	DK	
G1Perception of overall livelihood					
IC1	Many people's livelihood will be affected	100.0	0	0	1.0
IC2	My livelihood will be affected	100.0	0	0	1.0
IC3	Standard of living will decrease	92.0	7.2	0.8	0.956
Average perception score of impact on overall livelihood					**0.985**
G2 Perception of income					
IC4	Decrease in income	100.0	0	0	1.0
IC5	Income from agriculture has decreased	100.0	0	0	1.0
IC6	Darjeeling Mandarin, ginger, orchids are becoming less remunerative	100.0	0	0	1.0
IC7	There is no profit in large cardamom since the past decade	100.0	0	0	1.0
IC8	Income from cardamom is decreasing	98.4	1.6	0	0.992
IC9	Income from NTFP decreased	17.6	48.8	33.6	0.42
Average perception score of impact on income					**0.902**
G3 Perception of health-related expenditure					
IC10	Expenditure due to suffering from serious disease will increase	97.6	2.4	0	0.988
IC11	Malaria or dengue is increasing my hospital or inactive days	54.4	40.8	4.8	0.748
Average perception score of impact on expenditure					**0.868**
G4 Perception of occupation and migration					
IC12	There will be a large-scale migration or exodus of people and animals from here	96.0	1.6	2.4	0.968
IC13	People leaving agriculture	96.0	3.2	0.8	0.976
IC14	I am planning to abandon the traditional large cardamom cultivation	80.0	19.2	0.8	0.896
IC15	Dependency on forest is decreased	73.6	18.4	8.0	0.828
Average perception score of impact on changing occupation and migration					**0.917**
G5 Perception of poverty and starvation					
IC16	Starvation and food shortage will occur	99.2	0.8	0	0.996
IC17	Increased poverty	89.6	9.6	0.8	0.944
IC18	Widen the gap between rich and poor	91.2	6.4	2.4	0.944
Average perception score of impact on food security and poverty					**0.961**
Overall score					**0.927**

$N = 125$, *IC* indicator code, *Y* yes, *N* no, *DK* don't know, *MS* mean score

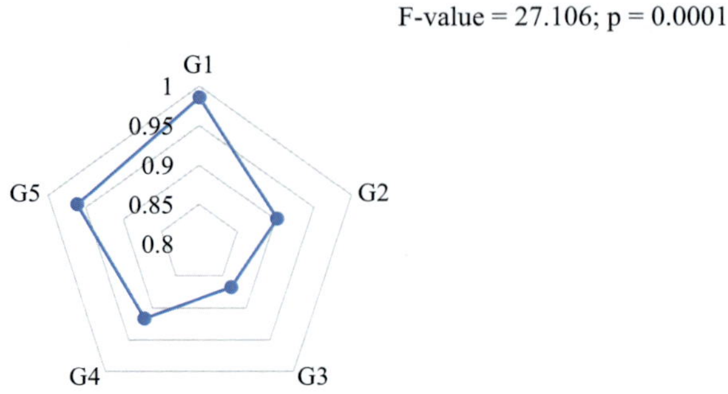

Fig. 4.3 Spider diagram depicting large cardamom grower's perception of different indicators related to climate change impact and risk on livelihoods. G1, perception of climate change impact on overall livelihood; G2-, perception of climate change impact on income; G3, perception of climate change impact on health-related expenditure; G4, perception of climate change impact on occupation and migration; G5, perception of climate change impact on poverty and starvation

suffering and expenditure from serious diseases, including anticipation of more than half of the respondents toward an increase in inactive days due to suffering from serious diseases like malaria or dengue.

Similar anticipation by locals in different parts of the Himalayan region toward the diverse aspects of life and livelihood has been documented (Dey et al. 2017a, b; Bhusal and Kafle 2020) and validated by earlier studies through agriculture and human well-being manifested through crop/plant phenological changes, new agricultural pests and weeds in agriculture (Yu et al. 2010; Rana et al. 2013), and changes in biodiversity (Willis and Bhagwat 2009).

3.4 Perception of Climate Change Mitigation Options

After assessing the perceptions related to climate change and its impact and risks of the large cardamom growers, it is imperative to understand their perceptions on climate change adaptation strategies for the large cardamom farming system that can provide valuable inputs for the design and implementation of adaptive responses and policies (Guariguata et al. 2012; Dey et al. 2017b; Vineeta et al. 2023).

The overall perception level on climate change mitigation of the large cardamom growers is medium, with an overall average score of 0.788 (Table 4.3). The overall perception level of large cardamom growers on conservation-based mitigation options was low. Mulching of large cardamom (MO3) was agreed by about 89% of the growers, particularly with weeds (MO6) by 75.2% of the growers to conserve soil moisture during rain-deficit periods. Other conservation-based mitigation options like rain water harvesting (MO9), soil conservation (MO7), water conservation (MO5), use of organic products (MO8), and use of traditional knowledge (MO4) were disagreed by 78.4–94.4% of the respondents.

Table 4.3 Distribution of respondents based on the perception score of climate change mitigation options for large cardamom farming

MOC	Climate change mitigation options for large cardamom farming	Distribution (% of N)			MS
		Y	N	DK	
G1 Changing the farming system					
MO1	You will abandon large cardamom farming	68.8	31.2	0	**0.844**
G2 Faith in climate change mitigations					
MO2	Belief in mitigation to improve large cardamom	90.4	8.0	1.6	**0.944**
G3 Conservation-based mitigations					
MO3	Mulching of cardamom with local material	88.8	10.4	0.8	0.940
MO4	ITK to control disease, insects, and pests	21.6	72.0	6.4	0.576
MO5	Use of water conservation techniques	14.4	85.6	0	0.572
MO6	Using weeds as mulch	75.2	24.8	0	0.876
MO7	Soil conservation techniques	9.6	90.4	0	0.548
MO8	Use of organic products to control disease, insects, and pests	20.8	78.4	0.8	0.600
MO9	Rain water harvesting	5.6	94.4	0	0.528
Average perception score of conservation-based mitigation					**0.663**
G4 Improved technology as mitigation					
MO10	Any methods for increasing pollinators	3.2	96.8	0	0.516
MO11	Use of improved/tolerant variety and quality planting material	94.4	5.6	0	0.972
MO12	Changing the planting time	36.8	62.4	0.8	0.680
Average perception score of improved technology as mitigation					**0.723**
G5 Institutional intervention as mitigation					
MO13	Accessed to improved technology, resource, and infrastructure	97.6	0	2.4	0.976
MO14	Use of insurance	8.8	90.4	0.8	0.540
MO15	Govt. schemes for coping climate change	11.2	88.0	0.8	0.552
MO16	Institutional interventions for capacity building	99.2	0.8	0	0.996
Average perception score of institutional intervention as mitigation					**0.766**
Overall score					**0.788**

N = 125, *MOC* mitigation option code, *MO* mitigation option, *Y* yes, *N* no, *DK* don't know, *MS* mean score, *ITK* indigenous traditional knowledge

Similarly, 3.2–94.4% of the growers were aware of improved technology as a mitigation option. Use of improved or tolerant variety (MO11) was agreed by 94.4% of the growers; thus, the option was estimated with a high perception level. However, changing the planting time of large cardamom (MO12) and increasing pollinators (MO10) was disagreed by 63.2% and 94.4% of the growers, respectively, thus documented with a low perception level. Institutional intervention as a mitigation option was known to 8.8–99.2% of the respondents. Among these options, almost all the respondents believed institutional intervention in terms of capacity building (MO16) and access to improved technology, resource, and infrastructure (MO13) by 99.2%

and 97.6% of the respondents, respectively, and thus were documented with a high perception level. However, crop insurance (MO14) and other government Schemes (MO15) were disagreed by 91.2% and 88.8% of the respondents, respectively, as institutional-based mitigation options.

Entire respondents of our study had perceived climate change. However, considering the total 16 mitigation options grouped into five categories, viz., changing the farming system (G1), faith in climate change mitigations (G2), conservation-based mitigations (G3), improved technology as mitigation (G4), and institutional intervention as mitigation (G5), the level of perception among these five mitigation option groups differed significantly (Fig. 4.4).

Similar mitigation options for large cardamom farming were also perceived by the farmers in Sikkim (Kumar et al. 2015; Sharma et al. 2016) and also validated by several studies from Darjeeling Himalayas (Tarafder et al. 2018; Vineeta et al. 2023). Diverse mitigation options adopted by the communities were reported from across Africa and Asia (Onyekuru and Marchant 2014; Dey et al. 2017b; Vineeta et al. 2023). The perception level of these large cardamom growers to mitigation options in response to changes in climatic variability events is medium, while to 16 different options was inconsistent and variable due to their dependency on natural resources, constraints in human and physical capital, and poor infrastructure (Shewmake 2008; Salau et al. 2012; Gukurume 2013).

It was also found that the level of perception score of those mitigation options was low, which was either unknown or not accessed and adopted by the growers. Increasing pollinators as a mitigation option was unknown to the growers. Rain water harvesting, soil and water conservation techniques, and use of organic products and traditional knowledge were not adopted by the growers though known due to personal beliefs while crop insurance and other government schemes were either unknown or not assessed by the growers. This indicates inadequate human–environment interactions in the study area where the large cardamom growers still lack robustness in livelihoods and depends on natural resources while neglected

F-value = 74.406; p = 0.0001

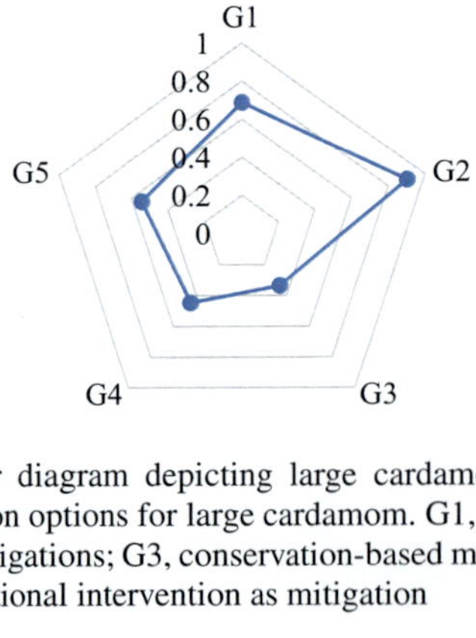

Fig. 4.4 Spider diagram depicting large cardamom grower's perception of different climate change mitigation options for large cardamom. G1, changing the farming system; G2, faith in climate change mitigations; G3, conservation-based mitigations; G4, improved technology as mitigation; G5, institutional intervention as mitigation

politically and institutionally with less institutional presence, negligible access to, and by the institutions for appropriate technology, skills, information, policies, microfinance, and equity, along with inadequate infrastructure development, particularly communication and market (Dey et al. 2017b).

The extent of perception level can be associated with a sustainable adaptation, which is influenced by the adaptive capacity of an individual or a community (Anonymous 2007; Dey et al. 2017b). The adaptive capacity of an individual or a community develops with adequate and proper human–environment interactions (Chapagain et al. 2009; Ericksen et al. 2011; Dey et al. 2017b) influenced by socio-economic, politico-cultural, ecological, and institutional-based in terms of robustness in livelihoods and alternatives, assets, access to appropriate technology, skills, infrastructure, information, policies, microfinance, equity, and perceptions inter alia (Dow et al. 2006; Hammill et al. 2008; Dey et al. 2017b). Perceiving climate change, its impact, and mitigations develops resilience, that is, the ability to cope with, adapt to, and recover from shocks into the future (Nyanga et al. 2011; Diyawadana et al. 2016; De Matos Carlos et al. 2020).

4 Conclusion

In conclusion, the study reveals that younger large cardamom growers in the Darjeeling Himalayas are more educated and economically better off, despite having smaller landholdings. Growers widely perceived climate change impacts such as rising temperatures, erratic rainfall, water scarcity, pest outbreaks, and reduced productivity. Recent climatic trends only partially validate these perceptions. The perception of climate change impacts and risks was consistent across socioeconomic groups, with Scheduled Tribe growers showing higher awareness. Most respondents expressed fear of declining livelihoods and food insecurity, leading many to consider crop diversification, alternate livelihoods, or migration. Although they believed that mitigation through improved planting materials, better curing techniques, mulching, and technology could enhance resilience, awareness about conservation practices, government schemes, and traditional knowledge remained low. The findings underscore the need for inclusive, need-based, and impact-oriented strategies to build resilience and ensure the sustainability of large cardamom-based agroforestry systems under changing climatic conditions.

5 Recommendations and Future Scope of Research

The present study on large cardamom-based TAF in the Darjeeling Himalayas highlights that there is a pressing need for scientists, policymakers, and NGOs to provide integrated technological, financial, and policy support for livelihood diversification and food security. Incorporating indigenous awareness alongside scientific understanding into climate policies can make adaptation strategies more participatory, cost-effective, and sustainable. Enhancing alternative livelihoods with financial, technical, and institutional backing is crucial, alongside strengthening local governance processes that ensure equity and participatory decision-making. Recognizing and empowering cardamom growers will foster successful people-centric initiatives for sustainable natural resource management. The study also recommends micro-financing, social safety nets, and public–private partnership (PPP) models for long-term resilience. Future research should explore the cultural and psychological dimensions influencing adaptation, which remain under-investigated due to their complex and sensitive nature. Additionally, documenting local climate knowledge and generating site-specific data on land degradation, biodiversity loss, and ecosystem services is essential to develop practical, science-based adaptation strategies and inform region-specific action plans.

References

Adger WN, Agrawala S, Mirza MMQ, Conde C, O'brien K, Pulhin J, Pulwarty R, Smit B, Takahashi K (2007) Assessment of adaptation practices, options, constraints and capacity. In: Parry ML, Canziani OF, Palutikof JP, Van Der Linden PJ, Hanson CE (eds) Climate change 2007: impacts, adaptation and vulnerability. Contribution of Working Group II to the fourth assessment report of the intergovernmental panel on climate change. Cambridge University Press, Cambridge, pp 717–743

Anonymous (2007) Climate change 2007: impacts, adaptation and vulnerability. IPCC, Cambridge University Press, Cambridge

Bagheri A, Teymouri A (2022) Farmers' intended and actual adoption of soil and water conservation practices. Agric Water Manag 259:107244

Bhusal S, Kafle S (2020) Impact of climate change on the livelihood of Allaichey Community Forest User Group (CFUG). Global Scientific Journals 8:1258–1273. https://doi.org/10.13140/RG.2.2.19783.32166

Brody SD, Zahran S, Vedlitz A, Grover H (2008) Examining the relationship between physical vulnerability public perceptions of global climate change in the United States. Environ Behav 40:72–95

Cajee L (2018) Physical aspects of the Darjeeling Himalaya: understanding from a geographical perspective. IOSR J Human Soc Sci 23:66–79

Chapagain BK, Subedi R, Paudel NS (2009) Exploring local knowledge of climate change: some reflections. J For Livelihood 8:106–110

Chhogyel N, Kumar L, Bajgai Y (2020) Consequences of climate change impacts and incidences of extreme weather events in relation to crop production in Bhutan. Sustainability 12:4319

Dahal KR, Dahal P, Adhikari RK, Naukkarinen V, Panday D, Bista N, Helenius J, Marambe B (2023) Climate change impacts and adaptation in a hill farming system of the Himalayan region: climatic trends, farmers' perceptions and practices. Climate 11:11. https://doi.org/10.3390/cli11010011

De Matos Carlos S, da Cunha DA, Pires MV, Do Couto-Santos FR (2020) Understanding farmers' perceptions and adaptation to climate change: the case of Rio das Contas basin, Brazil. GeoJournal 85:805–821. https://doi.org/10.1007/s10708-019-09993-1

De NK, Bera AK (1990) Soils of the Darjeeling Himalayas. In: Sah NK, Bhatt SD, Pande RK (eds) Himalaya: environment, resources and development. Shree Almora Book Depot, Almora, pp 151–163

Dey T, Pala NA, Shukla G, Pal PK, Chakravarty S (2017a) Perception on impact of climate change on forest ecosystem in protected area of West Bengal, India. J For Environ Sci 33:1–7. https://doi.org/10.7747/JFES.2017.33.1.1

Dey T, Pala NA, Shukla G, Pal PK, Das G, Chakravarty S (2017b) Climate change perceptions and response strategies of forest fringe communities in Indian Eastern Himalaya. Environ Dev Sustain. https://doi.org/10.1007/s10668-017-9920-1

Diallo A, Donkor E, Owusu V (2020) Climate change adaptation strategies, productivity and sustainable food security in southern Mali. Clim Change 159:309–327

Diyawadana DMN, Pathmarajah S, Gunawardena ERN (2016) Perception-driven coping strategies for climate change by smallholder farmers in Sri Lanka: a case study in Hakwatuna-oya major irrigation scheme. Trop Agric Res 27:336–349

Dow K, Kasperson RE, Bohn M (2006) Exploring the social justice implications of adaptation and vulnerability. In: Adger WN, Paavola J, Huq S, Mace MJ (eds) Fairness in adaptation to climate change. MIT Press, Cambridge, MA, pp 79–96

Ericksen S, Aldunce P, Bahinipati CS, Martins RD, Molefe JI, Nhemachena C, O'brien K, Olorunfem F, Park J, Sygna L, Ulsrud K (2011) When not every response to climate change is a good one: identifying principles for sustainable adaptation. Clim Dev 3:7–20. https://doi.org/10.3763/cdev.2010.0060

Frechtling J, Sharp L, Westat (1997) User-friendly handbook for mixed method valuations. National Science Foundation

Froehlich W, Sarkar S (2000) Environment: Soils. In: Starkel L, Basu S (eds) Rains, landslides and floods in the Darjeeling Himalaya. Indian National Science Academy, New Delhi, pp 16–19

Gillies RR, Wang SY, Sun Y, Chung OY (2013) Supportive empirical modelling for the forecast of monsoon precipitation in Nepal. Int J Climatol 33:3047–3054

Gioli G, Thapa G, Khan F, Dasgupta P, Nathan D, Chhetri N, Adhikari L, Mohanty SK, Aurino E, Mapstone Scott L (2019) Understanding and tackling poverty and vulnerability in mountain livelihoods in the Hindu Kush Himalayas. In: The Hindu Kush Himalaya assessment. Springer, Cham

Guariguata MR, Locatelli B, Haupt F (2012) Adapting tropical production forests to global climate change: risk perceptions and actions. Int For Rev 14:27–38

Gukurume S (2013) Climate change, variability and sustainable agriculture in Zimbabwe's rural communities. Russ J Agric Soc Econ Sci 2:89–100

Hammill P, Matthew R, Mc Carter E (2008) Microfinance and climate change adaptation. IDS Bull 39:113–122

Islam MA (2022) Farmers' perception to climate change: a review. Acad Lett:5419. https://doi.org/10.20935/al5419

Islam MDM, Jannat A, Dhar AR, Ahamed T (2020) Factors determining conversion of agricultural land use in Bangladesh: farmers' perceptions and perspectives of climate change. GeoJournal 85:343–362

Jagermeyr J (2020) Agriculture's historic twin-challenge toward sustainable water use and food supply for all. Front Sustain Food Syst 4:35

Karmacharya J, Levine RC, Jones R, Moufouma-Okia W, New M (2015) Sensitivity of systematic biases in South Asian summer-monsoon simulations to regional climate model domain size and implications for downscaled regional process studies. Clim Dyn 45:213–231

Kumar A, Shukla G, Pala NA, Chakravarty S (2015) Knowledge intensity and problem in ginger and large cardamom production technology of Lepcha tribes in Dzongu region of Sikkim. In: Munsi PS, Ghosh SK, Bhowmick N, Deb P (eds) Innovative horticulture: concepts for sustainable development, recent trends. New Delhi Publishers, New Delhi, pp 37–40

Lepcha LD, Shukla G, Pala NA, Vineeta PPK, Chakravarty S (2018) Contribution of NTFPs on livelihood of forest-fringe communities in Jaldapara National Park, India. J Sustain For. https://doi.org/10.1080/10549811.2018.1528158

Li W, Ruiz-Menjivar J, Zhang L, Zhang J (2021) Climate change perceptions and the adoption of low-carbon agricultural technologies: evidence from rice production systems in the Yangtze River Basin. Sci Total Environ 759:143554

Makondo CC, Thomas DSG (2018) Climate change adaptation: linking indigenous knowledge with western science for effective adaptation. Environ Sci Policy 88:83–91

Meena RK, Verma TP, Yadav RP, Mahapatra SK, Surya JN, Singh D, Singh SK (2019) Local perceptions and adaptation of indigenous communities to climate change: evidences from high mountain Pangi Valley of Indian Himalayas. Indian J Tradit Knowl 18:58–67

Moktan S, Das AP (2013) Diversity and distribution of invasive alien plants along the altitudinal gradient in Darjeeling Himalaya, India. Pleione 7:305–313

Mukherjee D (2013) Critical analysis of challenges of Darjeeling Himalaya: water, natural resources, hazards and the implications of climate change. Int J Agric Innov Res 2:13–22

Negi BK, Joshi RK, Pandey A (2018) Status of large cardamom (*Amomum subulatum*Roxb.) farming systems in the changing scenario of modern economics of Sikkim, Himalaya. Glob J Biosci Biotechnol 7:189–199

Nyanga PH, Johnsen FH, Aune JB, Kalinda TH (2011) Small holder farmer's perceptions of climate change and conservation agriculture: evidence from Zambia. J Sustain Dev 4:73–85

Onyekuru AN, Marchant R (2014) Climate change impact and adaptation pathways for forest dependent livelihood systems in Nigeria. Afr J Agric Res 9:1819–1832

Paudel B, Zhang Y, Yan J, Rai R, Li L, Wu X, Chapagain PS, Khanal NR (2020) Farmers' understanding of climate change in Nepal Himalayas: important determinants and implications for developing adaptation strategies. Clim Change 158:485–502

Pradhan R, Sarkar BC, K Abha M, Shukla G, Tamang M, Vineeta, Bhat JA, Kumar M, Chakravarty S (2022) Biomass carbon and soil nutrient status in urban green sites at foothills of eastern Himalayas: implication for carbon management. Curr Res Environ Sustain 4:100168. https://doi.org/10.1016/j.crsust.2022.100168

Rai S, Dahal B, Anup KC (2022) Climate change perceptions and adaptations by indigenous Chepang community of Dhading, Nepal. GeoJournal. https://doi.org/10.1007/s10708-022-10577-9

Rana RS, Bhagat RM, Kalia V, Lal H, Sen V (2013) Indigenous perceptions of climate change vis-`a-vis mountain agricultural activities in Himachal Pradesh, India. J Tradit Knowl 12:596–604

Rasul G, Saboor A, Tiwari PC, Hussain A, Ghosh N, Chettri GB (2019) Food and nutrition security in the Hindu Kush Himalaya: unique challenges and niche opportunities. In: Wester P, Mishra A, Mukherji A, Shrestha AB (eds) The Hindu Kush Himalaya assessment: mountains, climate change, sustainability and people. Springer, Cham, pp 301–338

Rathore M (2022) Households by annual income India FY 2021. https://www.statista.com/statistics/482584/india-households-by-annual-income/

Roy D, Datta A, Kuwornu JKM, Zulfiqar F (2021) Comparing farmers' perceptions of climate change with meteorological trends and examining farm adaptation measures in hazard-prone districts of northwest Bangladesh. Environ Dev Sustain 23:8699–8721

Salau ES, Onuk EG, Ibrahim A (2012) Knowledge, perception and adaptation strategies to climate change among farmers in southern agricultural zone of Nasarawa State, Nigeria. J Agric Ext 16:199–211

Saxena S (2005) Sustainable development in Darjeeling Hills, India; ecological and socio-economic aspects for small-scale farmers with supportive observations from Kanagawa, Japan. Ph. D. Dissertation. Am FachbereichGeowissenschaften, der Johannes Gutenberg-Universität Mainz. Unpubl

Sharma G, Partap U, Dahal DR, Sharma DP, Sharma E (2016) Declining large cardamom production systems in the Sikkim Himalayas- climate change impacts, agroeconomic potential and revival strategies. Mt Res Dev. https://doi.org/10.1659/MRD-JOURNAL-D-14-00122.1

Shewmake S (2008) Vulnerability and the impact of climate change in South Africa's Limpopo River Basin, IFPRI Discussion Paper 00804. Washington, DC

Shukla G, Kumar A, Pala NA, Chakravarty S (2016) Farmer's perception and awareness of climate change: a case study from Khangchendzonga Biosphere Reserve, India. Environ Dev Sustain 4. https://doi.org/10.1007/s10668-015-9694-2

Singh SP, Thadani R, Negi GCS, Singh RD, Gumber S (2020) The impact of climate change in Hindu Kush Himalayas: key sustainability issues. In: Dimri AP, Bookhagen B, Stoffel M, Yasunari T (eds) Himalayan weather and climate and their impact on the environment. Springer, Cham, pp 453–472

Tamang B, Pala NA, Shukla G, Rashid M, Rather MM, Bhat JA, Masoodi TH, Chakravarty S (2021) Trees outside forest (TOFs) aids in mitigating global climatic change through carbon sequestration: example from academic institutional landscapes. Acta Ecol Sin. https://doi.org/10.1016/j.chnaes.2021.06.007

Tarafder HK, Roy K, Rai B (2018) Large cardamom production scenario in mountainous region of West Bengal: constraints, issues and strategies. Asian J Soil Sci 13:153–156

Vineeta, Tamang B, Shukla G, Chakravarty S (2023) The urge of conserving tradition from climate change: a case study of Darjeeling Himalayan large cardamom-based traditional agroforestry farming system. Nat Based Sol 3:100064. https://doi.org/10.1016/j.nbsj.2023.100064

Vineeta, Tamang B, Siril S, Singh M, Das S, Shukla G, Chakravarty S (2021) Ecosystem services of traditional large cardamom-based agroforestry systems of Darjeeling and Sikkim Himalayas. J Tree Sci 40:78–91. https://doi.org/10.5958/2455-7129.2021.00006.6

Wangchuk K, Wangdi J (2018) Signs of climate warming through the eyes of Yak herders in northern Bhutan. Mt Res Dev 38:45–52

Willis KJ, Bhagwat SA (2009) Biodiversity and climate change. Science 326:806–807

Yu H, Luedeling E, Xu J (2010) Winter and spring warming result in delayed spring phenology on the Tibetan Plateau. Proc Natl Acad Sci 107:22151–22156

Chapter 5
Biodiversity Conservation and Livelihood Improvement Through Agroforestry Intervention in Marginal Farming Communities of the Indian Himalayan Region

Benatul Behar, Nazir. A. Pala, G. M. Bhat, Vaishnu Dutt, M. A. Islam, and T. A. Rather

Abstract The Indian Himalayan Region (IHR), a South Asian water tower and a global hotspot for biodiversity, is severely impacted by the vast loss of natural ecosystems that threatens marginal farming communities. These communities grapple with depleting biodiversity, lack of sustainable livelihoods, and ongoing ecological degradation. This chapter explores how agroforestry initiatives that combine targets for conservation with socioeconomic development can tackle these interrelated problems. Unlike traditional methods that prioritize protected areas, agroforestry systems (AFS) offer inclusive, landscape-level solutions that incorporate native tree species into farming activities. Homestead agroforestry, improved shifting cultivation, and diverse cropping are examples of models that boost ecological resilience while generating income from timber, fruits, fodder, and other commercial resources. Additionally, these systems enhance microclimatic conditions, soil fertility, and water retention. These benefits are particularly important for resource-constrained farmers in remote Himalayan settlements. The chapter assesses how well agroforestry technologies restore damaged landscapes and improve rural livelihoods using data from community-level studies and field research. Special focus is paid to the domestication of native fruit tree species and their integration into agricultural systems, which has shown promise in improving food security and biodiversity. The results reveal that agroforestry is a revolutionary strategy that combines ecological restoration with economic empowerment, rather than merely serving as a substitute for subsistence farming. The study emphasizes the value of agroforestry in fostering sustainable development and increasing adaptation ability throughout the IHR by

B. Behar · N. A. Pala (✉) · G. M. Bhat · V. Dutt · M. A. Islam · T. A. Rather
Division of Silviculture and Agroforestry, Faculty of Forestry, SK University of Agricultural Sciences & Technology of Kashmir, Srinagar, Jammu and Kashmir, India

B. Singh et al. (eds.), *Himalayan Agroforestry Systems*, Advances in Agroforestry 15,
https://doi.org/10.1007/978-3-032-25096-4_5

focusing on the experiences of marginal farming communities. Ultimately, it calls for a shift in policy and practice, recognizing that cultivated landscapes are essential for enhancing livelihoods and preserving biodiversity.

Keywords Agroforestry · Livelihood · Biodiversity · Himalayan Region · Sustainable Development · Marginal farming · communities

1 Introduction

The Indian Himalayan Region (IHR) stretches for over 2500 km 72°40′ to 97°25′ E from Jammu and Kashmir in the west to Arunachal Pradesh in the east with a width of 250–300 km 21°57′ N to 37°5′ N. Approximately 48.5 million people live in the region, which makes up 12.7% of India's total land area of 0.42 million km^2 (Bhatt et al. 2020). It has an extremely diverse range of climatic conditions, vegetation, and wildlife due to the significant variance in latitude, longitude, and elevation. The Indian Himalayan Region boasts exceptionally high plant biodiversity, making it one of the world's 12 hotspots for biodiversity (Sharma et al. 2023). In contrast to the country's plain areas, where agriculture accounts for 55.8% of land use, the Indian Himalaya supports a greater area covered by forest cover (52%), with wastelands (23.2%) and agriculture (12.4%) coming in second and third, respectively (Poonia and Sharma 2017). Nonetheless, the natural resources are under extreme strain due to the world's population growth and the resulting demand for land for construction and cultivation. The western Himalaya engages in settled agriculture on terraced slopes, while the northeastern Himalaya practices shifting agriculture/"slash and burn" agriculture, which is primarily rainfed. Though the IHR only has a small amount of land dedicated to agriculture—less than 14% of the region's total area—agriculture continues to be the main source of income for around 70% of the Himalaya's rural population (Negi et al. 2023). Because of the depletion of natural resources and shifting socioeconomic and policy conditions, traditional agriculture in the Himalaya is becoming less sustainable (Kumar et al. 2021).

In the western Himalaya, the primary agricultural problem is land abandonment and the resulting destruction of agro-biodiversity, while in the eastern Himalaya, the primary concern is the shortening of the jhum cycle and the ensuing degradation of land (Kumar et al. 2018). The local populace exploits forest resources for non-timber forest products (NTFPs), fuelwood, and fodder due to poverty and unemployment (Banday et al. 2021). Accordingly, the pervasive poverty in developing countries has led to overexploitation, slash-and-burn agriculture, farmland conversion, charcoal manufacture, bush fires, and wood harvesting, all of which have caused deforestation and biodiversity loss (Patra et al. 2022). This chapter seeks to highlight the vital role that agroforestry plays in supporting rural livelihoods, landscape biodiversity conservation, and direct enhancement of agro-biodiversity as an integrated land-use system. Agroforestry, a feasible substitute for maintaining ecological, social, and economic sustainability, is becoming more widely acknowledged as a productive ecological foundation for higher crop yields, more consistent financial returns, and greater diversity (McGunnigle et al. 2023). Because certain environmental circumstances and multiple micro-situations are present in the IHR, AFS components such as sericulture, apiculture, fisheries, tea, and medicinal plants are significant (Dagar et al. 2020). These systems have enormous potential to produce biomass and create jobs. If agroforestry is carried out over a 75,500 ha region, it may produce more than 5.763 million man-days of employment yearly (Kumar et al. 2018). This draws attention to the value of preserving biodiversity and the role that agroforestry as a form of land use may play in agro-biodiversity's ongoing preservation. The chapter also identifies some governance and policy issues that arise when agroforestry is promoted for biodiversity preservation and sustainable livelihoods.

2 Livelihood Dependency of Rural Communities on Traditional Agroforestry Systems (TAFS) across the Indian Himalayan Region

India has a large population living in close vicinity of forests with critical dependence on forest ecosystem for their livelihoods (Banday et al. 2021). Among the mountain ecosystems, the Indian Himalayan Region (IHR) has unparalleled significance. On one hand, despite having its own rich heritage of natural resources, people living in IHR are marginalized and exploit forest resources extensively for fuelwood collection, grazing, and other subsistence needs (Singh et al. 2017). On the other hand, the indigenous agroforestry systems of Himalayan region form an integral part of the traditional communities and local environment where in the crops, cattle rearing and forestry constitute an interlinked system (Negi et al. 2018). Such systems are indigenously managed, with best practices that have evolved through trial and error over time by the farmers (Phondani et al. 2020). Naturally regenerating tree species are maintained by the farmers without any significant external inputs or manpower on the margins of rainfed terraced agriculture, which in turn play an important role in adaptation to climate change and by reducing pressure on natural forests (Kumar et al. 2018; Phondani et al. 2020). Agroforestry is the most competent land-use option for ensured sustainable development, increased productivity of land, enhanced livelihoods, and eased environmental stress (Rajput et al. 2023). Therefore, it becomes imperative to understand agroforestry as an intervention in mitigating livelihood crisis by providing adequate supply of fuelwood, fodder, timber, and fruits (Table 5.1) to marginal farming communities in the Himalayan region, thereby reducing pressure on natural forests and boosting biodiversity management in the region.

Table 5.1 Regional pattern of agroforestry/forestry-based livelihoods among marginal farming communities of India

Forest/agroforestry products	Involvement in collection (%)	Collection/ annum	Consumption/ annum	Involvement in marketing (%)	Sale/ annum	Income (Rs./ annum)	References
Central plateau and hill region							Islam et al. (2015a)
Fuelwood	100	326.08$^{\Delta}$	598.60$^{\Delta}$	17.07	17.92$^{\Delta}$	89,600	
Fodder	98.78	1396.86$^{\Delta}$	2824.55$^{\Delta}$	12.20	12.80 $^{\Delta}$	64,000	
Timber	100	58.41$^{£}$	56.78$^{£}$	8.54	31.00 $^{£}$	358,400	
Fruits	29.77	1135.87$^{\Delta}$	345.37$^{\Delta}$	7.72	790.50$^{\Delta}$	8597	
NTFPs	36.59	1717.94$^{\sharp}$	430.40$^{\sharp}$	21.10	1287.53$^{\sharp}$	34,985	
Western Himalayan region							Behar (2021)
Fuelwood	100	251.46$^{\Delta}$	188.65$^{\Delta}$	35.57	62.81$^{\Delta}$	439,670	
Fodder	93.26	579.21$^{\Delta}$	527.78$^{\Delta}$	29.81	51.43$^{\Delta}$	565,730	
Timber	87.50	132.00$^{£}$	62.44$^{£}$	43.27	69.56$^{£}$	859,579	
Fruits	51.78	141.74$^{\Delta}$	22.96$^{\Delta}$	53.03	118.78$^{\Delta}$	7,064,720	
Trans Himalayan region							Bhat et al. (2022)
Fuelwood	100	381.00$^{\Delta}$	358.45$^{\Delta}$	9.40	22.55$^{\Delta}$	383,350	
Fodder	100	612.4$^{\Delta}$	547.2$^{\Delta}$	17.07	65.20$^{\Delta}$	847,600	
Timber	98.17	59.82$^{£}$	55.41$^{£}$	38.41	4.400$^{£}$	264,366	
Fruits	64.66	38.68$^{\sharp}$	21.67$^{\sharp}$	23.33	17.01$^{\sharp}$	3,591,000	
NTFPs	50.61	2.25$^{\sharp}$	0.31$^{\sharp}$	53.04	1.83$^{\sharp}$	109,980	

Source: Authors' summary from this chapter and review of literature (Δ = tons, $\sharp$ = kg, £ = m^3/Volume of pole = 0.0346 m^3)

2.1 Fuelwood

The world's renewable energy consumption, along with traditional uses of biomass, occupies a share (85%) of total domestic energy (Mahapatra et al. 2021), which is increasing by 2.1% annually tuning to 64.2 exajoules (EJ), with a sheer decline in traditional uses of biomass reaching up to 7% between 2008 and 2018 (Sikkema et al. 2021). Regardless of the decrease in traditional use of biomass, still 660 million people in India lack access to clean cooking fuels (Sreeja et al. 2023). In India, the energy demand has tripled during the last three decades, where traditional biomass like fuelwood and charcoal, along with animal waste, constitute 12% of energy mix. The demand for fuelwood purpose was 228 million ton and the sustainable supply accounts for 128 million ton (Ranjan 2021), which is being met from forests, community lands, and farm lands. The total fuelwood extraction from forests in India is 85.29 million tons, with an annual per capita consumption of 278.21 kg in 2019 in comparison to 294.28 kg in 2011 due to the promotion of clean fuel in the country (Nagar et al. 2022). Banday et al. (2021) revealed that the consumption of fuelwood in the Central Himalayan village is 418.86 MT, with an annual fuel availability of 211.03 MT; thus, a deficit of 207.83 MT. The average fuelwood consumption in Garhwal Himalayas observed by Paul et al (2019) was highest (2.52 kg per capita per day) in the tropical region and lowest (1.63 kg per capita per day) in the subtropical region, with labor energy expenditure for fuelwood collection greatest (146.62 MJ per capita per day) in the temperate region and lowest (88.80 MJ per capita per day) in the tropical region. While evaluating the status of fuelwood extraction and consumption in Sopore, Kashmir (Islam et al. 2011), revealed that the per capita fuelwood extraction and consumption in the surveyed population was recorded to be 0.28 t/year and 0.42 t/year, respectively. The total extraction of fuelwood from diverse sources was 696.28 t/year. The total fuelwood consumption was estimated to be 1044.41 t/year. Out of total consumption, 438.65 t was consumed in cooking, followed by 426.74 t in heating, cottage industries 136.00 t, community function 32.36 t, and the rest 10.66 t for other purposes in a study in Bandipora (Banyal et al. 2013).

Agroforestry on farms facilitates meeting the energy demand for fuel as the contribution of natural forests is approximately 40% of the country's energy needs. The main source of cooking fuel is biomass supplying 76% for all rural communities, and this proportion is likely to increase by a further 5.5% by 2030 (Nagar et al. 2021). In India, the prevalent agroforestry systems are agrisilviculture, agrihorticulture, agrihortisilviculture, and silvipasture (NRCAF 2007), spread over 25.32 million hectare land, that is, 8.2% of the country's total geographical area (Dhyani 2014). A total of 517.25 tons were obtained from silvihorticulture plantation, followed by homestead farming 97.48 tons, 62.67 tons from forest, and 18.88 tons from community forestry. Silvihorticulture plantation contributed about 74.29% of the fuelwood requirement, and the rest (25.71%) people use to rely on other sources in Kashmir (Islam et al. 2011). The fuelwood consumption rate, however, oscillates significantly across different seasons as well as along different altitudes throughout the Indian Himalayan Region (IHR). In summer, the consumption of fuelwood per household was estimated to be 7.5 kg/day and 10.7 kg/day in winter. The overall fuelwood consumption was 40.8% from agroforestry, 29.5% by LPG, 14.1% crop residue, 10.4% dung cake, and 5.2% kerosene (Singh et al. 2019).

2.2 *Fodder*

India ranks first in milk production in the world (Oh 2021; FAO 2022). Despite such a good ranking, the per capita production of milk is too low due to the relatively low quality of feed provided to the cattle and huge population. Availability of fodder, litter, and fuelwood in the forests is an important source of biomass, contributing about 90% of energy demands for various household purposes for the marginal farming communities of the Himalayan region (Shaheen et al. 2022). The demand of biomass for fodder in India was 1594 million tons and the sustainable supply accounts for 741 million tons only (Aggarwal et al. 2009), which is being met from forests, community lands, and farmlands. The total fodder extraction from forests in India is 1053.04 million tons (ISFR 2019). However, depletion of natural resources and population pressure gives way to fodder and fuelwood shortages of Himalayan communities but natural resources, along with mixed farming systems, are the lifeline for livelihoods of Himalayan communities (Pandey et al. 2021). Fodder trees, shrubs, crop residues, and grazing in forests are the main sources of feed for livestock in the Himalayas (Lata and Mondal 2021). Under such circumstances as well as strict enforcement of conservation policies for natural resources, it has encouraged the cultivation of trees on farmlands of rural smallholders with agroforestry on farms as an aid in fulfillment of the energy demands for fodder across India. Nagar et al. (2021) revealed that the biomass consumption from agroforestry and other sources fluctuated considerably during different seasons.

Contribution of agroforestry toward fodder has an estimated value of 12.96, 10.35, and 7.65 kg/adult cattle unit/day in winter, summer, and rainy seasons, respectively, while other sources contributed 3.53, 4.76, and 11.58 kg/ACU/day, respectively. The preferred agroforestry tree species for fodder purpose were *Grewia optiva*, *Ficus roxburghii*, *Celtis australis*, *Ficus palmata*, *Ficus cunea*, *Bauhinia variegata*, *Mangifera indica*, *Leucaena leucocephala*, *Morus alba*, *Ougeinia oojeinensis*, *Pennisetum purpureum*, *Andropogon munroi*, *Crysopogon fulvus*, *Rhus parviflora*, and *Woodfordia fruticosa*. Islam et al. (2022) investigated the dependence of livestock on natural forests in the Ganderbal district of the Kashmir Himalayas, aiming to develop a strategy for revitalizing fodder security through silvopastoral agroforestry. Their study emphasized that implementing silvopastoral systems on cultivable wastelands could yield substantial benefits, including an estimated 11,847.73 tons of green fodder per year, 3846.67 tons of tree fodder per year, 307.73 tons of fuelwood annually, and 320.43 cubic meters of small timber per year. Additionally, the system could produce 8.46 tons of grass seed annually, generate an average income of $66.13 per household per year, and create 9.98 man-days of employment per household annually. Timber Himalayan rural communities are mainly dependent on forests for deriving their day-to-day requirements as well as their livelihood support. The lower and higher altitude villages of district Bagh, Central Himalayan region, depicted the wood consumption pattern to be 2.19 kg and 3.76 kg/capita/day, respectively, and an average of 2.97 kg/capita/day. In India, the requirement of timber was 92 million m^3, or 3.5cft/capita/year (Rai and Chakrabarti 2001; Banday et al. 2021). However, from natural forests, timber is less available due to conservation, social, and environmental concerns. Industrial plantations account for only 5 percent of overall forest area but support 35 percent of global supply of wood (Rawat and Tekleyohannes 2021). Competition from alternative land-use practices has limited the scope of expansion of plantation forestry. Besides, the demand for timber is booming not only locally but also at regional and international levels. In response to this, many small-scale agroforestry systems have evolved, where the production of timber on farms either as small-scale woodlots or as scattered trees is potentially an important element of farm livelihoods (Baig et al. 2021).

Islam et al. (2022) reported that the dependency of rural households on the forestry resources for the procurement of small timber and their multiple use patterns has become an important implication in developing economies, braced economic upswing, and aided in bringing down poverty. Palit (2021) remarked that small timber plays a notable role in supporting the livelihoods of the rural poor. Collection and sale of small timber are the principal form of income for the rural populace. Baba et al. (2015) reported that the total extraction of small timber from various sources in selected villages amounted to 39.46 tons per annum. This timber was primarily used for housing and roofing (39.63%), followed by cattle sheds and storehouses (15.25%), rural furniture and packing cases (14.75%), agricultural implements (13.25%), fencing (12.50%), and other uses (4.62%). Islam et al. (2015b) further estimated the average household timber requirement to be 0.346 cubic meters per annum, resulting in a total demand of 282.49 cubic meters per year across the study area. The supply of small timber was met through multiple sources, including agroforestry and natural forests (69.09 m^3/year), community forestry (41.33 m^3/year), and homestead forestry (35.71 m^3/year).

2.3 *Fruits*

Fruit trees are one of the significant components of traditional agroforestry systems (AFSs); thus, they are grown with or without intercrops and provide long-lasting returns to the farmers. Fruit-based land uses (FBLUs) have numerous advantages in decreasing landslides, conserving moisture, soil erosion, and carbon sequestration, along with livelihood, income security, and nutritional security (Rathore et al. 2020). Commercially important fruit species (mango, citrus, peach, litchi, guava, etc.) are grown in the region with great dependence on edaphic and climatic conditions, but in comparison to national average, their productivity is low (Acrhm et al. 2018). Gangoo et al. (2017) studied the dependence of the people on anardana in the livelihood of Ramban district and had a share amounting to 48.78% in the annual income of every household. Vibhuti et al. (2019), while analyzing the diversity, status, and utilization of plants in homegardens of Kumaun Himalaya, recorded a total number of 70 plant species belonging to 35 families, tree density was highest (825 ind./ha) in large-sized homegardens in comparison to medium (750 ind./ha) and (317 ind./ha) in small-sized homegardens. Among different components, vegetables shared half (50.78%) of the plant diversity in total, followed by fruits, medicinal plants, fodder, fuelwood, and ornamental species of the order of 22.57%, 18.03%, 12.21%, 7.35%, and 1.43%, respectively. *Artocarpus integrifolia, Mangifera indica, Bohermaria olerosa,* and *Litchi chinensis* were the most dominant and frequent trees in almost all-sized homegardens.

3 Link Between the Biodiversity and Livelihood Strategies of Rural Communities

Nearly 1.6 billion people throughout the world are dependent on the forests for deriving their livelihoods and sustenance (Dean et al. 2016; Pandey et al. 2021) and depend on forests for different products and services (Taye et al. 2021). Income that is derived from the forest resources plays a significant role in operating households at the subsistence level, which support to fill their gaps in long- or short-term income and serve as a safety net for income crises (Pereira et al. 2014). The traditional agroforestry systems in mountainous regions mimic the natural forests as well as offer ecosystem services that are similar to those under the natural forests (Vineeta et al. 2022). The ecosystem services provisioned from the Himalayan ecosystems in the form of timber, fiber, food, medicine, drinking, and irrigation water are the livelihood source to people living there (Sharma and Singh 2021). Some traditional agroforestry systems in the tropical regions show high diversity of flora compared to other agroecosystems (Reang et al. 2022). Thus, harvesting of products from woodlands is broadly encouraged as an integral constituent of the rural livelihoods across the developing nations of the world, thereby offering goods for consumption at the household level as well as income generation.

3.1 Food Security

Agriculture provides the majority of livelihoods for rural residents in IHR; people also take advantage of natural resources to increase their earnings (Singh et al. 2018). The collection, processing, and sale of various NTFPs are the primary sources of income for rural populations. Bamboos, Lac, colors, edible mushrooms, tannins, fragrant oils, wax, etc., are examples of these nontraditional food products (Islam et al. 2017). The value and application of woodland resources in mitigating the risk linked with food supply at critical times have been emphasized by numerous writers. These times include the start of the rainy season, which is when food shortages are most severe because households have typically used up all of the previous year's harvest but have not started harvesting for the upcoming season. NTFP availability acts as a source of revenue as well as a vital gap-filler during periods of low food stocks (Tangjang et al. 2014). The indigenous populations of the Indian Himalayan Region rely on a wide variety of wild plants for food and other uses (Devi et al. 2022). More than 4000 kinds of flowering plants, or wild fruit species, can be found in large quantities growing along the altitudinal gradient in the western portion of the Central Himalayan region or the Garhwal Himalayas (Negi et al. 2011).

A summary of the findings revealed that the average income earned by NTFPs in the central Himalayas was ₹4791.16/household/year. This income was broken down into different categories: 36.26% cottage industry, 13.49% fruit, 11.40% fuelwood, 8.55% toothbrush, 8.15% mahua (*Madhuca latifolia*), 7.22% flowers, 6.74% oilseeds, 4.31% vegetables, and 0.88% ethno medicine (Islam et al. 2017). As reported by Sarmah and Arunachalam 2011, the forests of Arunachal Pradesh sustain a rich diversity of nontimber species, which means that the Miao circle had the highest total contribution of NTFPs to annual household income (23% of total income), followed by the Diyun circle (21% of total income), the Nampong circle (19% of total income), the Vijayanagar circle (18% of total income), and the Bordumsa circle (11% of total income), in that order. India produces less food now that its natural resources are being depleted (UNEP 2011; Selaiyur. and Nadu n.d.). The primary causes of the depletion of natural resources are increased pressure from livestock and population growth (Maja and Ayano 2021). Since an agroforestry system consists of more than two parts, it produces food grains, vegetables, fruits, spices, and livestock. The most important agroforestry systems that will be vital to the nation's food security are agrisilviculture, silvipastoral, agrihorticulture, and other systems like apiculture, aqua forestry, and tree species (Sahoo and Wani 2020).

Agroforestry components or their interactions that enhance or increase food quantity and quality, farmer income, environmental aspects, health conditions, and year-round food and income were all viewed as beneficial to farmers' food security (Duffy et al. 2021); on the other hand, any negative effects of agroforestry on these factors were viewed as disbenefits (Table 5.2).

Table 5.2 The agroforestry handouts to farmers' food security

Basis of food security	Agroforestry systems' (pros and cons) contribution to food security Aspect (growing/decreasing) overview	
Availability	Amount and standard of food	Enhancing the variety of foods available and producing crops and animal feed
Access	Income	Creating and varying sources of revenue
Utilization	Health	Provision of fire wood, management of pests and diseases, provision of micronutrients, and satisfaction of medical needs
Stability	Preserving access to and supply of food, environmental services	Constant revenue creation, food supply, water conservation, weed suppression, soil enrichment, soil erosion control, refuge from extreme weather events, microclimate alteration, preservation of biodiversity, and sequestration of carbon

Source: Datta and Behera (2024)

3.2 Health and Nutrition

Presently, intense food insecurity impinges 300 million people, much as malnutrition sway a large proportion of children less than 5 years old (FAO 2022). Moreover, disturbance in the food supply network can cause price hikes and economic uncertainty, leading to social and political altercation (Datta and Behera 2022). Therefore, food and nutritional security is crucial to support economic strength and social well-being. In order to acquire food and nutritional security, a multitude of interconnected strategies are needed, such as enhancing crop production and soil fertility, adjoining biofortification to principal foods, and cultivating a manifold span of plants that produce fruits, vegetables, and nuts to diversify the dietary regime of the general public (Sahoo et al. 2020). Nonetheless, farming communities around the world tend to favor growing exotic crops over native ones, which are frequently higher in fiber, protein, and nutrients than conventional crops. These natural variants are what the surrounding populations have harvested from woods as nontimber forest products (NTFPs). Around 85% of rural Indians resorted to using forest resources for medical purposes; the reason for this higher reliance on traditional medicine based on forests is the remoteness of the area (Kumar and Saikia 2020). Utilizing forest-based species, traditional medicine has been practiced in Ladakh, Tamil Nadu, and West Bengal; these regions have been documented by Ballabh and Chaurasia 2007, Muralidharan and Narasimhan 2012, and Sinhababu and Banerjee 2013, respectively. The use of biodiversity has produced many nutritional and health benefits; nevertheless, overuse of resources and significant habitat loss have negatively impacted certain populations' dietary diversity, nutrition, and general health. Apart from this, sustainable agriculture operations are pivotal in the Himalayan region, as unsustainable practices have by this time aggravated soil erosion, water exhaustion, and biodiversity damage and made the agricultural sector extra receptive to climate change bang (Aryal et al. 2019; Datta and Das 2019). Agroforestry systems have the prospective to raise land output and enhance food accessibility (Datta and Behera 2024). Trees planted in such systems can provide shade and fend off soil moisture evaporation, and compatible plants such as legumes can hold up nutrient cycling, yielding increased outputs (Duffy et al. 2021). The plentitude of insect-eating birds in the wooded surroundings of agroforestry systems can assist in prevention against pests, leading to enhanced yields; nevertheless, the benefaction of trees to spontaneous pest control is species-reliant on as several trees can also spoon out as asylum for pests (Sow et al. 2020) on the top of that, including livestock in agroforestry systems that can furnish supplementary protein foods and improve nutritional security (Fig. 5.1).

An agroforestry system's cultivation of nutrient-dense fruit trees alongside another crop offers the benefit of enhancing rural people's health and nutrition. Fruit plants are the most beneficial part of agri-horticulture systems because they are rich sources of essential minerals, alkaloids, and vitamins. Nutritive values of most prevalent commercial fruits of the Himalayan region as depicted in Table 5.3

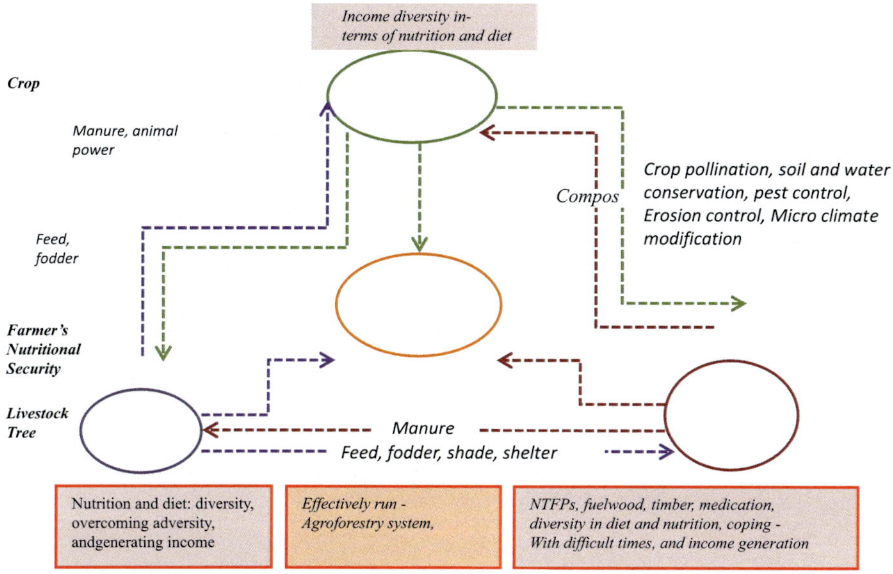

Fig. 5.1 Integrated agroforestry framework: synergistic roles of crops, trees, and livestock in enhancing farmers' nutritional security

(Sundriyal and Sundriyal 2004; Suresh et al. 2013). It is commonly known that animals, a variety of crops, and the collection of wild plants have all continuously and significantly improved human meals, for example, *Moringa oleifera* is a very nutritious food crop with certain promising traits including drought resilience and quick growth (Devkota and Bhusal 2020). *E. coracana* and *E. frumentaceae* are part of the renowned Garhwal "Barahnaja" crop and are mainly grown for the diet of the rural populace (Tariyal et al. 2022). Traditional agroforestry promotes a variety of indigenous medicinal plants in Uttarakhand. There are 68 species of plants in this region, mostly used for ethno-medical like herbs (36; 52.94%), followed by trees, shrubs, and climbers (19; 27.94%), (11; 16.17%), and (2; 2.94%), respectively, and the most often utilized plant parts among the known medicinal plants were bark (42%), rhizome, and flower (16%) (Monika et al. 2020).

3.3 Energy and Biomass Resource Use and Ramifications on Biodiversity

Traditional agroforestry methods mirror natural forests and promote wildlife--friendly land use, while also providing an enhanced supply of fuelwood and feed (Saraev et al. 2022). There are numerous conventional agroforestry systems that provide chances for soil fertility enhancement, reduced poverty, enhanced livelihoods,

Table 5.3 Nutritive values of the most prevalent commercial fruits of the Himalayan region

Name of fruit	Carbo hydrate (%)	Protein (%)	Fiber (%)	Mineral matter (%)	Fat (%)	Moisture (%)	Ca (%)	Fe (%)	P (%)	Vitamin C (mg/100 g)
Mango	11.80	0.60	1.10	0.30	0.10	86.10	0.01	0.02	0.03	13.00
Walnut	11.0	15.60	2.60	1.80	64.50	4.50	0.10	4.80	0.38	–
Apple	13.40	0.30	–	0.30	0.10	85.90	0.01	1.70	0.02	2.00
Orange	10.60	0.90	–	0.40	0.30	87.80	0.05	0.10	0.02	13.00
Almond	10.50	20.80	1.70	2.90	58.90	5.20	0.23	3.50	0.49	–
Plum	8.90	0.70	–	0.40	0.20	89.80	0.02	0.50	0.02	1.00
Banana	36.40	1.30	–	0.70	0.20	61.40	0.01	0.40	0.05	1.00
Pine apple	12.00	0.60	0.30	0.50	0.10	86.50	0.20	0.90	0.01	63.00
Avocado	0.80	1.70	–	1.10	22.80	73.60	0.01	0.70	0.08	13.00
Bel	30.60	1.30	2.20	1.50	0.20	64.20	0.09	0.30	0.05	15.00
Pear	11.50	0.20	1.00	0.30	0.10	86.90	0.01	Trace	0.70	–
Ber	12.80	0.80	–	0.40	0.10	85.90	0.03	0.80	0.03	–
Peach	7.60	1.50	–	0.60	0.20	90.10	0.01	1.70	0.03	1.00
Cashewnut	22.30	21.20	1.30	2.40	46.90	5.90	0.05	5.00	0.45	–
Passion fruit	22.00	0.90	–	0.70	0.10	76.30	0.01	2.00	0.06	–
Fig	17.10	1.30	–	0.60	0.20	80.80	0.06	1.20	0.03	2.00
Papaya	9.50	0.50	–	0.40	0.10	89.60	0.01	0.40	0.01	46.00
Grape	10.10	0.90	3.00	0.40	0.10	86.70	0.03	0.30	0.02	17.00
Litchi	9.40	0.70	2.25	0.70	0.30	84.30	0.21	0.03	0.31	Trace
Guava	8.10	0.10	4.80	0.60	0.20	85.30	0.05	1.20	0.02	16.00
Lime	10.90	1.50	1.30	0.70	1.00	84.60	0.09	0.30	0.02	63.00
Jackfruit	18.90	1.90	1.10	0.80	0.10	77.20	0.02	0.50	0.03	–
Lemon	11.10	1.00	1.70	0.30	0.90	85.00	0.07	0.30	0.01	39.00
Jamun	19.70	0.70	0.90	0.40	0.10	28.20	0.02	0.00	0.01	–
Karonda	67.10	2.30	–	2.80	9.60	18.20	0.16	39.10	0.06	–

Source: Sundriyal and Sundriyal (2004)

biodiversity preservation, and carbon sequestration (Gupta et al. 2020). The following characteristics were primarily employed by rural communities to sift through species proposals: stable burning flame, low smoke production, high calorific value, and, most importantly, usefulness as animal feed. *Grewia optiva* was the fodder of choice for 81% of smallholder agroforestry farmers in Uttarakhand, according to a survey by Nagar et al. 2022. Next in preference were *Celtis australis*, *Bauhinia variegate*, and *Morus alba*, in that order 74%, 66%, and 65%. Approximately 77% of farmers stated that *Celtis australis* was the best tree species for fuelwood, with *Grewia optiva*, *Mallotus philippensis*, and *Morus alba* coming in second and third, respectively, at 56%, 57%, and 53%. *Emblica officinalis*, *Psidium guajava*, *Citrus limon*, and *Mangifera indica* were ranked as the most preferred tree species, according to 92% of the farmers, among other uses (such as nutrient supplements, food items, lumber, and fiber).

Previous studies have recorded consumption of fuelwood ranging from 1.63 to 2.52 kg/capita/day in temperate, subtropical, and tropical regions of Garhwal Himalaya (Kumar and Sharma 2009) and between 1.29–1.98 kg/capita/day (Nagar et al. 2018) for rural communities in Garhwal Himalaya; from 2.83 to 5.81 kg/capita/day in tribal areas of Nanda Devi biosphere reserve (Rajwar and Kumar 2011); from 1.1 to 2.8 kg/capita/day (Bhatt and Sachan 2004); between 0.455–2.388 kg/capita/day (Khuman et al. 2011) and between 1.53–2.91 kg/capita/day (Rawat et al. 2018). Consumption of fuelwood varies between 3.1 and 10.4 kg/capita/day in Arunachal Pradesh and is 5.0 kg/capita/day in Garos (Maikhuri 1991). According to reports, the average daily use of fuelwood in the western Himalayas is 1.49 kg/capita (Bhatt et al. 1994) while in the Garhwal Himalaya it is 3 kg/capita (Awasthi et al. 2003). According to estimates, agroforestry produced 7.65 kg/ACU/day of fodder during the rainy season, 10.35 kg/ACU/day in the summer, and 12.96 kg/ACU/day in the winter, respectively. By contrast, fodder consumption from other sources was 11.58, 4.76, and 3.53 kg/ACU/day, respectively (Nagar et al. 2021). Average day-to-day consumption of fodder was about 22 kg/ACU with 9 kg/ACU from grasses, tree leaves, and market concentrate and 13 kg/ACU from forest in the Himalayan region (Pandey 2011). Dhanai et al. (2014) reported that the consumption of fodder ranges between 12.37 and 22.16 at lower, 16.40 and 17.38 at middle, and 18.33 and 26.04 kg/animal/day at high altitude, respectively. Agroforestry practices helped to meet the requirement of fodder and fuelwood in the mountainous areas (Nagar et al. 2021). Additionally, by lowering the need for people to purchase fruits and vegetables on a daily basis, they improved livelihood possibilities and offered more support. In addition to lowering the labor needed for other cultural practices on the farm, income is produced by selling milk and fodder to neighbors and other communities (Fig. 5.2). Agroforestry practices also assist in lessening the problems experienced during resource collection, save energy, and time spent obtaining items from outside the farm, and lower the incidence of human–wildlife conflict.

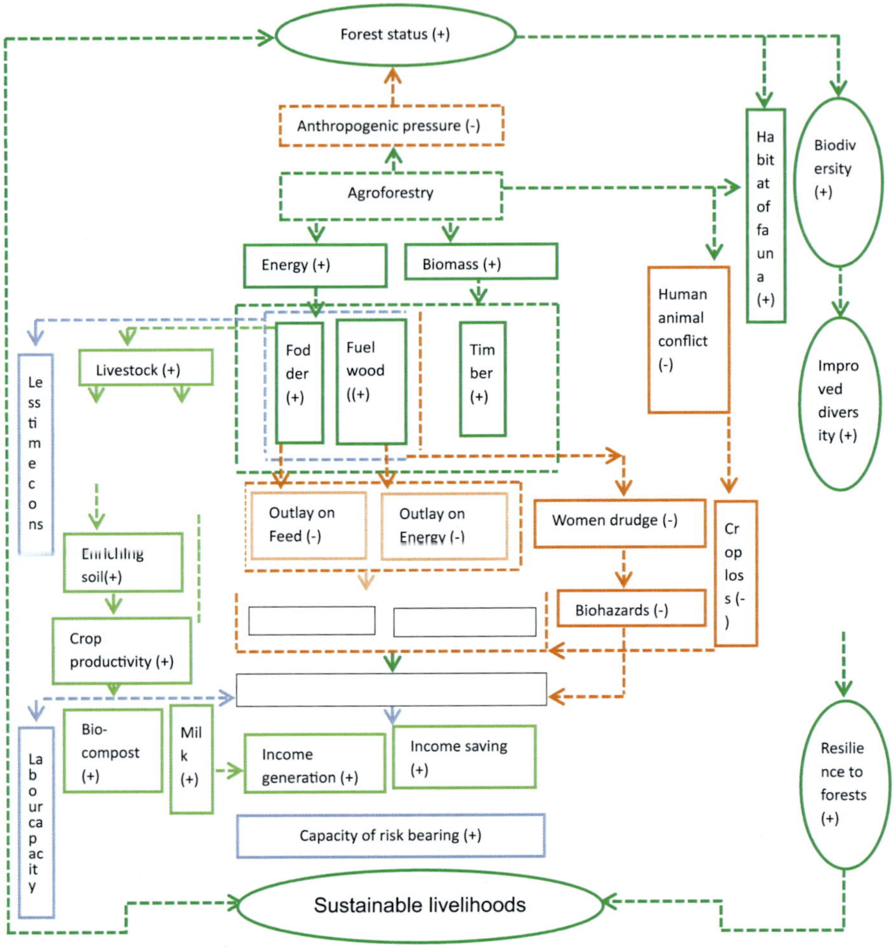

Fig. 5.2 Agroforestry-led energy supply and resilience: a schematic representation of potential roles

3.4 *Livelihood Strategies and Impact on Biodiversity*

The welfare of forests and specific species thriving there was impacted by livelihood practices in a variety of ways (Zafar et al. 2023). Regeneration, whether artificial or natural, has never been able to keep up with the rate of extraction (Chouhan and Sarma 2020). In developing nations, the use of fuelwood and slash-and-burn agriculture has precipitously increased the destruction of forests (Sati 2022). Large-scale cash crop cultivation is operated by a sociocultural shift from subsistence farming to a market economy, which is bolstered by improved accessibility and government-subsidized prices for staple food grains (Mundoli et al. 2016). The entire situation has led to a loss of agriculture biodiversity, which implies increased risks for local livelihoods in the event of things like a decline in market price or demand for commercial crops, the closure of providing principal food grains at subsidy rates, pest outbreaks in cash crops, and homogenous landscapes (Behera et al. 2016).

In the meantime, indigenous innovations for agro-biodiversity conservation do exist, but they are limited (Ghosh et al. 2020). Natural woods play a major role in the traditional livestock-crop mixed agricultural practices found in the Himalayas (Shekhar et al. 2022). The 115 million people who live in the mountains as well as those who live in the plains that border the forests depend on the various concrete and intangible benefits that the region's forests give (Vejre et al. 2010). Widespread changes include the increased use of land for agriculture, the trend toward producing cash crops rather than traditional staple food crops, and the replacement of multipurpose agroforestry trees with fruit trees. *Fagopyrum tataricum, Fagopyrum esculentum, Pisum arvense, Panicum miliaceum,* and *Setaria italica* cultivation have all but stopped (Maikhuri et al. 2015).

Nomadic settlements have a significant impact on species composition, fire incidence, and diversity. Livestock settlements adjacent to forests influence species richness and diversity by encouraging the growth of weedy species. The dense growth of invasive plants such as *Polygonum aviculare*, *Urtica dioica*, and *Cirsium falconeri* in the vicinity of cattle settlement has resulted in changes to the composition of the plant community. Such weeds, which are incredibly allelopathic, have outcompeted tree recovery, altered species composition, and caused compaction in the soil (Haq et al. 2022). The Himalayan range, which contains a broad diversity of flora, wildlife, culture, human populations, and social systems, is significant from a geodynamic standpoint (Sati 2020). The region's species are under jeopardy due to ongoing human pressure and the notable changes in climate that have happened recently. Continuous depletion of forest resources results in low agricultural yield to the population because farm production is connected with inputs derived from forests (Singh et al. 2021). The likelihood of a fall in agro-biodiversity and hazards it poses to livelihoods and Himalayan ecosystems could be lessened by taking steps like enhancing the use of traditional manure, limiting the number of trees planted on farms, encouraging the development of agroforestry in degraded forests, and passing laws that support the sale of nontimber forest products to rural communities.

4 Agroforestry Technologies and Biodiversity Managements

The protection of biodiversity could be greatly aided by traditional agroforestry techniques. By utilizing agroforestry technology, rural households can increase their diversity and range of livelihood possibilities while reducing the loss of biodiversity (Gupta et al. 2020). In order to address ecological, social, and economic demands, agroforestry technologies emphasize the significance of trees on farms as well as in agricultural landscapes (Zerihun 2014). Both agroforestry and conservation scientists have given much attention to the benefits that agroforestry practices offer to biodiversity protection and rural livelihoods (Nair et al. 2021a).

4.1 Multifunctional Agroforestry

AF replenishes soil organic matter and nutrient levels, and provides a variety of goods and services, including food, fuel wood, fodder, timber, and medicines. It also helps to preserve biodiversity, maintain healthy ecosystems, and safeguard soil and water resources, and storage of carbon on land (Kumar et al. 2018). Additionally, AF provides low-income farmers with less expensive options than pricey traditional conservation methods (Bijalwan et al. 2020; Mehta et al. 2024). SALT (sloping agricultural land technology) or CHIAT [contour hedgerow intercropping (agroforestry) technology] is one such system. It involves planting dense hedgerows of perennial, quickly growing trees or shrubs that affix nitrogen along contour lines, forming a living barrier that collects sediments and slowly turns sloping land into terraced land (www.icimod.org, accessed 20 December, 2023). The traditional farmers in the Central Himalaya do, in fact, use a variety of multipurpose tree species as sporadic trees on terraced hills. The Himalayan alder, Indian fig tree (*Ficus glomerata*), shisham, wild Himalayan cherry, bihul, Chinese tallow tree (*Sapium sebiferum*), and wild Himalayan pear (*Pyrus pashia*), are a few examples. However, there is slight information available about the growth and ecological effects of such tree species (Raj et al. 2020).

4.2 Shifting Cultivation or Jhum Improvement

Jhuming has been improved over time by experience and regional requirements (such as horticultural intervention or agroforestry), particularly in light of the notion that this traditional method is environmentally detrimental (Jinger et al. 2023). Two externally funded development projects, viz., Nagaland Environment Protection and Economic Development Project, Kohima, funded by India Canada Environment Facility (NEPED-ICEF: http://cigrasp.pik-potsdam. de/adaptations/Nagaland--environment-protection and-economic-development, accessed 22 Nov 2023) and North Eastern Region Community Resource Management Project (NERCORMP), funded by International Fund for Agricultural Development (IFAD) and North Eastern Council (NEC) in Manipur and Meghalaya (http://mdoner.gov.in/sites/default/files/silo2_content/NERCORMP/NERCORMP%20%20Project%20Details.pdf, accessed 22 Nov 2023) were put into effect over the course of the previous 10 years with the intention of enhancing vulnerable populations' sustainable means of subsistence through better resource base management, helping to protect and restore the environment (Kumar et al. 2018). Jhuming cycles, which used to last up to 20–30 years, now only last 3–6 years or less (Marak et al. 2023). Such brief fallowcycles result in a decline in soil quality because the biological processes connected to the site's vegetation are unable to restore soil fertility (Ahirwal et al. 2023).

It is essential to provide incentives for the traditional jhum producers to engage in agroforestry (Paul et al. 2023). This could support a balance between economic expansion and ecological conservation, without negatively affecting local feelings a farmer-led extension program might be able to aid in the rehabilitation of the jhum lands (Mathur and Bhattacharya 2023; Mandal and Lonchung 2023).

4.3 Hippophae rhamnoides-*Based Agroforestry Systems*

Hippophae rhamnoides (Sea buckthorn) is an indigenous plant that has persevered in extreme weather, with an average annual temperature of 10–15 °C and occasionally dropping as low as −40 to −50 °C. Its typical rainfall is 350–1200 mm (Hussain et al. 2023). One of the few possible resources in high mountain regions that is both ecologically and ethno botanically sustainable, this food crop offers the mountain people several advantages (Raghuvanshi et al. 2022). A propitious tool for the preservation of biodiversity in the Trans-Himalayas (Acharya et al. 2010). Sea buckthorn is used for many purposes, such as fuel, having a high calorific value of 4785 kcal (Devi et al. 2023), fodder during thin times, and it can affix atmospheric nitrogen in root nodules by forming consortium with Frankia (Negi and Sharma 1998; Kumar et al. 2018). Despite the fact that the plant grows organically, its benefits to the environment and economy have not received enough attention (Jubayer et al. 2023). There are many chances to work on the features of sea buck thorn's genetic variety, propagation, root nodulation, and therapeutic qualities, all of which will contribute to the high mountain people's improved quality of life (Malik et al. 2020; Trivedi et al. 2020).

4.4 Alder-Based Agroforestry Systems

The large cardamom-based system in eastern Himalayas based on alder is perceived to accelerate the cycling of nutrients, enhance productivity and soil fertility, conserve biodiversity, decrease soil erosion, soil and water act as a carbon sink, raise the standard of living of communities through increased income from farms, and offer aesthetic values to mountain communities (Tamang et al. 2021). The tribes of *Angami, Chakhasang, Chang, Yimchaunger,* and *Konyak* benefit from the increased soil fertility caused by alder trees (Ray et al. 2018). *Khonoma*, Nagaland, the alder-based production system is a remarkable illustration of long-lasting sustainable land usage (Nath et al. 2023). According to this method, alder seedlings are planted on sloping terrain, and the trees take 6–8 years to grow to full maturity (Fibers et al. 2020). In order to prepare the land for the cultivation of arable crops, the trees are then pollarded, their residue is burned, and the ash that remains is mixed with the soil. A 2- to 4-year fallow period follows cropping, and subsequent pollarding is carried out once every 4–6 years. All coppices, with the exception of the five or six on top of the main trunk, are chopped down (Sarkar et al. 2023). In addition to supplying phosphorus, potassium, calcium, and other nutrients to the soil through root nodules, the litter given to soil fixes atmospheric nitrogen (Choudhury et al. 2022; Sharma and Singh 1994).

4.5 Homestead Agroforestry System

The *Adi* tribal people of the Siang valley in northeastern India have used the *Bari* traditional homestead agroforestry system for many generations (Das et al. 2023). It is a model of productivity, food security, sustainability, biodiversity conservation, environmental preservation, and wise management of natural resources (Misra and Ghosh 2024). An agroforestry system that is representative of the *Bari* community is very important for biodiversity management and conservation (Goswami et al. 2022). Through a complex arrangement of different living forms in space and time, as well as niche diversification, it has been developed to allow for the best possible harvesting of solar energy. In terms of the crop component's horizontal arrangement, (Nair et al. 2021b), the inner zone—which is closest to the home—often contains aromatic plants, condiments, and spices as well as vegetables such as giant taro (*Alocasiaindica*), chillies, pumpkin (*Cucurbita maxima*), ginger, sponge gourd (*Luffa* spp.), turmeric, and medicinal plants (George and Christopher 2020). Subsequently, there is a second zone that comprises Citrus spp., banana (*Musa* spp.), rattan (*Calamus* spp.), jackfruit (*Artocarpus heterophyllus*), areca nut (*Areca catechu*), and bamboo, A third zone is occupied by trees, including mango, *Terminalia chebula* (myrobalan), *Garcinia cowa* (garcinia), *Spondias pinnata* (wild mango), Indian gooseberry, and *Dillenia indica* (elephant apple) (Upadhyay et al. 2002; Kumar et al. 2018). It is imperative to revive these historic land-use systems, which have withstood the test of time and are able to offer a variety of goods and services.

5 Policy and Governance

The main purpose of policies is to create an institutional and economic environment that facilitates the relatively unhindered implementation of private local initiatives. Farmers may be able to engage in agroforestry in an environment where control is limited with the aid of long-term or short-term incentives, like subsidies, meant to develop markets and gradually lower economic risk (Foresta 2013).

6 Major Policy Issues Responsible for Weakening the Himalayan Mountain Farming System

The execution of the mountain/hill agriculture stratum is until now declining in spite of multiple strategies and policies (Bhatt et al. 2022). The marginalization of mountain farmers has been exacerbated by an overemphasis on some of the crops that respond well to high levels of external inputs, plains with more "operable" field sizes, and a dearth of experimentation, technological expansion, and policy systematics on hill farming (Khanduri et al. 2023). For example, in mountain farming, the use of irrigation and fertilizer inputs is severely limited and is necessary for the promotion of high-yielding crop varieties. However, there are still too few and inadequate market and technology connections for agriculture to modernize (Bhatt et al. 2022). The prevalent policies are not conducive to the subsistence agricultural practices used in the mountains, nor are they economically feasible for large-scale farming (Ullah and Kim 2020). This has led to decreased growth in producing food grains, resulting in weakening the economic base and food security of marginal hill farmers (Maikhuri et al. 2010). These policies have put the region's traditional agriculture, with its variety of crops and cropping practices, in grave danger (Kumar et al. 2020). Some topics of immediate concern in this sector with reference to smallholders' farming systems are provided in Table 5.4.

Table 5.4 Policy areas of immediate concern and their strategic solutions across (IHR)

Areas of concern	References	Way forward	References
Land-use policies: There has been a decline in agricultural diversity and food security as a result of large holdings for fruit and hybrid crops and a lack of support for traditional crops.	Bhatt et al. (2022), Ullah and Kim (2020), Chaudhary et al. (2020)	*Improved land-use policies*: Land-use policies targeting ubiquitous hilly areas encouraging agro-biodiversity through traditional agroforestry methods maintained by family farmers.	Kumar et al. (2023) and Darlong (2017)
Forest/wildlife conservation policies: Crop damage due to wild animals, Forest (Conservation) Act (1986), Wildlife (Protection) Act (1972), restrict farmers from killing such animals' compensation for damage less (Watershed Management Directorate 2010) leading to lessen cropping intensity/ waive farm lands entirely.	Bhatt et al. (2022)	*Eliminate legal and institutional constraints*: A successful policy should encourage both public and private efforts to sustainably enhance the welfare of the country. Regulations should not impede relevant, empirically based development models that are at hand, such as the effective agroforestry techniques and approaches.	Foresta (2013) and Bhatt et al. (2022)

(continued)

Table 5.4 (continued)

Areas of concern	References	Way forward	References
Policy credit and subsidies for food imports: The government's policy of importing food and subsidizing its cost to those below the poverty line (BPL) through public distribution systems (PDs) has caused the hill people to alter their eating habits and give up on traditional crop varieties that are more varied and nutritious.	Maikhuri et al. (2010)	*Support positive policy outcomes*: Farmers are compensated for environmental services (PES) by receiving financial support for growing trees on their lands, which lowers financial and climatic risks while creating public ecosystem services (PES). For PES, these economic and environmental services are valued. Therefore, encouraging traditional agricultural kinds that are more varied and nutritious.	Bhatt et al. (2020), Foresta (2013), and Ranjan (2024)
Farm input subsidies: Input costs have been lowered to support exotic, high-yielding crop varieties at the expense of conventional hill agriculture, including water, herbicides, chemical fertilizers, and seeds. These subsidies might have helped support a variety of local crops.	Bhatt et al. (2022) and Barthel et al. (2013)	*Farmers for the delay in returns*: Changes in production systems need investment from producers since it takes time for the new system to yield the same amount as the old one. A few examples of policy interventions are regulatory instruments like state forest programs and legislation, economic tools like taxing schemes and in-kind or financial incentives (PES), and informational tools like education and technical help.	Foresta (2013) and Bhatt et al. (2020)
Low investment in research and development: Agriculture policy, research and development focused on handful of crops, negligence toward traditional/mountain food crops have deprived the development of improved varieties, leaving them economically less competitive and farmers increasingly replace such crops with hybrid crops, thereby eroding the agro-biodiversity of the region.	Bhatt et al. (2022) and Kumar et al. (2021)	*Higher levels of investment in quality training*: Concern for and dedication to the goal of agricultural development must be reflected in the agricultural education process. Learner-centered methods that rely on group learning must emphasize critical evaluation of issues, solutions, and reactions. Since these action-oriented approaches are more difficult than conventional chalk-and-talk techniques, agricultural education approaches must be restructured, quality must be raised, and suitable training and capacity-building support are needed.	Bhatt et al. (2022) and Gulati et al. (2020)

Source: The authors' summary from this chapter and the literature examination

7 Way Forward

A system for a diverse team of scientists and extension workers to work on the challenges with regular feedback from the farmers is required, as well as diversified mandates and strong inter-sectoral and interdisciplinary linkages within the agricultural educational institutions in the Himalayan region (Gulati et al. 2020). This requires a “paradigm shift” since the objectives and scope of agriculture must be revised and reoriented in light of a sustainable future (Singh et al. 2020). India’s National Agroforestry Policy of 2014 aims to improve productivity and sustainable environmental practices by bringing together trees, crops, and livestock on one site (Kumar et al. 2023). It was created as a countermeasure to the limited agricultural output caused by the country’s rapid population growth and reliance on seasonal rainfall for agriculture, which has resulted in a steady drop in farmers’ land ownership percentage (Chaudhary et al. 2020). Adoption has to be supported by favorable policies as well as local and national institutional arrangements because of its long-term nature and the process may not occur in a policy vacuum to enable (Table 5.4). In order to accomplish the UN Sustainable Development Goals-2 (SDGs-2), agricultural development and programs must be created in compliance with state and federal policies while taking local and regional objectives into consideration, attain food security, eliminate hunger, improve nutrition, and facilitate sustainable agriculture.

8 Governance

There are major ramifications for on-farm tree planting and management as a result of the changing nature of environmental governance, which encompasses all institutions and policies influencing the condition of the environment in the developing world (Garrity 2006). Local user organizations and decentralized multistakeholder committees are gaining formal authority from national forestry agencies (Leimona 2011). Market- and incentive-based methods to environmental management are gradually replacing bans and rules (Tomich et al. 2004). More attention is being paid to integrated methods of managing ecosystems and landscapes that involve the local community as vital partners (Shirvani Dastgerdi and Kheyroddin 2022). Agroforestry is facing both new opportunities and challenges as a result of these shifts. Numerous laws and rural organizations influence farmers’ incentives to plant and maintain trees in their agricultural landscapes, even though there are not many that are specifically focused on agroforestry. The five policy concerns (Table 5.5) that are most important to agroforestry are global environmental governance, biodiversity and forest protection, land and tree tenure, forest classification, and environmental service payment systems (Swallow et al. 2006). Focused applied research and participation in regional policy processes boost the positive effects of agroforestry growth in regional policy domains and support national and international policy reform.

Table 5.5 Delineation of links between AFS and five elements of environmental governance

	Relation: intentional tree management to farmer incentives	Policymaking location in relation to the farmer	Trends in relation to governance and policy
Ownership of land and trees	Assurance to farmers of future returns on current investments; encouragement to farmers to acquire tree products for their own farm or elsewhere in the landscape	National policies; local standards; and dispersed government agencies	Progressive personalization; decentralized government organizations are becoming more significant overall
Forest classification and governance	Farmer incentives to preserve and manage surrounding forests; farmer incentives to harvest tree products from forest areas	Both national and decentralized forest agencies	State agency decentralization and a shift away from the command and control paradigm
Buffer zone and landscape approaches to conservation	Rewards and penalties for maintaining trees close to protected areas; kinds of trees that are permitted and encouraged in various landscape areas	Worldwide conservation demands; national forest agencies; decentralized conservation/forest agencies	Integrated conservation and development programs have had varying degrees of success; landscape approaches are still mostly experimental.
Environmental service mechanisms	Tree and vegetation management by individual farmers, organizations, and/ or municipal governments is a major component of environmental service mechanisms	National policy and international frameworks for carbon and biodiversity; regional dialogue for watershed services	Integrating into government initiatives in other Latin American nations; conducting small-scale trials in other areas
Global environmental governance	There are substantial forestry components in UNCBD[1], UNCCD[2], UNFCCC[3]and GEF[4], but insufficient support is given to smallholder agroforestry; there has not been a dedicated forestry treaty since the Rio meeting in 1992	Global agreements and funding opportunities: National ratification, domestication, and funding	The UNFCCC has made the greatest headway in clearly taking smallholders' and agroforestry's potential into account

Source: The authors' synopsis from this chapter and the literature analysis:
1. United Nations Convention on Biological Diversity
2. United Nations Convention on Combating Desertification
3. United Nations Framework Conventions on Climate Change
4. Global Environment Facility

9 Conclusion

The abundance of biodiversity found in forests signifies that the natural forests still play an important role in enhancing the standards of living by providing food, nutrition, and energy to rural inhabitants in the Indian Himalayan Region. The region has a high population density, which leads to higher rates of resource extraction, a widespread subsistence practice throughout the belt. It is crucial to create measures to lessen livelihood-based extraction as Himalayan forests play a major role in regulating the country's climate. The area's forest-dependent populations would benefit from the measures having more diverse sources of income and jobs, and must take into account socioeconomic variables grounded in community-specific characteristics that define the boundaries of forest resource extraction. They enable rural communities to sustain their production systems, and better agroforestry systems have the potential to reduce overreliance of these populations on natural forest resources, thereby contributing to the maintenance of biodiversity in natural systems. Because AFS incorporate many different components, they have unintentionally boosted agro-biodiversity. In certain situations, they have even made it easier to implement so-called new AF technologies by emulating older systems. Nonetheless, more germplasm distribution and lobbying of legislators are still necessary to advance AFS.

References

Acharya SOMEN, Stobdan T, Singh SB (2010) Seabuckthorn (Hippophae sp. L.): New crop opportunity for biodiversity conservation in cold arid Trans-Himalayas. J Soil Water Conserv 9(3):201–204

Acrhm NK, Gupta SAK, Dogra NASIP (2018) Performance of litchi (Litchi chinensis Sonn.) based Agri-horticultural land uses in rainfed condition on degraded lands in North Western Himalayas, India

Aggarwal A, Paul V, Das S (2009) Forest resources degradation, livelihoods and climate change in green India: looking back to change track. The Energy and Resources Institute, Delhi

Ahirwal J, Thangjam U, Sahoo UK (2023) Land use change and its impacts on soil carbon dynamics in Mizoram, Northeast India. In: Soil carbon dynamics in Indian Himalayan region. Springer Nature Singapore, Singapore, pp 217–234

Aryal JP, Sapkota TB, Rahut DB, Jat ML (2019) Agricultural sustainability under emerging climatic variability: role of climate smart agriculture and relevant policies in India. Int J Innov Sustainable Dev 14(2):219–245. https://doi.org/10.1504/IJISD.2019.10020869

Awasthi A, Uniyal SK, Rawat GS, Rajvanshi A (2003) Forest resource availability and its use by the migratory villages of Uttarkashi, Garhwal Himalaya (India). For Ecol Manag 174:13–24

Baba MY, Islam MA, Qaisar KN (2015) Assessing the household fuelwood extraction and consumption situation in rural Kashmir, India. Int J For Crop Improv 6(1):55–63

Baig MB, Burgess PJ, Fike JH (2021) Agroforestry for healthy ecosystems: constraints, improvement strategies and extension in Pakistan. Agrofor Syst 95:995–1013

Ballabh B, Chaurasia OP (2007) Traditional medicinal plants of cold desert Ladakh used in treatment of cold, cough and fever. J Ethnopharmacol 112:341–349

Banday M, Islam MA, Pala NA, Rashid M, Ahmad PI, Rather MM, Raja R (2021) Livelihood security and forest resource extraction by forest fringe communities in Indian Himalayan region. In: Diversity and dynamics in forest ecosystems. Apple Academic Press, pp 163–194

Banyal R, Islam MA, Masoodi TH, Gangoo SA (2013) Energy status and consumption pattern in rural temperate zone of western Himalayas: a case study. Indian Forester 139(8):683–687

Barthel S, Crumley CL, Svedin U (2013) Bioculturalrefugia: combating the erosion of diversity in landscapes of food production. Ecol Soc 18(4):71

Behar B (2021) Livelihood dependency on prominent agroforestry systems in Kulgam district of U.T. Jammu and Kashmir. MSc thesis. Faculty of forestry, SK University of Agricultural Sciences & Technology of Kashmir, India

Behera RN, Nayak DK, Andersen P, Måren IE (2016) From jhum to broom: agricultural land-use change and food security implications on the Meghalaya plateau, India. Ambio 45:63–77

Bhat GM, Fatima S, Islam MA, Rather TA, Dutt V, Pala NA, Rashid M, Dar MUD (2022) Livelihood dependency on agroforestry systems in trans Himalayan region of Ladakh. Indian J Agrofor 24(2):103–110

Bhatt BP, Negi AK, Todaria NP (1994) Fuelwood consumption pattern at different altitudes in Garhwal Himalaya. Energy 19(4):465–468

Bhatt BP, Sachan MS (2004) Firewood consumption along an altitudinal gradient in mountain villages of India. Biomass Bioenergy 27:69–75

Bhatt ID, Negi VS, Rawal RS (2020) Promoting nature-based solution (NbS) through restoration of degraded landscapes in the Indian Himalayan region. In: Nature based solutions for resilient ecosystems and societies, pp 197–211

Bhatt GC, Singh R, Baduni P, Kandari LS, Rawat LS, Prasad C, Phondani PC and Maikhuri RK (2022) Education, research and policy issues in Indian Himalayan Mountain farming system: priorities for sustainable development

Bijalwan A, Sharma C, Sah V, Raj A (2020) Structure and composition of trees and their effect on crop yield in existing agrihortisilvicultural and agrihorticultural systems of Garhwal Himalaya. Indian J Agrofor 11(2):64–70

Chaudhary S, Wang Y, Dixit AM, Khanal NR, Xu P, Fu B, Yan K, Liu Q, Lu Y, Li M (2020) A Synopsis of farmland abandonment and its driving factors in Nepal. Land 9(3):84

Choudhury BU, Nengzouzam G, Ansari MA, Islam A (2022) Causes and consequences of soil erosion in northeastern Himalaya, India. Curr Sci 122(7):772–789

Chouhan SF, Sarma C (2020) Disadvantaged neighborhoods: causes, dimensions, and impacts. In: Zero Hunger. Springer, Cham, pp 211–219

Dagar JC, Gangaiah B, Gupta SR (2020) Agroforestry to sustain Island and coastal agriculture in the scenario of climate change: Indian perspective. In: Agroforestry for degraded landscapes: recent advances and emerging challenges-Vol. 1. Springer Singapore, Singapore, pp 367–424

Darlong V (2017) Reflections on the impacts of state policies on shifting cultivators in Northeast India. In: Shifting cultivation policies: balancing environmental and social sustainability. CABI, Wallingford UK, pp 344–378

Das A, Gujre N, Devi RJ, Mitra S (2023) A review on traditional ecological knowledge and its role in natural resources management: Northeast India, a cultural paradise. Environ Manag 72(1):113–134

Datta P, Behera B (2022) Assessment of adaptive capacity and adaptation to climate change in the farming households of Eastern Himalayan foothills of West Bengal, India. Environ Challenges 7:100462. https://doi.org/10.1016/j.envc.2022.100462

Datta P, Behera B (2024) Assessing the role of agriculture-forestry-livestock nexus in improving farmers' food security in South Asia: a systematic literature review. Agric Syst 213:103807

Datta P, Das S (2019) Analysis of long-term precipitation changes in West Bengal, India: an approach to detect monotonic trends influenced by autocorrelations. Dyn Atmos Oceans 88:101118. https://doi.org/10.1016/j.dynatmoce.2019.101118

Dean C, Young J, Ashley-Cantello W, Ellis K, Jeffries B (2016) Cover photo

Devi SG, Abraham S, Babu P, Vasanthi HR (2022) Traditional Manipuri cuisine and prospects of promoting medicinal plant-based food tourism post-COVID-19. International Journal of Hospitality and Tourism Systems. Special Issue on CoVID-19, pp 114–130

Devi NB, Lepcha NT, Bhutia PT, Rocky P, Sahoo UK, Pandey R, Nath AJ (2023) Biodiversity and ecosystems services of the agroforestry systems of the Himalayan region: an overview. In: Agroforestry for sustainable intensification of agriculture in Asia and Africa, pp 487–513

Devkota S, Bhusal KK (2020) Moringaoleifera: a miracle multipurpose tree for agroforestry and climate change mitigation from the Himalayas–a review. Cogent Food Agric 6(1):1805951

Dhanai R, Negi RS, Parmar MK, Singh S (2014) Fuelwood and fodder consumption pattern in Himalayan watershed. Int J Environ Biol 4(1):35–40

Dhyani SK (2014) National Agroforestry Policy 2014 and the need for area estimation under agroforestry. Curr Sci 107:9–10

Duffy C, Toth GG, Hagan RP, McKeown PC, Rahman SA, Widyaningsih Y, Sunderland TC, Spillane C (2021) Agroforestry contributions to smallholder farmer food security in Indonesia. Agrofor Syst 95(6):1109–1124

FAO (2022) Food security and nutrition in the world the state of. https://www.fao.org/

Fibers A, Barrackpore K, Bengal W (2020) Rehabilitation of old river bed lands by an intensively managed silvi-pastoral system in the North-West Himalayas. Indian J Soil Conserv 48(1):70–79

Foresta HD (2013) Advancing agroforestry on the policy agenda–a guide for decision-makers

Gangoo SA, Islam MA, Tahir M (2017) Wealth of non timber forest products and their trade in Jammu and Kashmir. Indian Forester 143(9):737–744

Garrity DP (2006) World agroforestry into the future. World Agroforestry Centre

George MV, Christopher G (2020) Structure, diversity and utilization of plant species in tribal homegardens of Kerala, India. Agrofor Syst 94(1):297–307

Ghosh PK, Nath CP, Hazra KK, Kumar P, Das A, Mandal KG (2020) Sustainability concern in Indian agriculture: needs science-led innovation and structural reforms. Indian J Agron 65(2):131–143

Goswami, S., Bijalwan, A., Singh, B., Ahmed, T. and Kumar, S., 2022. Floristic diversity, community structure and species richness of Malta (Citrus sinensis) based traditional homegardens of Tehri Garhwal, Uttarakhand, India

Gulati MP, Priya S, Bresnyan EW (2020) Grow solar, save water, double farmer income

Gupta SR, Dagar JC, Teketay D (2020) Agroforestry for rehabilitation of degraded landscapes: achieving livelihood and environmental security. In: Agroforestry for degraded landscapes: recent advances and emerging challenges-Vol 1, pp 23–68

Haq SM, Yaqoob U, Hamid S, Hassan M, Bashir F, Waheed M, Bussmann RW (2022) Localized impact of livestock settlements on vegetation patterns in fir forests of Kashmir Himalaya. Acta Ecol Sin 42(4):407–416

Hussain A, Abidi SH, Syed Q, Saeed A, Naib N, Bano A, Malik M, Nadeem AA (2023) A supernatural multipurpose plant sea buckthorn (*Hippophaerhamnoides* L.): an updated overview of its folk traditional uses, phytochemical profile and biological activities. J Berry Res:1–46. (Preprint)

ISFR (2019) India State of Forest Report 2013, Forest Survey of India. Ministry of Environment & Forests, Dehradun

Islam MA, Banyal R, Masoodi NA, Masoodi TH, Gangoo SA, Sharma LK (2011) Status of fuelwood extraction and consumption in rural North Kashmir: a case study. Indian Forester 137(11):1265–1268

Islam MA, Quli SMS, Sofi PA, Bhat GM, Malik AR (2015a) Livelihood dependency of indigenous people on forest in Jharkhand, India. Vegetos 28(3):106–118

Islam MA, Masoodi TH, Gangoo SA, Sofi PA, Bhat GM, Wani AA, Gatoo AA, Singh A, Malik AR (2015b) Perceptions, attitudes and preferences in agroforestry among rural societies of Kashmir, India. J Appl Nat Sci 7(2):976–983

Islam MA, Quli SMS, Mushtaq T (2017) Wasteland reclamation strategy for household timber security of tribes in Jharkhand, India. J Appl Nat Sci 9(4):2264–2271

Islam MA, Mugloo JA, Raj A, Bhat GM, Wani AA, Gatoo AA, Malik AR, Pala NA, Shah M (2022) Agroforestry strategy for revitalizing fodder security in Kashmir Himalaya, India. Agric Res 11(3):528–538

Jinger D, Kaushal R, Kumar R, Paramesh V, Verma A, Shukla M, Chavan SB, Kakade V, Dobhal S, Uthappa AR, Roy T (2023) Degraded land rehabilitation through agroforestry in India: achievements, current understanding, and future prospectives. Front Ecol Evol 11:1088796

Jubayer MF, Mazumder MAR, Nayik GA, Ansari MJ, Ranganathan TV (2023) Hippophae Rhamnoides L.: sea buckthorn. In: Immunity boosting medicinal plants of the Western Himalayas. Springer Nature Singapore, Singapore, pp 463–491

Khanduri S, Saklani RD, Chetry BM (2023) Increasing risk of silent disaster in Uttarakhand Himalaya: an example from higher Himalaya. Afetve Risk Dergisi 6(3):870–890

Khuman YSC, Pandey R, Rao KS (2011) Fuelwood consumption patterns in Fakot watershed, Garhwal Himalaya, Uttarakhand. Energy 36(8):4769–4776

Kumar R, Saikia P (2020) Forest resources of Jharkhand, Eastern India: socio-economic and bio-ecological perspectives. In: Socio-economic and eco-biological dimensions in resource use and conservation: strategies for sustainability. Springer, Cham, pp 61–101

Kumar M, Sharma CM (2009) Fuelwood consumption pattern at different altitudes in rural areas of Garhwal Himalaya. Biomass Bioenergy 33:1413–1418

Kumar BM, Handa AK, Dhyani SK, Arunachalam A (2018) Agroforestry in the Indian Himalayan region: an overview. In: Temperate agroforestry systems, pp 153–172

Kumar R, Verma TS, Chandra A, Kumar M, Rawat L (2020) Temporal and spatial assessment of physico-chemical properties in surface and groundwater in the Himalayan foothill (Uttarakhand), India

Kumar P, Sharma PK, Kumar P, Sharma M, Butail NP (2021) Agricultural sustainability in Indian Himalayan region: constraints and potentials. Indian J Ecol 48(3):649–661

Kumar S, Dev I, Kumar N (2023) Forestry and agroforestry. In: Trajectory of 75 years of Indian Agriculture after Independence. Springer Nature Singapore, Singapore, pp 497–519

Lata M, Mondal BC (2021) Uses of tree leaves as alternative feed resources for ruminant animals

Leimona B (2011) Fairly efficient or efficiently fair: success factors and constraints of payment and reward schemes for environmental services in Asia. Wageningen University and Research

Mahapatra S, Kumar D, Singh B, Sachan PK (2021) Biofuels and their sources of production: a review on cleaner sustainable alternative against conventional fuel, in the framework of the food and energy nexus. Energy Nexus 4:100036

Maikhuri RK (1991) Fuelwood consumption pattern of different tribal communities living in Arunachal Pradesh in Northeast India. Bioresour Technol 35(3):291–296

Maikhuri RK, Rawat L, Negi V, Bahuguna A, Agarwal SK, Farooquee NA, Phondani P, Negi CS (2010) Demonstration dissemination of simple eco-friendly technologies for natural resource management in central Himalaya. Indian J Sci Technol 3(7):823–831

Maikhuri RK, Rawat LS, Semwal RL, Rao KS, Saxena KG (2015) Organic farming in Uttarakhand Himalaya, India. Int J Ecol Environ Sci 41(3–4):161–176

Maja MM, Ayano SF (2021) The impact of population growth on natural resources and farmers' capacity to adapt to climate change in low-income countries. Earth Syst Environ 5:271–283

Malik N, Saxena S, Nangyal H, Zohra T, Ikram H (2020) Sea buckthorn: its sustainable utilization for biodiversity conservation. In: Environmental pollution, biodiversity, and sustainable development: issues and remediation, pp 147–158

Mandal RK, Lonchung K (2023) A Review of environmental issues due to shifting cultivation in Arunachal Pradesh, India. J Exp Agric Int 45(11):83–91

Marak SR, Kumar M, Kundu A, Dutta D, Pebam R, Chattoraj KK, Lal D (2023) Shifting cultivation in the East Garo Hills, Meghalaya (India): an earth observation perspective. In: Water, land, and forest susceptibility and sustainability. Academic, pp 439–458

Mathur I, Bhattacharya P (2023) Perception of agroforestry practices and factors influencing adoption among shifting cultivators in Tripura, India. For Trees Livelihoods 33(1):23–19

McGunnigle N, Bardsley D, Nuberg I, Cedamon E, Pandit BH (2023) The succession of farmers' perceptions of transitioning landscapes–a case study of agroforestry in the middle hills of Nepal. Hum Ecol 51(4):699–717

Mehta H, Rathore AC, Tomar JMS, Mandal D, Kumar P, Kumar S, Sharma SK, Kaushal R, Singh C, Chaturvedi OP, Madhu M (2024) Minor millets based agroforestry of multipurpose tree species of Bhimal (Grewiaoptiva Drummond JR ex Burret) and Mulberry (Morus alba L.) for resource conservation and production in north western Himalayas–10-year study. Agric Ecosyst Environ 359:108761

Misra S, Ghosh A (2024) Agriculture paradigm shift: a journey from traditional to modern agriculture. In: Biodiversity and bioeconomy. Elsevier, pp 113–141

Monika DM, Bisht PS, Chaturvedi P (2020) Medicinal uses of traditionally used plants in Bhatwari block, district Uttarkashi, Uttarakhand, India. J Sci Res 64(1):119–126

Mundoli S, Joseph G, Setty S (2016) "Shifting agriculture": the changing dynamics of Adivasi farming in the forest-fringes of a tiger reserve in South India. Agroecol Sustain Food Syst 40(8):759–782

Muralidharan R, Narasimhan D (2012) Ethnomedicinal plants used against gastrointestinal problem in Gingee hills of Villupuram district, Tamil Nadu. J Appl Pharm Sci 2(10):123–125

Nagar B, Kumar M, Pandey R, Rawat S (2018) Fuelwood and fodder consumption from agroforestry at different altitudinal zones of Garhwal Himalaya. J Plant Dev Sci 10(12):659–668

Nagar B, Rawat S, Pandey R, Kumar M (2021) Variation in specific gravity and carbon proportion of agroforestry tree species of Himalaya. Environ Chall 4:100156

Nagar B, Rawat S, Pandey R, Kumar M, Alatalo JM (2022) Fuelwood and fodder consumption patterns among agroforestry-practicing smallholder farmers of the lower Himalayas, India. Environ Dev Sustain 24(4):5594–5613

Nair PR, Kumar BM, Nair VD, Nair PR, Kumar BM, Nair VD (2021a) Agroforestry for biodiversity conservation. In: An introduction to agroforestry: four decades of scientific developments. Springer, Cham, pp 539–562

Nair PR, Kumar BM, Nair VD, Nair PR, Kumar BM, Nair VD (2021b) Agroforestry systems in the temperate zone. In: An introduction to agroforestry: four decades of scientific developments. Springer, Cham, pp 195–232

Nath U, Puzari A, Changmai S (2023) Use of traditional plant-based pesticides in the Jhum agroforestry system of Angami tribe in Kohima district of Nagaland, India. Agrofor Syst 98:129–149

Negi, A. and Sharma, P.N. (1998) Indigenous technology knowledge for watershed management in Upper North-West Himalayas of India. Published by Participatory Watershed Management Training in Asia (PWMTA) Program, GCP/RAS/161/NET, FAO (UN), Kathmandu, Nepal

Negi VS, Maikhuri RK, Rawat LS (2011) Non-timber forest products (NTFPs): a viable option for biodiversity conservation and livelihood enhancement in central Himalaya. Biodivers Conserv 20(3):545–559

Negi VS, Maikhuri RK, Chandra A, Maletha A, Dhyani PP (2018) Assessing sustainability of farming systems in mountain agroecosystems of Western Himalaya, India. Agroecol Sustainable Food Syst 42(7):751–776

Negi VS, Pathak R, Thakur S, Joshi RK, Bhatt ID, Rawal RS (2023) Scoping the need of mainstreaming indigenous knowledge for sustainable use of bioresources in the Indian Himalayan region. Environ Manag 72(1):135–146

NRCAF (2007) Vision 2025. National Research Centre for Agroforestry, Jhansi

Oh S (2021) IDF global dairy conference 2021. J Dairy Sci Biotechnol 39(4):166–171

Palit PN (2021) Non-timber forest products and their role in catalyzing the rural economy in the Barjora Forest range, Bankura District, West Bengal. Adv Sustainable Dev Manag Environ Nat Resour Econ Outlook Opin 2:401

Pandey R (2011) Forest biomass extraction for livestock feed and associated carbon analysis in lower Himalayas, India. Mitig Adapt Strateg Glob Chang 16(8):879–888

Pandey DK, Adhiguru P, De HK, Upadhyay AD, Radhakrishnan K (2021) Income inequality among indigenous people dependent on traditional agroforestry system in Indian Himalayas. Indian J Agric Sci 91:847–851

Patra S, Kumar A, Saikia P (2022) Deforestation and forests degradation impacts on livelihood security and climate change: Indian initiatives towards its mitigation. In: Environmental degradation: challenges and strategies for mitigation. Springer, Cham, pp 371–392

Paul A, Dutta PK, Khan ML, Das AK (2019) Rhododendrons: a major resource of fuelwood in high altitude region of Arunachal Himalaya, India. Biodiversitas J Biol Divers 20(9):2628–2635

Paul S, Chakraborty D, Padaria R, Tripathi AK (2023) A comparison between climate change perceptions and meteorological observations to improve the understanding of adaptation decisions in shifting cultivation. Theor Appl Climatol 153(3):1139–1155

Pereira LM, Cuneo CN, Twine WC (2014) Food and cash: understanding the role of the retail sector in rural food security in South Africa. Food Secur 6:339–357

Phondani PC, Maikhuri RK, Rawat LS, Negi VS (2020) Assessing farmers' perception on criteria and indicators for sustainable management of indigenous agroforestry systems in Uttarakhand, India. Environ Sustainability Indic 5:100018

Poonia MP, Sharma SC (2017) Environmental Studies. KHANNA PUBLISHING

Raghuvanshi MS, Sawal RK, Arunachalam A, Moharana PC, Landol S, Stanzin J, Dorjay N, Spalbar E, Pandey L (2022) Traditional resource of food and nutritional security through sea buckthorn (Hipphophaerhamnoides) in cold arid Ladakh, India. Climate Change Environ Sustainability 10(2):142–150

Rai SN, Chakrabarti SK (2001) Demand and supply of fuelwood and timber in India. Indian Forester 127(3):23–29

Raj A, Jhariya MK, Yadav DK, Banerjee A (2020) Agroforestry systems in the hills and their ecosystem services. In: Environmental and Sustainable Development Through Forestry and Other Resources. Apple Academic Press, pp 25–50

Rajput A, Roy S, Waghmare MN, Singh S, Shukla VK, Singh V, Singh P (2023) Exploring sustainable practices in modern agronomy and their environmental impact-a review. Int J Environ Climate Chang 13(11):3146–3161

Rajwar GS, Kumar M (2011) Fuelwood consumption in two tribal villages of the Nanda Devi biosphere Reserve of the Indian Himalaya and strategies for fuelwood sustainability. Environ Dev Sustain 13(4):727–741

Ranjan R (2021) Can combining two environmental services under a single PES program result in better environmental outcomes and lower costs? Nat Resour Model 34(2):12303

Ranjan R (2024) Creating synergies between payments for ecosystem services, green bonds, and catastrophe insurance markets for enhanced environmental resilience. Land Use Policy 136:106970

Rathore AC, Mehta H, Sharma NK, Gupta AK, Alam NM, Islam S, Dogra P (2020) Performance of litchi (Litchi chinensisSonn.) based agri-horticultural land uses in rainfed condition on degraded lands in North Western Himalayas, India. Agrofor Syst 94:2225–2236

Rawat YS, Tekleyohannes AT (2021) Sustainable forest management and forest products industry development in Ethiopia. Int For Rev 23(2):197–218

Rawat S, Kumar M, Pandey R, Nagar B (2018) Forest resource utilization pattern and heterogeneity in rural household characteristics in Garhwal Himalaya. Indian Forester 144(2):150–158

Ray SK, Das A, Deka BC (2018) Soil conservation measures for arresting soil and nutrient erosion under shifting cultivation. Conserv Agric Adv Food Secur Changing Climate 2:695–709

Reang D, Sahoo UK, Giri K, Hazarika A, Nath AJ (2022) Ethnic homestead forests of North-East India revealed as diverse land-use systems. Agrofor Syst 96(3):465–478

Sahoo G, Wani AM (2020) Agroforestry in food security. In: Recent research in agriculture for doubling of farmer's income, p 73

Sahoo G, Wani AM, Sharma A, Rout S (2020) Agroforestry for forest and landscape restoration. Int J Adv Study Res Work 9:536–542

Saraev V, Low WH, Beauchamp K, Perks M (2022) The potential for agroforestry to reduce net GHG emissions in Scotland through the Woodland Carbon Code. Forest Research
Sarkar BC, Tamang M, Shukla G, Shahina NN, Nath AJ, Chakravarty S (2023) Variation in tree density, biomass and carbon stock across an altitudinal gradient under large cardamom agroforestry system of Darjeeling Himalaya
Sarmah R, Arunachalam A (2011) Contribution of non-timber forest products (NTFPs) to livelihood economy of people living in forest fringes in Changlang district of Arunachal Pradesh, India. Indian J Fundam Appl Life Sci 1(2):157–169
Sati VP (2020) Himalaya on the threshold of change. Springer
Sati VP (2022) Assessment of natural capital for sustainable rural livelihoods: an empirical study of Mizoram, India. Asia-Pacific J Rural Dev 32(1):7–25
Selaiyur C and Nadu T (n.d.) A study on depletion of natural resources in India-a view
Shaheen SM, Antoniadis V, Shahid M, Yang Y, Abdelrahman H, Zhang T, Hassan NE, Bibi I, Niazi NK, Younis SA, Almazroui M (2022) Sustainable applications of rice feedstock in agro-environmental and construction sectors: a global perspective. Renew Sustainable Energy Rev 153:111791
Sharma UC, Singh KA (1994) Agriculture with Alder-A potential indigenous farming system of Nagaland. In: Proceedings of the national seminar on developments in soil science. ICAR, New Delhi, pp 570–571
Sharma S, Singh P (eds) (2021) Wetlands conservation: current challenges and future strategies. Wiley
Sharma UC, Datta M, Sharma V (2023) Climate and biodiversity. In: Soils in the Hindu Kush Himalayas: management for agricultural land use. Springer, Cham, pp 27–57
Shekhar C, Verma S, Mishra AK, Singh V, Prakash S, Lal P (2022) Fodder diversity, availability and utilization pattern in Western Himalayas. Int J Agric Environ Biotechnol 15(2):225–228
Shirvani Dastgerdi A, Kheyroddin R (2022) Policy recommendations for integrating resilience into the management of cultural landscapes. Sustainability 14(14):8500
Sikkema R, Proskurina S, Banja M, Vakkilainen E (2021) How can solid biomass contribute to the EU's renewable energy targets in 2020, 2030 and what are the GHG drivers and safeguards in energy-and forestry sectors? Renew Energy 165:758–772
Singh RK, Zander KK, Kumar S, Singh A, Sheoran P, Kumar A, Hussain SM, Riba T, Rallen O, Lego YJ, Padung E (2017) Perceptions of climate variability and livelihood adaptations relating to gender and wealth among the Adi community of the Eastern Indian Himalayas. Appl Geogr 86:41–52
Singh RD, Tewari P, Dhaila P, Tamta KK (2018) Diversifying livelihood options of timberline resource dependent communities in Uttarakhand Himalayas: conservation and development implications. Trop Ecol 59(2):327–338
Singh AP, Kumar M, Nagar B, Pala NA, Bussman W (2019) Ethnomedicinal use of plant resources in Kirtinagar block of Tehri Garhwal in Western Himalayas. Ethnobot Res Appl 18:1–11
Singh RB, Paroda RS, Dadlani M (2020) Indian agriculture towards 2030. World 821(51):151
Singh C, Chauhan N, Upadhyay SK, Singh R, Rani A (2021) The Himalayan natural resources: Challenges and conservation for sustainable development. J Pharmacogn Phytochem 10(1):1643–1648
Sinhababu A, Banerjee A (2013) Ethno-botanical study of medicinal plants used by tribal of Bankura district, West Bengal, India. J Med Plants Stud 1:98–104
Sow A, Seye D, Faye E, Benoit L, Galan M, Haran J, Brévault T (2020) Birds and bats contribute to natural regulation of the millet head miner in tree-crop agroforestry systems. Crop Prot 132:105127
Sreeja AT, Dhengle A, Kumar D, Pradhan AK (2023) Does access to clean cooking fuels reduce environmental degradation? Evidence from BRICS nations. Environ Sci Pollut Res 30(32):78948–78958
Sundriyal M, Sundriyal RC (2004) Wild edible plants of the Sikkim Himalaya: nutritive values of selected species. Econ Bot 58(2):286–299

Suresh CP, Bhutia KD, Shukla G, Pradhan K, Chakravarty S (2013) Free listed wild edible fruit species of Sikkim Himalayas and their uses

Swallow BM, Russell D, Fay C (2006) Agroforestry and environmental governance, World agroforestry into the future. World Agroforestry Centre, pp 85–94

Tamang B, Saril S, Singh M, Das S, Shukla G, Chakravarty S (2021) Ecosystem services of traditional large cardamom based agroforestry systems of Darjeeling and Sikkim Himalayas. J Tree Sci 40(1):78–91

Tangjang S, Borang A, Arunachalam A (2014) Improving sustenance of small and marginal (Adi) farmers through traditional vegetable crops in East Siang District of Arunachal Himalaya Northeast India. Indian J Hill Farm 27:90–94

Tariyal N, Bijalwan A, Chaudhary S, Singh B, Dhanai CS, Tewari S, Kumar M, Kumar S, Cabral Pinto MM, Thakur TK (2022) Crop production and carbon sequestration potential of *Grewiaoppositifolia*-based traditional agroforestry systems in Indian Himalayan Region. Land 11(6):839

Taye FA, Folkersen MV, Fleming CM, Buckwell A, Mackey B, Diwakar KC, Le D, Hasan S, Saint Ange C (2021) The economic values of global forest ecosystem services: a meta-analysis. Ecol Econ 189:107145

Tomich TP, Thomas D, Van Noordwijk M (2004) Environmental services and land use change in Southeast Asia: from recognition to regulation or reward. Agric Eco Syst Environ 104(1):229–244

Trivedi VL, Nautiyal MC, Sati J, Attri DC (2020) In vitro propagation of male and female Hippophaesalicifolia D. Don. In Vitro Cell Dev Biol Plant 56:98–110

Ullah I, Kim DY (2020) A model of collaborative governance for community-based trophy-hunting programs in developing countries. Perspect Ecol Conserv 18(3):145–160

United Nations Environment Programme. International Resource Panel, 2011. Decoupling natural resource use and environmental impacts from economic growth. UNEP/Earthprint

Upadhyay K, Arunachalam A, Khan ML (2002) Designing proper agroforestry models for Arunachal: scope and challenges. In: Sundriyal RC, Singh T, Sinha GN (eds) Arunachal Pradesh: environmental planning and sustainable development. HIMAVIKAS Occasional Publication No. 16. G.B. Pant Institute of Himalayan Environment and Development, Almora, India, pp 349–355

Vejre H, Jensen FS, Thorsen BJ (2010) Demonstrating the importance of intangible ecosystem services from peri-urban landscapes. Ecol Complex 7(3):338–348

Vibhuti BK, Bargali SS, Bargali SA (2019) Species composition, diversity and traditional uses of homegarden in Kumaun Himalaya, India. Indian J Agric Sci 89(9):1415–1418

Vineeta, Sarkar BC, Tamang M, Shukla G, Debnath MK, Nath AJ, Chakravarty S (2022) Floristic diversity, and conservation status of large cardamom based traditional agroforestry system along an altitudinal gradient in the Darjeeling Himalaya, India. Agrofor Syst 96(8):1199–1210

Zafar S, Zafar Khan M, Mehmood T, Begum F, Sadiq M (2023) Role of community-based conservation and natural resource management in building climate resilience among vulnerable mountain societies. Clim Dev 15(7):608–621

Zerihun MF (2014) A socioeconomic analysis of factors that affect the adoption of agroforestry technologies in the Eastern Cape Province of South Africa (Doctoral dissertation, Tshwane University of Technology)

Chapter 6
Apple-Based Agroforestry Systems: A Pathway to Economic Growth and Environmental Resilience

D. R. Bhardwaj, Alisha Keprate, Jatin Kumar, and Prashant Sharma

Abstract The threat of climate change, decreasing soil health, and declining productivity necessitates the development of an agroecologically sound and climate--resilient land-use system. Apple-based agroforestry systems (ABAFSs) represent a desirable long-term option for temperate ecosystems that are economically and environmentally resilient. They reduce ecological exposure and increase the flexibility of mountain ecosystems, thereby fitting into sustainable development in sensitive ecosystems. ABAFSs comprise apple trees (*Malus domestica* Borkh.) with intercrops such as peas, soybeans, and medicinal crops, or animals, in multifunctional agroecosystems to maximize productivity and resilience. Apple trees suited to temperate climates have a medium canopy and deep roots, efficiently utilizing resources and minimizing competition. Economically, ABAFSs diversify revenue with apples, intercrops, and value-added products like cider and herbal extracts, with 30–50% higher returns than monocultures, with 1.39–4.06 benefit–cost ratios and land equivalent ratios of 1.27–1.45. Ecologically, they improve soil health, increase organic carbon (0.84–1.43%), and enhance nutrient levels (e.g., N up to 557.20 kg/ha), while improving water use efficiency by 3–5% through complementary root stratification. Carbon sequestration ranges from 48 to 166 t ha^{-1}, supporting climate mitigation under the Paris Agreement. ABAFSs also stabilize slopes, promote biodiversity, and regulate microclimate. According to Sustainable Development Goals (SDGs 1, 2, 13, 15), they enhance ecosystem recovery and food security. Challenges related to its suitability and adoptability require strategic intervention, but ABAFSs remain a climate-resilient model for temperate agriculture and SDGs.

Keywords Apple-based agroforestry · Economic resilience · Carbon sequestration · Water use efficiency · Sustainability

---

D. R. Bhardwaj (✉) · A. Keprate · J. Kumar
Dr. Yashwant Singh Parmar University of Horticulture and Forestry, Nauni, Solan, Himachal Pradesh, India

P. Sharma
ICAR-Indian Institute of Soil and Water Conservation, Research Centre, Datia, India

B. Singh et al. (eds.), *Himalayan Agroforestry Systems*, Advances in Agroforestry 15,
https://doi.org/10.1007/978-3-032-25096-4_6

1 Introduction

The temperate region has evolved a diverse array of land-use systems, each meticulously adapted to fulfill specific human needs while aligning with the site-specific ecological variability inherent to its heterogeneous landscapes. This region is notably susceptible to rising temperatures and increasing atmospheric CO_2 concentrations, adversely affecting the sustainability of higher-altitude populations (Dmuchowski et al. 2022). These environmental problems are intensified by topographical features characterized by terraced slopes, resulting in a fragile ecosystem that impedes farming activities and often makes it challenging to meet yearly household food requirements (Bijalwan 2012). In response to these challenges, agroforestry offers a promising approach by incorporating tree-based systems, such as fruit cultivation, into agricultural frameworks to improve resource efficiency, enhance ecosystem resilience, and support the livelihoods of local communities under these increasingly adverse conditions (Kashyap et al. 2014; Zhao et al. 2022).

Among the different agroforestry systems (AFSs), fruit-tree-based AF, the deliberate integration of annual or perennial crops with perennial fruit-producing trees on the same land unit (Bellow et al. 2008), stands out as a globally recognized land--use system. Small landholders often favor fruit trees because of their relatively short juvenile phases, high market value, and contributions to household nutrition (Bellow et al. 2008). ABAFSs represent an innovative and sustainable approach to land use, combining the cultivation of apple trees (*Malus domestica* Borkh.) with various crops, livestock, or forest species to optimize productivity, ecological balance, and economic returns. These systems are particularly prominent in temperate regions across the globe, including parts of Europe, North America, Central Asia, and the Himalayan belt, where climatic conditions are favorable for apple cultivation. As a leading temperate fruit crop, apples account for approximately 50% of global deciduous fruit tree production, which reflects their commercial significance (Nigussie et al. 2019; Admasu and Jenberu 2022). China dominates as the top apple-producing country, followed by the United States and India (Joshi et al. 2021). Globally, the Loess Plateau is the main apple-producing area (Liu et al. 2020); the cover and yield of apple trees from this region amount to 1.31 million hectares and 23 million tons, accounting for 25.2% and 26.3% of global coverage and production, respectively (Wang et al. 2020). It is the primary source of income for 3.4 million rural households and 10 million smallholder farmers (Gao et al. 2021). In the northwestern Himalayan region of India, such as Himachal Pradesh, Jammu and Kashmir, and Uttarakhand, fruit-based AF has become a foundation of rural economies. While posing challenges for monocropping, the steep, terraced, sloped topography provides a favorable microclimate for apple-based agroforestry systems (ABAFSs) in high altitudes, offering optimal temperature and chilling hours for fruit development.

The multilayered structure of AF not only maximizes land productivity but also offers numerous ecological benefits, including improved soil fertility, enhanced biodiversity, carbon sequestration, and microclimate regulation (Agroforestry network 2019). In apple-based systems, apple trees are the primary component, often intercropped with vegetables, cereals, legumes, or fodder species. Additionally, integration with livestock such as sheep, poultry, or bees (apiculture) is standard, contributing to soil health, build-up of carbon content, diversified farm income, and enhanced resource use efficiency (Yadav et al. 2019; Verma et al. 2021). The popularity of apple-based agroforestry is primarily driven by its multiple advantages. Apple is a high-value fruit crop with a strong market demand, offering significant income potential for smallholder and commercial farmers. Combined with intercrops or livestock, these systems ensure diversified farm outputs, improve food and nutritional security, and reduce economic risks (Sharma et al. 2021; Lauri and Gautier 2024). Furthermore, ABAFSs are environmentally beneficial, supporting soil conservation, carbon sequestration, biodiversity enhancement, and microclimate moderation—particularly important in fragile and hilly ecosystems (Kumar et al. 2024).

Many tree characteristics make apples highly suitable for agroforestry adoption. Apple trees have a moderate canopy structure, allowing sufficient light penetration for understorey crops (Volk 2017). Their relatively deep root system minimizes competition with shallow-rooted intercrops, promoting complementary resource use. In addition, apple trees have a well-defined phenological cycle, enabling strategic crop planning and seasonal management (Kumar 2024). The deep root systems of apple trees help in stabilizing the fragile slopes of the Himalayan region, while their canopy mitigates the impact of heavy monsoon rains important for soil conservation. Their long productive life and perennial nature provide long-term economic returns and ecological services, contributing to system sustainability.

These systems align strongly with international climate and development frameworks. They directly contribute to several Sustainable Development Goals (SDGs), including SDG 1 (No Poverty), SDG 2 (Zero Hunger), SDG 13 (Climate Action), and SDG 15 (Life on Land) as well as national goal of doubling the farmers' income by promoting sustainable livelihoods, food security, climate resilience, and ecosystem restoration (United Nations SDGs 2024). Moreover, apple-based agroforestry supports global climate commitments under the Kyoto Protocol and the Paris Agreement by acting as a nature-based solution for C-sequestration and land-based mitigation. Enhancing above- and belowground biomass captures atmospheric carbon dioxide, helping countries meet their Nationally Determined Contributions (NDCs) toward emissions reduction (Verma et al. 2023). These systems are gaining increasing recognition under the climate-smart agriculture framework, providing climate resilience while ensuring food and nutritional security (Sharma et al. 2017a, b). Apple-based agroforestry also plays a vital role in temperate hilly and mountainous regions by preventing soil erosion and improving land use on marginal terrains (Chisanga et al. 2018; Bhardwaj et al. 2023; Kumar et al. 2024).

Despite their benefits, challenges such as market accessibility, pest management, and the need for technical know-how persist (Racsko and Schrader 2012; Simon et al. 2017; Pitchers et al. 2021). Therefore, ongoing research and policy support are essential to optimize and promote these systems worldwide. Overall, apple-based agroforestry holds immense potential for sustainable development, particularly in temperate and mountainous regions. To fully realize the importance of these AFSs, it becomes necessary to research and develop appropriate tree–crop combinations with optimal tree density and climate-smart methodologies for different regions. As we know, temperate areas are some of the most delicate ecosystems on earth, and apple-based agroforestry can be an example of sustainable land use, ecological integrity, and socioeconomic advancement.

1.1 Apple-Based Agroforestry Systems (ABAFSs)

ABAFSs offer a sustainable and balanced approach to land utilization that supports both environmental sustainability and economic viability. Apple trees are integrated into the temperate ecosystem because of their adaptability and multifunctionality. Beyond producing fruit, apple trees actively contribute to soil health, climatic moderation, and climatic resilience in the agroecosystem. To truly exploit the value of apple in these agroecosystems, it becomes necessary to understand the biological and agronomic traits that make them such a strong fit for integrated, sustainable farming systems.

2 Tree Traits Powering Agroforestry Success

2.1 Climatic Suitability and Adaptability

Apple trees are cultivated in temperate and sub-temperate climate areas of the world, above all in hilly and mid-altitude regions. Their chilling hour requirement, typically 1200–1500 hours below 7 °C depending upon the type of cultivar (Rai et al. 2015), matches well with the climatic conditions of temperate ecosystems such as the northwestern Himalayas, making them an appropriate option to be included in agroforestry. Moreover, apple trees show flexibility toward different soil and moisture conditions, given the availability of adequate drainage, topographic feature, and organic matter. This further endorses their scope for incorporation into these multifunctional systems.

2.2 Canopy Architecture

One of the traits that makes apple trees appropriate for use in agroforestry is their manageable canopy structure. The canopy is kept open and well-ventilated with good training and pruning practices, like central leader or spindle systems (Dallabetta 2014). This helps sunlight reach the lower strata, producing adequate light for intercrops or companion species. In the last few decades, research in tree architecture management has focused on improving canopy porosity to increase light and air movement through precision pruning (Robinson et al. 2019). Techniques like "artificial extinction" have positively affected fruit size, return bloom, and pest reduction (Breen et al. 2012). Additionally, pruning helps mitigate shading effects, especially when combined with overstory species. After plantation, training and pruning procedures must be adapted to optimize light penetration in agroforestry settings (Lauri et al. 2020). The production of dwarf and semi-dwarf rootstocks has further improved the structural harmony of trees by controlling their height and lateral spread (Verma et al. 2024). This not only optimizes space but also minimizes interspecific competition so that intercrops can thrive. Additionally, canopy management has the potential to offer supplementary benefits, like fodder and fuelwood, to farmers. This combination of managements renders apple trees a key feature in diversified, multifunctional AFSs, increasing productivity, soil health, and sustainability.

2.3 *Livelihood Security and Economic Contribution*

Apple production has high economic benefits and is thus among the most commercially significant fruit trees of temperate AFSs. The trees start producing within a few years after planting, particularly when propagated on dwarf rootstocks, and can continue to produce for many decades. This long-term productivity also provides a good source of income for households on rural farms (Tajebe et al. 2013). Coupled with crops or livestock, apple trees increase farm-level resilience through diversified incomes and decreased reliance on individual commodities. Their high market value and high demand also increase their worth in commercial agroforestry enterprises (Vasylieva and Harvey 2021).

2.4 *Compatibility with Companion Species*

Due to their structural and functional traits, apple trees are highly compatible with a broad suite of agroforestry species. Being deciduous, apple trees shed leaves in winter, and thus complete sun reaches the ground cover—a blessing to winter crop production in alleyways unencumbered by shading restriction. Leaf fall avoids light competition and facilitates natural cycling of nutrients, leaving organic residues on the ground. Apple tree canopies are easily open to pruning, which also helps penetration of light (Dallabetta 2014) and competition minimization with interspecific companion crops, facilitating spatial flexibility in management.

Because of their dense canopy profile and low root competition, apple trees may be a reliable mid-stratum crop in advanced AFSs. It has been found that while apple tree structure is impacted by reduced light intensity, fruit production is not affected until it drops as much as 65% (Pitchers et al. 2021). This information can be utilized in optimizing the design of ABAFSs. Apple trees in traditional agroforestry designs take up the topmost stratum, that is, full sun, with shrubs, grasses, or herbaceous crops in lower strata. More complex systems, however, can have overstory plants such as timber or nut trees casting shade over the apple canopy, with understory shrubs (such as black currant) and annual crops (such as pea–maize or corn–soybean rotation) in lower strata (Chisanga et al. 2013; Liu et al. 2020; Zheng et al. 2021). These vertically stacked, multifunctional arrangements, often termed multifunctional woody polycultures (Lovell et al. 2018), maximize vertical and horizontal space, increasing productivity and system resilience. Apple trees can also be incorporated into agrisilvicultural and agrosilvopastoral systems (Lauri et al. 2020) with planning, diversifying their function and ecological role.

2.5 Ecological and Soil Health Benefits

Apart from their structural and economic advantages, the integration of apple trees also yields benefits to the agroecosystem. At temperate elevations, apple orchards act as climatic moderators and buffers against climatic extremes. Canopies of trees reduce diurnal temperature variability, decrease wind speed, and establish stable microclimatic conditions that protect intercrops from frost and heat stress (Liu et al. 2020). Additionally, seasonal leaf drop contributes organic matter to the soil (Tagliavini et al. 2007), increasing its carbon status and physical properties such as structure and water-holding capacity. With mulch management and organic fertilizer application, apple-based systems encourage soil microbial processes and nutrient cycling, increasing soil fertility (Mia et al. 2021; Tagliavini et al. 2007). Furthermore, apple trees harbor many pollinators and beneficial insects, increasing the diversity of the farming systems and facilitating natural pest management.

2.6 Sustainable and Climate-Resilient Land Use

Apple trees make essential contributions to climate-resilient agroforestry options. Their long-term existence is helpful in the sequestration of carbon both above and underground (Kumar et al. 2024; Zahoor et al. 2021a). By being included within a diversified structure, they counter the climate-related risks involved in variability by steadying production levels and augmenting ecosystem services. By promoting apples, using supportive species and appropriate farming practices, apple-based agroforestry options are sustainable land-use models combining ecological stewardship with economic returns.

2.7 System Design and Components

AFSs are widely regarded as an effective means of bridging the gap between the world's increasing human population and the limited availability of agricultural land. By incorporating trees, crops, and sometimes livestock into the same unit of land or the farming system, agroforestry is an excellent way to enhance land-use efficiency. This land-use system has become very popular, where it has been taken up on a large scale for its role in ecological and economic gains, such as increased biodiversity, soil conservation and protection, and diversified income sources for farmers (Pantera et al. 2021). Designing ecologically intensive agroecosystems requires a deep knowledge of biological processes in ecosystems and the integration of local traditional agricultural knowledge (Malézieux 2012). Any agroforestry system in the area requires a thorough understanding of system design and the inclusion of agricultural components to minimize resource competition (Jamar et al. 2015). This competition occurs where trees and crops overlap, leading to decreased agricultural productivity. These losses pose a grave danger to the functionality and sustainability of agroforestry ecosystems (Gao et al. 2013) (Fig. 6.1).

The productivity and sustainability of ABAFSs also rely on a well-planned system that includes compatible species selection, optimum spatial arrangement, and effective management practices. In these systems, apple trees are integrated as the main woody perennial species, along with suitable understory crops and, occasionally, livestock to develop a multifunctional agroforestry system. Apple cultivars such as "Red Delicious," "Golden Delicious," "Gala," and "Fuji" are popular in the northwestern Himalayas, for their compatibility with the area's temperate climatic conditions and consumer demands (Ahmed et al. 2008). These apple trees are often interplanted with suitable understory crops such as red clover, peas, or beans, which do well in the shade under the trees (Fig. 6.2). Various apple tree and crop combinations being practiced in different regions/countries are tabulated in Table 6.1. It can

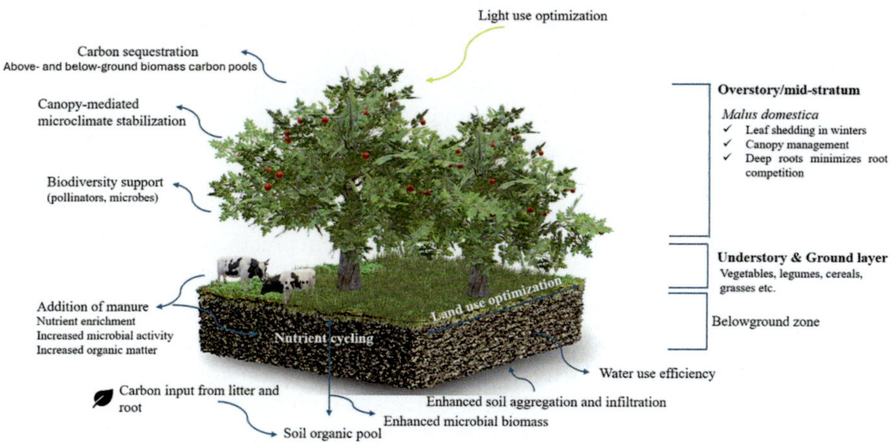

Fig. 6.1 Schematic representation of ecological and functional dynamics of ABAFSs

Fig. 6.2 Different apple-based agroforestry system in dry temperate Himalayas

be observed that in India, multi-strata designs integrate annuals and companion woody perennials. At the same time, China uses wider spacing with linear apple tree arrangements, intercropped with legumes and cereals. The USA combines fruit and nut shrubs with perennial grasses in a silvopastoral system, while France adopts a simplified silvoarable approach. Legumes in this group are essential in improving soil fertility by adding nitrogen and conserving moisture, creating a beneficial environment that promotes apple tree development (Pavek and Granatstein 2014). Research reveals that apple trees with legumes yield more fruit than those under monocropping or heavy feeder intercrops. Bhat et al. (2018) observed that high yields and better quality of fruits were obtained with pea intercrops followed by red clover and French beans. Heavy feeder crops like maize and oats were observed to diminish the quality of the apple fruit. Conversely, legumes enhance the soil's water-holding capacity and supply nutrients, especially in the fertile topsoil, to facilitate root uptake. This minimizes water loss, keeps the soil slightly warmer, and enhances healthy root growth, increasing yields.

In AFSs, closely spaced trees also attenuate photosynthetically active radiation (PAR), affecting understory crops' photosynthetic ability and yield (Keprate et al. 2024). Maintaining proper tree architecture and optimizing spatial arrangement of trees are among the many effective canopy management practices that help mitigate aboveground competition by enhancing light penetration to the understory (Birla et al. 2024). Planting apple trees in wider rows with regular pruning helps reduce competition for sunlight. Competition between trees and crops goes beyond aboveground resources, such as light interception, to subsurface interactions where root systems vie for soil moisture and nutrients. Tree and crop roots' spatial arrangement and morphological traits are primary determinants of belowground competitive dynamics (Sun et al. 2018). In various AFSs, it has been observed that the delicate

Table 6.1 Spatial arrangements and crop combinations in ABAFSs across global regions

Location	Apple cultivar	Spacing	Tree component	Crop component	
India	Royal Delicious Red Gold	4 m × 4 m	*Populus ciliata, Morus alba, Prunus armeniaca, Juglans regia, Celtis australis, Pinus gerardiana,* Robinia sp., Ailanthus altissima, *Salix tetrasperma*	Pea, maize, French bean, Rajmash, barley, wheat, tomato, brinjal, cauliflower, cabbage, buckwheat, capsicum, onion, turnip, radish, green gram, garlic Grasses: oats, *Festuca pratense*, *Trifolium* spp., *Dactylis glomerata*	Anjulo (2009), Rajput (2010), Chisanga et al. (2013), Kashyap et al. (2014), Salve (2018), Ahmad et al. (2018), Zahoor et al. (2021a), Kumar et al. (2024)
China	Apple-Freedom, Liberty, Priscilla Red Fuji	5 m × 4 m	–	Soybean, peanut, corn, canola	Gao et al. (2013), Liu et al. (2020), Zheng et al. (2021), Gao et al. (2022), Zhao et al. (2022), Shen et al. (2023)
USA	Apple-Freedom, Liberty, Priscilla	4 m × 4 m	*Castanea sp, Ribes sps (red and black currant), Rubus idaeus (Raspberry),Corylus* spp. *(hazelnut)*	*Bromus inermis, Lolium X Festuca, Dactylis glomerata, Sche donorus phoenix, Phleum pratense, Trifolium pra tense, and Trifolium hybridum*	Kranz et al. (2019)
France	Dalinette grafted on Geneva® G202 C.O.V. rootstock	4 m × 4 m	*Juglans nigra, Juglans regia*	*Medicago sativa* L.	Pitchers et al. (2021)
Ethiopia	Ana Jona Gold Crispin Princisa Dorset Golden CP92	2.95 × 9 m 4.2 × 4.2 m	–	Cabbage (*Brassica oleracea*), potatoes (*Solanum tuberosum*), tomato *(Lycopersicon esculentum*), Enset (Ensete ventricosum), potatoes (*Solanum tuberosum*)	Tajebe and Gelan (2018), Amare et al. (2023)

roots of plants in the system possess certain morphological and physiological plasticity, allowing adaptation to heterogeneous soil resource distributions (Zamora et al. 2007). This interspecific competition reduces fine root biomass, with tree fine roots often redistributing to deeper soil strata while crop fine roots shift toward shallower soil horizons (Sun et al. 2018). The key is to optimize resource use, minimize competition, enhance productivity, and strengthen ecosystem resilience by encouraging complementary interactions between trees and crops while maintaining a balanced resource distribution.

Strategic management practices are vital for minimizing competition and maximizing resource use efficiency in AFSs. Zheng et al. (2021) examined the apple–soybean AFSs in China and demonstrated that belowground competition reduced with increasing age of trees. With rising levels of irrigation, there was also increased competition belowground. Still, the application of plastic mulch increased water use efficiency, soybean yield, and total water use, emphasizing the need to choose effective mulching and irrigation methods before the establishment of an agroforestry system. With careful planning of spatial arrangements, water management, and competition control, AFSs can be optimized to be more productive, resilient, and sustainable in the long term.

3 Role of Apple-Based Agroforestry Systems

ABAFSs represent a scientifically grounded, multifunctional approach to sustainable agriculture, particularly in temperate and hilly regions like the northwestern Himalayas. Integrating apple trees with intercrops, medicinal herbs, or livestock creates synergistic agroecosystems that enhance economic viability, soil health, water use efficiency, and C-sequestration. ABAFSs leverage ecological interactions to optimize resource use, improve land productivity, and mitigate climate challenges, while providing diversified income streams and bolstering food security. Supported by empirical research, these systems demonstrate superior land-use efficiency, nutrient cycling, and resilience compared to monocultures, making them a cornerstone for sustainable land management and rural development in challenging environments.

3.1 Economics of Apple-Based Agroforestry Systems

ABAFSs are an economically and ecologically sound land-use option, well adapted to temperate and hilly areas where climatic conditions are conducive to apple production. Socioeconomically, the systems promote a transition from subsistence to market agriculture. In contrast to the conventional cereal- or vegetable-based farming, apple production avails relatively high regional and foreign market value, generating secure and sure returns that augment domestic fiscal security (O'Rourke

2021). Integrating apple trees with crops, vegetables, medicinal herbs, or livestock promotes diversified farming, increasing incomes and efficient resource utilization (Bhardwaj et al. 2023). Diversification reduces the economic risks of monoculture systems (Shahane 2023), which are vulnerable to market price fluctuation, climatic catastrophes, and pest or disease infestations. By combining different productive elements within one holding, apple-based agroforestry provides a cushion against income uncertainty, allowing farmers to earn other forms of income when one venture collapses. In the early years, as tree crops become commercially productive only after 3–5 years, farmers use inter-row spacings for rapidly growing species like legumes, vegetables, or cereals, which provide interim income. As the canopy spreads with the trees, intercrops are progressively replaced by shade-tolerant crops such as ginger, turmeric, basil, vegetables, or grasses to maintain productivity. When animals are included, they introduce another dimension with organic manure, reducing dependence on chemical fertilizers, and recirculating nutrients in the system (Lemaire et al. 2023). Some agroforestry interventions integrate livestock, which contribute to additional income of the farm along with addition of nutrients through manure into the system, thereby reducing the use of chemical fertilizers. Such integration promotes circular nutrient flows, minimizes wastage, and reduces input costs, which helps improve the farming system's overall benefit–cost ratio.

3.2 Land-Use Productivity and Efficiency

The economic benefit of ABAFSs includes increased land-use efficiency (Wei et al. 2024). Land in monoculture apple orchards is wasted, particularly in the juvenile, nonproductive years of apple trees. In agroforestry, vertical and horizontal space dimensions are utilized to the maximum by incorporating several complementary layers of production (Halli et al. 2016). Apple trees form the upper canopy, and the understory is innovatively utilized to grow intercrops such as legumes, grasses, vegetables, or medicinal plants based on the orchard age and planting arrangement. Such multi-strata configurations are very effective in optimizing land productivity per unit area and for year-round utilization of available space. The increased biomass production and diversified yield streams in ABAFSs are a key factor in improving economic returns, increasing income stability, and strengthening food and nutritional security at the household level (Zahoor et al. 2021a). In Xinjiang, China, Wei et al. (2024) proved that apple orchards intercropped with ryegrass had an LER of 1.27–1.30, equating to considerably higher land-use efficiency than monoculture orchards. In another study, Shen et al. (2023) proved that apple trees intercropped with soybean possessed an LER value of 1.45, which reinforces the assertion that ABAFSs have potential for maximizing land productivity and resource use efficiency.

3.3 Economic Efficiency and Resource Utilization

Economic efficiency of ABAFSs has been reported a few times in temperate conditions, with higher net return and higher benefit–cost ratio than the farming system in isolation under standard conditions. The BC ratio, a key indicator of economic sustainability, effectively reflects the profitability of apple-based agroforestry. As shown in Table 6.2, in the Kinnaur region of India, ABAFSs exhibit an impressive BC ratio range of 3.10–4.06, indicating exceptional returns on investment. Farmers adopting apple agroforestry models in the northwestern Himalayas, for example, have been earning 30–50% higher income annually (Bhat et al. 2018; Zahoor et al. 2021b), varying with design, cropping pattern, and market opportunities. Besides, the system ensures greater financial efficiency by reducing the cost of inputs such as chemical fertilizers, pesticides, and supplemental feed. Incorporation of leaf litter and animal manure contributes to the buildup of soil organic matter, reducing the long-term cost of expenditure on soil amendments. Likewise, wind and shade protection by apple trees improves the region's microclimate, leading to increased produce yields and quality, hence greater economic returns. In line with India's national goal of doubling farmers' income, ABAFSs present an effective option by raising productivity, significantly improving profitability and maximizing utilization of resources. Table 6.2 also displays gross returns of INR 243080.7–5,959,580, highlighting the economic advantage of combining intercrops with apple trees. These figures indicate that ABAFSs offer long-term environmental stewardship and significantly enhance economic resilience. In the Kashmir valley, Zahoor et al. (2021b) evaluated the economic benefits of ABAFSs for improving livelihood. Their study revealed greater returns from AFSs than monocropping. Likewise, Bhat et al. (2018) reported for Kashmir region that intercropping apple with peas (BC ratio: 1.71), red clover (1.40), cabbage (1.41), and French beans (1.39) exhibited greater benefit–cost ratios than monoculture (1.34), demonstrating the economic benefit of ABAFSs in different temperate regions.

3.4 Postharvest Value Addition

Postharvest value addition is essential to improve the financial sustainability of ABAFSs (Mole 2018). Apart from direct consumption of the fresh fruit, apples can be subjected to additional processing to make high-value products such as vinegar, jam, juice, cider, and dried chips (Fernandes et al. 2022; Ullah et al. 2018), which earn premium prices in markets. These products have more shelf life, reducing the urge to sell immediately and better navigating markets with limited access. Further, value-added intercrops such as turmeric, ginger, ashwagandha, and Tulsi can be converted into powders, oils, and extracts, meeting more demand in the pharmaceutical and cosmetic industries. Organic or conventionally processed status also commands a higher market value. Such enterprises, usually owned and run by

Table 6.2 Economic performance of apple-based agroforestry systems across different regions

Region	System	Component	BC ratio	Cost of cultivation	Gross returns	Net returns	References
Kinnaur	Agrihorticulture	Pea, bean, buckwheat, barley	3.79–4.06	793,780.5–1,006,444	3,022,097–3,813,289	2,228,318–2,806,845	Kumar (2024)
Solan	Agrihorticulture	Radish, pea, maize	3.17	272,199.2	861,136.6	588,937.4	Sinha (2023)
Solan	Agrihorticulture	Pea+coriander	3.24	266,117.6	861,266.6	595,149	Dogra (2023)
Kullu, H.P	Agrihorticulture	Tomato, brinjal, capsicum, black gram	2.19–2.33	468,846–535,588.2	1,026,684–1,249,883	557,838.4–714,295.2	Acharya (2022)
Jammu and Kashmir	Apple-based agroforestry system	Maize, pea, strawberry, cabbage, red clover, French bean, and oats	1.06–1.71	300,551–531,495.6	864,673.6–1,243,759	486,927.7–784,979.7	Bhat et al. (2018)
Kinnaur, H.P	Agrihorticulture	Almond, maize, cabbage, beans, capsicum, barley, persimmon, pear, beetroot	2.05–2.12	635,640.79–1,381,907	243,080.7–1,634,4.132.10	998,491.31–3,092,591	Salve (2018)

(continued)

Table 6.2 (continued)

Region	System	Component	BC ratio	Cost of cultivation	Gross returns	Net returns	References
Kinnaur, H.P	Agrihortisilviculture	Apricot, oak, Himalayan poplar, Chilgoza pine, maize, cabbage, common bean, barley, willow, almond, celtis, mulberry, beetroot, walnut, tomato	2.75–2.93	1,206,446–1,733,656	3,101,583–4,738,031	1,895,137–3,004,375	Salve (2018)
Kinnaur, H.P	Apple-based agroforestry system/ agrihorticulture	Apple+ peas, maize, wheat, rajmash, barley, cabbage	3.10–3.78	890,214–1,577,028	2,757,200–5,950,580	1,867,890–4,375,360	Chisanga et al. (2013)
Kinnaur, H.P	Agrihortisilviculture	Robinia, pea, maize, Ailanthus, Salix, cabbage, poplar, wheat, rajmash, deodar, and Chilgoza pine.	1.8–3.25	329,734–890,214	1,133,842–2,757,200	804,108–1,867,890	
Kullu, H.P	Agrihorticulture	Wheat, pea, tomato, garlic, maize	2.64–3.44	875,299.1–1,414,902	3,012,800–3,752,281	2,137,501–2,607,132	Rajput (2010)

Compounded @ 6% interest rate

cooperatives or self-help groups (Alemu et al. 2018), leave room for branding, direct marketing, agri-tourism, diversifying returns, and market risk reduction. Overall, value addition adds to the viability, profitability, and durability of ABAFSs, making them a competitive option among smallholders and rural entrepreneurs.

3.5 Income Resilience to Climate Change

Another economic benefit is the increased resilience of family income. Apple cultivation can be profitable, but is highly vulnerable to unstable weather patterns, hailstorms, and volatile market prices (Singh et al. 2016; Hou et al. 2022). However, when combined with crops and animals, the monetary effect of such uncertainties is significantly reduced (Wilson and Lovell 2016; Mihrete and Mihretu 2025). A failed apple orchard because of frost or poor pollination can be compensated for by selling intercrops, livestock products, or value-added products. Such resilience is particularly crucial under the climate change regime, which would probably increase the intensity and occurrence of extreme climate conditions (Nyong et al. 2020). Therefore, apple agroforestry is a resilient farming practice by climate standards and yields more certain and stable long-term income.

3.6 Market Integration and Value Chains

Market entry and value chain construction also contribute to the economic multiplier impact of ABAFSs. Having multiple products on offer raises the engagement of a farmer in more than one value chain, like fresh fruits and vegetables, processed food, herbs, and animal products (Raj et al. 2014). Agroforestry farmers benefit from reduced dependence on single buyers and greater flexibility to stagger sales across multiple markets. Having various products also provides the potential to cater to niche markets like organic, health, or ethnic consumer segments paying premium prices. Farmer producer organizations and cooperatives have appeared in numerous places to organize collective marketing, value addition, and brand creation for agroforestry produce, further enhancing farm incomes and saving transaction costs (Escribano et al. 2020).

3.7 Labor Generation and Inclusive Employment

ABAFSs provide more on-farm employment because they incorporate multiple farm enterprises (Smith et al. 2022; Mukhlis et al. 2022). Round-the-year activities—from orchard maintenance to intercropping, livestock management, harvesting, processing, and marketing—create year-round employment demand for unskilled and skilled labor. ABAFSs stabilize on-farm labor requirements, alleviate seasonal migration, and retain rural labor force (Admasu and Jenberu 2022). The system provides inclusive opportunities for women, youth, and marginal groups through crop management and value addition, boosting household income and socioeconomic empowerment (Mukhlis et al. 2022).

3.8 Soil Health and Sustainability

AFSs improve soil's physical, chemical, and biological characteristics, significantly impacting soil health (Sharma et al. 2023). ABAFSs are one of the most popular agroforestry models because of their capacity to improve soil fertility, sequester carbon, and promote sustainable land management. The data from Table 6.3 reveal that ABAFSs show variability in soil nutrient and organic carbon levels. These apple-based agrihorticulture and hortipastoral systems demonstrate significant potential for enhancing soil fertility and sustainability through region-specific management practices, which rely on effective species selection and management strategies. The interactions between apple trees and related crops ensure long-term productivity and ecological sustainability, improving soil structure, nutrient availability, and microbial activity. In AFSs, trees are essential for altering the physico-chemical characteristics of the soil. Improving soil structure and lowering bulk density are two of ABAFS's main advantages (Eddy and Yang 2022). Apple tree root systems improve water infiltration and aeration by forming pore holes in the soil. Furthermore, the steady input of organic matter through root biomass and leaf litter encourages the development of soil aggregates, which improves soil stability (Das 2024). Increased soil moisture retention is another one of ABAFS's significant effects (Fahad et al. 2022). Apple trees' canopy cover lowers evaporation rates, and the soil's buildup in organic matter increases its ability to retain water (Sharma et al. 2022). This is especially advantageous in temperate and semi-arid areas where agricultural productivity is constrained by water availability (Kumar et al. 2024).

In ABAFSs, tree–crop interactions also positively impact nutrient dynamics. Apple leaf litter and root exudates can change the pH of the soil, increasing the availability of nutrients for crops. Additionally, the buildup of organic matter improves nutrition availability and retention by increasing cation exchange capacity (CEC) (Rinady et al. 2023). Because of the breakdown of organic residues and enhanced microbial activity, apple trees help to raise levels of micronutrients and key macronutrients, including potassium (K), phosphorus (P), and nitrogen (N) (Ramananjatovo et al. 2024). Additionally, ABAFSs play a significant role in soil organic carbon (SOC) sequestration. Increased SOC levels result from the ongoing addition of organic wastes from microbial activity, tree biomass, and root turnover. This enhances soil fertility and general soil health.

AFSs based on apples also significantly improve the biological qualities of the soil. According to Sharma et al. (2022), one of the most noticeable consequences is the rise in soil microbial biomass and activity in the dry temperate Himalayas.

Table 6.3 The soil enrichment potential of apple-based agroforestry systems

Region	Type of agroforestry system	Nutrients	Organic carbon (OC)	Others	References
Solan, India	Agrihorticulture	N-327.15 P-57.20 K-254.93	OC-1.25	BD-1.31	Dogra (2023)
Solan, India	Agrihorticulture	N-352.38 P-57.39 K-251.40	OC-1.26	BD-1.27	Sinha (2023)
Solan, Himachal Pradesh	Agrihorticulture	N-348.59 P-53.75 K-228.09	OC-1.43%	BD-1.23 gcm^{-3}	Kumari et al. (2025)
Kullu, Himachal Pradesh	Agrihorticulture (apple+tomato)	N-372.88 P-27.56 K-277.57	–	MBC-- 469.70 mg/kg	Acharya (2022)
Srinagar, India	Hortipastoral system	N-475.06 P-17.38 K-450.31	OC-0.84%	BD-1.24-1.25 gcm^{-3} MBC-- 1382 mg/kg	Dand (2019)
Kinnaur, Himachal Pradesh	Agrihorticulture	N-0.19% P-0.79 K-1.21	OC-1.24%	Ca-5.52 Mg-4.0	Salve (2018)
Kinnaur, Himachal Pradesh	Agrihortisilviculture	N-0.15% P-0.79 K-1.21	OC-1.26%	Ca-5.19 Mg-4.06	
Kinnaur, Himachal Pradesh	Agrihorticulture	N- 557.20	OC-1.38%	BD-1.01 (gcm^{-3})	Chisanga (2013)
Kinnaur, Himachal Pradesh	Agrihortisilviculture	N-540.80	OC-1.40%	BD-0.89 gcm^{-3}	
Kullu, Himachal Pradesh	Agrihorticulture	N-542.5	–	BD-1.213	Rajput (2010)

Apple trees' constant supply of organic matter promotes microbial growth, which raises respiration rates and microbial biomass carbon (MBC) (Mungai et al. 2005). This is essential for soil nutrient mineralization and decomposition processes. Additionally, agroforestry supports advantageous mycorrhizal associations—especially those involving arbuscular mycorrhizal fungi (de Carvalho et al. 2010). These fungi develop symbiotic interactions with apple roots through their hyphal networks, improving soil structure and facilitating phosphate uptake. Higher enzymatic activity, such as dehydrogenase, phosphatase, and urease, which are critical for nutrient cycling and the breakdown of organic matter, is also a result of the greater microbial diversity (Vallejo et al. 2010). Therefore, ABAFSs are an environmentally and financially sound choice for temperate agroecosystems in the northwest Himalayas that can substantially contribute to soil conservation, C-sequestration, and sustainable land-use practices over time. It is a crucial tactic in sustainable agriculture since it increases productivity and guarantees improved environmental resilience when apple trees are integrated with appropriate intercrops.

3.9 Water Use Efficiency

In ABAFSs, water use efficiency (WUE) is affected by complex interactions of soil water availability, temperature, and interspecific competition. These factors all directly impact crop nutrient uptake, growth, and development (Yinglong et al. 2019). One key element influencing such systems' stability and productivity is the water dynamics between coexisting species (Chen et al. 2023). AFSs are bound to have complex root and canopy structures that result in intense competition for water resources compared to sole cropping, which shows relatively uniform resource allocation (Kumar and Jose 2018; Zhou et al. 2019). Because of this complexity, these systems have unique physiological adaptations and patterns of water use that determine how they operate in different environments.

Competition between species for water availability is expected to severely reduce the effectiveness of agroforestry, especially in areas with poor soil quality or frequent droughts. The high transpiration rates of AFSs can exacerbate water scarcity in desert or seasonally dry conditions, making it difficult to maintain balanced development among various plant species. Also, introducing tree species with high water-uptake capacities, like apple trees, can lower soil moisture levels and possibly jeopardize crop growth, productivity, and total yield (Yang et al. 2020). However, in regions prone to excessive soil water retention, this high-water uptake capacity can be beneficial in mitigating waterlogging and improving soil aeration, thereby creating more favorable conditions for species adapted to well-drained soils. Despite these challenges, AFSs also offer notable benefits in regulating microclimatic conditions and improving soil water dynamics. By decreasing evaporation, boosting infiltration, and altering the microenvironment in ways that affect plant water supplies, trees can improve soil water concentration (SWC). Additionally, tree canopies offer shade, which helps mitigate temperature extremes and reduces evapotranspiration, improving water use efficiency (WUE) across the system (Stocker et al. 2018).

The complex interactions in agroforestry system highlight the need for careful species selection and management strategies that optimize water resources while maximizing agroforestry's ecological and agricultural benefits. Liu et al. (2020) determined that apple trees predominantly utilize water from the 60–80 cm (20.9–25.9%) and 80–100 cm (22.8–24.7%) soil layers, whereas corn accesses water from shallower depths of 20–40 cm (18.8–33.1%) and 40–60 cm (20.6–33.7%) during the growing season. This complementary water use pattern elevates WUE in intercropped systems, with corn exhibiting a WUE increase of 3.03–5.26% compared to monoculture. At the same time, apple trees surpass their monoculture WUE only under conditions of reduced soil moisture. In a separate study, Shen et al. (2023) observed a 30.54% increase in SWC in fields intercropped with apple trees and soybeans, with this enhancement concentrated in the 0–40 cm soil layer, where the root systems of both species intersect. Soybeans demonstrate greater competitive ability near the soil surface and at distances from the trees, whereas apple trees predominate in deeper soil horizons. These observations emphasize the necessity of strategic water and nutrient supplementation during periods and locations of intense interspecific competition to alleviate stress and optimize yields.

3.10 C-Sequestration Potential of Apple-Based Agroforestry Systems

Fruit-based AFSs, integrating fruit trees like apple trees with agricultural crops and/or livestock, provide multiple ecological and economic benefits while acting as remarkable carbon sinks. In addition to their primary function in fruit production, research on apple orchards demonstrates their significant contribution to C-sequestration (Kumar et al. 2024; Gelaye and Getahun 2024). The C-sequestration potential of apple trees rivals many other tree species. This C-sequestration process involves absorbing and storing carbon from the atmosphere in plant biomass, that is, roots, stems, fruits, branches, and foliage (Hurd et al. 2022). When incorporated into the farms along with intercrops, the apple trees not only mitigate CO_2 but also yield high-value fruit and support numerous value-added products, making them economically viable for smallholder farmers. Carbon stocks in ABAFSs vary widely across spatial and temporal scales due to differences in age, climate, edaphic conditions, and management regimes. It is found that the apple trees reach their maximum potential for sequestering carbon around the age of 18, after which a gradual decrease starts, mainly because of physiological senescence and slower growth rates (Wu et al. 2012). These temporal dynamic underscores the necessity of strategic orchard management to maintain optimal carbon uptake.

The total amount of carbon stored in aboveground and belowground biomass in China, a central apple-producing region, was estimated to be between 230 and 475 Tg C. Between 1990 and 2010, net C-sequestration in apple orchards varied between 14 and 32 Tg C (Wu et al. 2012). These apple orchards are significant in China's national carbon budgets, accounting for around 4.5% of the net carbon sink across China's terrestrial ecosystems. In another study of the northwestern region of India, apple sole crop attained a maximum total biomass of 26.48 tons per hectare (t ha^{-1}). In contrast, agricultural crops ranged from 8.95 to 9.48 t ha^{-1}. Concurrently, AFSs exhibited superior biomass accumulation, storing between 36 and 36.85 t ha^{-1}, reflecting their enhanced capacity for C-sequestration and ecological stability (Zahoor et al. 2021a). Numerous factors such as tree age, genetic constitution, planting density, fruit output, pruning and training methods and other orchard management strategies (Wang et al. 2023) affect how the carbon that is fixed will be distributed across the different parts of apple trees, including the roots, stems, branches, leaves, and fruits. Table 6.4 presents the carbon stock studies in ABAFSs,

Table 6.4 Biomass and carbon storage capacity of apple-based agroforestry systems

Location	Crop combination	Age of the tree component	Total biomass of agroforestry system (t ha^{-1}) (Tree + Crop)	Vegetation carbon density (t ha^{-1})	SOC (t ha^{-1})	Total carbon (t ha^{-1}) (plants + soil)	C sequestration (t ha^{-1} yr^{-1})	References
Himachal Pradesh, India	Apple–tomato	11	44.05	–	30.62	–	79.75	Acharya (2022)
	Apple–brinjal	11	43.29	–	25.74	–	78.50	
	Apple–capsicum	11	42.68	–	27.97	–	77.51	
	Apple–black gram	11	40.09	–	23.61	–	73.22	
Jammu and Kashmir, India	Apple–rajmash–oats	11	36.85	–	45.76	64.18	–	Zahoor et al. (2021a)
	Apple–French beans–oats	11	35.92	–	43.67	61.62	–	
	Apple–green gram–oats	11	36.00	–	46.02	64.02	–	
Jammu and Kashmir, India	White clover+apple	–	45.81	22.90	30.60	53.50	192.11	Dand (2019)
	Red clover+ apple	–	47.59	23.80	31.15	54.95	195.59	
	Tall fescue+ apple	–	44.72	22.36	26.18	48.54	177.23	
	Orchard grass+ apple	–	44.13	22.06	26.44	48.50	177.57	
Kinnaur, Himachal Pradesh	Agrihorticulture	–	81.70	40.85	56.34	97.58	–	Salve (2018)
	Agrihortisilviculture	–	112.40	56.20	58.75	115.37	–	
Kinnaur, Himachal Pradesh	Agrihorticulture	–	29.18	14.59	151.77	166.36	–	Chisanga et al. (2013)
	Agrihortisilviculture	–	76.75	38.37	125.37	163.94	–	

Kullu, Himachal Pradesh	Orchard–cereal–cereal	20	99.23	49.61	19.89	90.88	–	Rajput (2010)
	Orchard–cereal–vegetable	20	95.14	47.57	20.51	89.39	–	
	Orchard+ vegetable+ vegetable	20	91.27	45.62	22.50	88.14	–	
	Agrihortisilviculture	–	112.40	56.20	58.75	115.37	–	

emphasizing the importance of apple-based land-use systems in carbon dynamics, offering a strong basis for optimizing management practices essential in the changing climate scenarios.

In addition to the carbon stored in the biomass, ABAFSs also improve soil organic carbon (SOC) pools by adding organic matter from leaf litter, root turnover, etc. These additions enhance the soil's stability and structure, promoting subsurface carbon retention for a long duration. This potential is further evidenced by a study in northwestern India by Zahoor et al. (2021a), which determined that ABAFSs achieved an ecosystem carbon stock of 64.18 t ha^{-1}—approximately 1.5–2 times higher than that of traditional agriculture in the region. According to their research, apple trees planted with oats and green gram as understory crops had the highest carbon stock (46.02 t ha^{-1}), followed by apple trees planted with French bean-oats (43.67 t ha^{-1}) and rajmash-oats (45.76 t ha^{-1}). On the contrary, the soil carbon stock in apple tree sole cropping was 39.99 t ha^{-1}, whereas the control agricultural systems ranged from 33.23 to 36.17 t ha^{-1}. The proportionate contributions to total organic carbon (TOC) in soil across various land-use systems were assessed using carbon fractionation studies in the dry temperate regions of Himachal Pradesh, India. It revealed horticulture systems having the greatest TOC, measuring 15.55 mg g^{-1}, followed by agriculture (13.42 mg g^{-1}), agrihorticulture (15.13 mg g^{-1}), and grass fallow (11.23 mg g^{-1}). While horticulture's passive carbon pool peaked at 10.81 mg g^{-1}, followed closely by agrihorticulture at 9.91 mg g^{-1}, this suggests a striking potential for agroforestry as a measure for increased CO_2 levels. Agrihorticulture systems also showed the largest active carbon pool at 5.22 mg g^{-1}with a carbon management index (CMI) reaching its highest value of 167.89. These results demonstrate how combining horticulture and agricultural methods can support soil carbon stocks and foster long-term ecological resilience, offering a viable route for sustainable land management under harsh weather conditions.

3.11 *Bottlenecks in Apple-Based Agroforestry*

AFSs based on apples have encouraging prospects for sustainable land use, particularly in hilly and temperate areas. However, several obstacles prevent them from operating at their best and being widely used (Nigussie et al. 2019). The struggle between apple trees and intercrops for light, water, and nutrients is one of the main obstacles. When apple trees get older, their thick canopies drastically reduce the amount of light that can reach the understory, resulting in low crop yields (Kumar 2024). The competition between the two components for soil moisture and nutrients also impacts their total production, especially in rainfed or marginal locations. Another issue is pest and disease management. Numerous pests and diseases may affect apple orchards, and adding intercrops can make integrated pest control techniques more difficult and occasionally necessitate additional chemicals. Due to restricted access to training and extension services, many smallholder farmers lack the technical competence and labor-intensive skills to manage such a complex system (Tamirat and Muluken 2018). Furthermore, diversification choices within these systems are limited by the absence of well-researched and location-specific intercrop suggestions (Girmay et al. 2014; Alemu et al. 2017). Economic limitations, including limited market access and price swings for apples and intercrops, further diminish these systems' appeal. Farmers' potential revenue is limited since they frequently do not receive enough assistance with marketing or value addition (Ali and Kachroo 2023). Institutional and policy inadequacies also present significant obstacles. Apple-based agroforestry receives little financial or technical aid since it is often poorly represented in government programs or land-use planning (Admasu and Jenberu 2022). Climate change introduces a further degree of complexity.

In conclusion, ABAFSs confront several issues that need to be resolved, even if they have the potential to increase ecological sustainability, boost land productivity, and offer a variety of revenue sources. It is essential to address problems with market access, pest and disease control, resource competitiveness, and institutional support. These systems have the potential to develop into robust and successful models of sustainable agriculture with the right kind of research, policy support, farmer education, and infrastructural development.

4 Future Directions

Apple-based AFSs offer a viable option for improving ecological sustainability, farm productivity, and climate resilience in temperate climates (Zahoor et al. 2021a; Mehta et al. 2024). Strategic interventions are required to enhance further their productivity, environmental services, and socioeconomic viability. One of the key directions is the focus on integrated nutrient management (INM), combining organic and inorganic sources of nutrients to enhance soil fertility, crop yields, and system health (Liu et al. 2025). The application of farmyard manure, biofertilizers, compost, and moderate quantities of chemical fertilizers in a prudent manner not only ensures the sustainability of nutrient supply but also creates long-term soil organic matter buildup (Mehta et al. 2024; Zhang et al. 2024). Amelioration of INM application modalities on a scale that matches various agro-climatic zones and intercropping combinations in apple cultivation could improve nutrient-use efficiency and crop yields.

Concurrently, applying natural farming techniques like bijamrit, jeevamrit, and mulching can enhance the ecological resilience of these systems. Natural farming helps minimizing dependence on external input, thereby creating a system of self-reliant ecosystem working in synergy through beneficial soil microorganisms, enhanced moisture retention, and natural pest control mechanisms (Choudhary et al. 2022). Promoting farmer experimentation and community-driven extension services can encourage faster adoption of low-cost regenerative techniques in apple-based agroforestry. Technological innovations are also essential to re-engineer these apple-based systems into future models and advance agroforestry toward climate-smart agriculture (Ntawuruhunga et al. 2023). Technologies like automation of drip irrigation, remote sensing for disease monitoring, precision delivery of nutrients, and the use of climate-resilient apple varieties through molecular breeding can maximize the use of resources and reduce risks due to climate (Zhang et al. 2021; Gonzalez 2023; Meng et al. 2023). In addition, technologies such as mobile-based advisory services, GIS-based land assessment, and robotic pollination can assist farmers in better managing their orchards, particularly during labor shortages and unpredictable weather conditions (Kansiime et al. 2019).

Another new frontier is the incorporation of apple-based agroforestry into carbon credit systems. Such systems hold significant potential for C-sequestration through above- and belowground biomass and enhanced soil organic carbon stocks (Kumar et al. 2024). Through carbon storage quantification and the adoption of monitoring procedures, farmers can sell into voluntary carbon markets and gain financial rewards for ecosystem services (Keerthika and Parthiban 2022). Creating standardized methodologies for carbon accounting in a range of agroforestry designs and ease of access to carbon credit schemes will be important in rendering this opportunity scalable and more inclusive (Nath et al. 2021). In essence, the future trajectory of ABAFSs should align ecological stewardship with economic resilience. A holistic strategy that fuses sustainable input management, agroecological innovations, digital tools, and climate finance mechanisms will be essential to unlock their full potential. Strengthening institutional support, farmer capacity-building, and region-specific research will be critical levers in realizing ABAFS as transformative models for sustainable land management in the Himalayan and other temperate landscapes.

5 Conclusions

ABAFSs constitute a climate-resilient and economically viable land-use pattern, especially tailored to temperate regions like the northwestern Himalayas. Their capability to combine food, fodder, and ecosystem services makes them the most essential tools in response to the twin imperatives of environmental degradation and rural livelihood insecurity. The future direction of such systems lies in synergizing existing agroecological wisdom and new technologies. Integrated nutrient management and natural farming will enhance soil health and input sustainability, whereas technological innovation, ranging from precision agriculture machinery to climate-resilient varieties, will ensure resilience in the face of climate change. Additionally, identifying and pricing the C-sequestration value of these systems through carbon credit mechanisms can unlock new revenue streams for farmers, thus encouraging sustainable practices. A comprehensive approach involving research, supportive policies, institutional support, and community-driven extension will be required to mainstream apple-based agroforestry in regional and national development agendas. Through concerted efforts, these systems can gradually transform into strong, sustainable agriculture models, simultaneously strengthening ecological integrity and rural prosperity.

References

Acharya M (2022) Ecological and economical appraisal of apple based agroforestry systems in Kullu Valley of Himachal Pradesh (Doctoral dissertation, College of Horticulture and Forestry Neri, Hamirpur (HP))

Admasu TG, Jenberu AA (2022) The impacts of apple-based agroforestry practices on the livelihoods of smallholder farmers in Southern Ethiopia. Trees, Forests and People 7:100205. https://doi.org/10.1016/j.tfp.2022.100205

Agroforestry network (2019). https://siwi.org/wp-content/uploads/2020/02/agroforestry-for-adapation-and-mitigation-to-climate-change_web.pdf (accessed 15 Mar 2025)

Ahmad S, Khan PA, Verma DK, Mir NH, Sharma A, Wani SA (2018) Forage production and orchard floor management through grass/legume intercropping in apple-based agroforestry systems. Int J Chem Stud 6(1):953–958

Ahmed N, Das B, Verma MK, Verma RK (2008) Lead lecture Indian apple industry and challenges from imported apple. Growth 54(75.07):12–80

Alemu SH, van Kempen L, Ruben R (2017) Explaining technical inefficiency and the variation in income from apple adoption in highland Ethiopia: the role of unequal endowments and knowledge asymmetries. J Agric Rural Dev Trop Subtrop 118:31–43

Alemu SH, Van Kempen L, Ruben R (2018) Women empowerment through self-help groups: the bittersweet fruits of collective apple cultivation in highland Ethiopia. J Hum Dev Capab 19(3):308–330

Ali J, Kachroo J (2023) Bottlenecks in production and marketing of apple in mountainous and inaccessible areas of Chenab Valley. Agric Sci Dig 43(5):675–680. https://doi.org/10.18805/ag.D-5313

Amare G, Desta B, Tena N, Muluneh K (2023) Apple production systems and post-harvest management across se-lected locations of North Shewa Zone, Ethiopia. Trends Hortic 6(1):2575. https://doi.org/10.24294/th.v6i1.2575

Anjulo A (2009) Component interactions and their influence on the production of apple based agroforestry system in wet temperate zone of Himachal Himalayas (Doctoral dissertation)

Bellow JG, Hudson RF, Nair PKR (2008) Adoption potential of fruit-tree-based agroforestry on small farms in the subtropical highlands. Agrofor Syst 73:23–36. https://doi.org/10.1007/s10457-008-9105-x

Bhardwaj DR, Salve A, Kumar J, Kumar A, Sharma P, Kumar D (2023) Biomass production and carbon storage potential of agroforestry land use systems in high hills of north-western Himalaya: an approach towards natural based climatic solution. Biomass Convers Biorefinery 1:–14. https://doi.org/10.1007/s13399-023-03952-0

Bhat R, Hussain S, Akhter M, Sharma MK, Nazir N, Khalid A, Bangaroo SA, Malik RA (2018) Effect of intercrops on apple production cv. Red delicious. Int J Curr Microbiol Appl Sci 7(6):1543–1550. https://doi.org/10.20546/ijcmas.2018.706.184

Bijalwan A (2012) Structure, composition and diversity of horticulture trees and agricultural crops productivity under traditional agri-horticulture system in Mid Hill situation of Garhwal Himalaya, India. Am J Plant Sci 3:480–488. https://doi.org/10.4236/ajps.2012.34057

Birla D, Yadav SL, Gajanand, Patel RA, Sanodiya P (2024) Agronomic techniques to improve environmental restoration and climatic resilience in the agroforestry system. In: Agroforestry solutions for climate change and environmental restoration. Springer Nature Singapore, Singapore, pp 437–462

Breen KC, van Hooijdonk BM, Tustin DS, Wilkie JD, Bound SA, Middleton SG, Close DC (2012) Changes in fruit set of 'Gala' apple in response to environment and artificial spur extinction. In: X international symposium on integrating canopy, rootstock and environmental physiology in orchard systems, vol 1058, pp 77–83

Chen Z, Wang G, Yang X, Li Z, Shen Y (2023) Water competition among the coexisting Platycladusorientalis, Prunus davidiana and Medicago sativa in a semi-arid agroforestry system. Agric Water Manag 279:108206

Chisanga K, Bhardwaj DR, Sharma S (2013) Bio-economic appraisal of agroforestry systems in dry temperate Western Himalayas. J Tree Sci 32(1&2):43–48

Chisanga K, Bhardwaj DR, Pala NA, Thakur CL (2018) Biomass production and carbon stock inventory of high-altitude dry temperate land use systems in North Western Himalaya. Ecol Process 7:1–13

Choudhary R, Kumar R, Sharma GD, Sharma RP, Rana N, Dev P (2022) Effect of natural farming on yield performances, soil health and nutrient uptake in wheat + gram inter cropping system in sub-temperate regions of Himachal Pradesh. J Crop Weed 18:01–08. https://doi.org/10.22271/09746315.2022.v18.i2.1566

Dallabetta N, Forno F, Mattedi L, Giordan M, Wehrens R (2014) Effect of training systems and pruning methods on fruit quality in apple. Agric For 60(4):173–179

Dand SA (2019) Forage yield, quality and carbon stock in an apple-based hortipastoral system (Doctoral dissertation, SKUAST Kashmir)

Das SK (2024) Soil organic carbon fractionation, carbon index, and microbial activity under different agroforestry systems in North Eastern Himalayas, India. Environ Monit Assess 196:469. https://doi.org/10.1007/s10661-024-12652-9

de Carvalho AMX, de Castro Tavares R, Cardoso IM, Kuyper TW (2010) Mycorrhizal associations in agroforestry systems. In: Dion P (ed) Soil biology and agriculture in the tropics. Soil biology, vol 21. Springer, Berlin, Heidelberg. https://doi.org/10.1007/978-3-642-05076-3_9

Dmuchowski W, Baczewska-Dąbrowska AH, Gworek B (2022) Agronomy in the temperate zone and threats or mitigation from climate change: a review. Catena 212:106089

Dogra K (2023) Response of pea (Pisum sativum L.) to natural farming practices under high density apple based agroforestry system (Doctoral dissertation, UHF, NAUNI)

Eddy WC, Yang WH (2022) Improvements in soil health and soil carbon sequestration by an agroforestry for food production system. AgrEcosyst Environ 333:107945. https://doi.org/10.1016/j.agee.2022.107945

Escribano M, Gaspar P, Mesias FJ (2020) Creating market opportunities in rural areas through the development of a brand that conveys sustainable and environmental values. J Rural Stud 75:206–215

Fahad S, Chavan SB, Chichaghare AR, Uthappa AR, Kumar M, Kakade V, Pradhan A, Jinger D, Rawale G, Yadav DK, Kumar V, Farooq TH, Ali B, Sawant AV, Saud S, Chen S, Poczai P (2022) Agroforestry systems for soil health improvement and maintenance. Sustainability 14:14877. https://doi.org/10.3390/su142214877

Fernandes PA, Wessel DF, Coimbra MA, Cardoso SM (2022) Apple (Malus domestica) by-products: chemistry, functionality and industrial applications. In: Mediterranean fruits bio-wastes: chemistry, functionality and technological applications. Springer International Publishing, Cham, pp 349–373

Gao L, Xu H, Bi H, Xi W, Bao B, Wang X, Bi C, Chang Y (2013) Intercropping competition between apple trees and crops in agroforestry systems on the Loess Plateau of China. PLoS One 8(7):e70739

Gao X, Zhao X, Wu P, Yang M, Ye M, Tian L, Zou Y, Wu Y, Zhang F, Siddique KHM (2021) The economic–environmental trade-off of growing apple trees in the drylands of China: a conceptual framework for sustainable intensification. J Clean Prod 296:0959–6526. https://doi.org/10.1016/j.jclepro.2021.126497

Gao X, He N, Jia R, Hu P, Zhao X (2022) Redesign of dryland apple orchards by intercropping the bioenergy crop canola (Brassica napus L.): achieving sustainable intensification. GCB Bioenergy 14(3):378–392

Gelaye Y, Getahun S (2024) A review of the carbon sequestration potential of fruit trees and their implications for climate change mitigation: the case of Ethiopia. Cogent Food Agric 10(1):2294544

Girmay G, Menza M, Mada M, Abebe T (2014) Empirical study on apple production, marketing and its contribution to household income in Chencha district of Southern Ethiopia. Scholarly J Agric Sci 4(3):166–175

Gonzalez MFS (2023) The role of technology adoption in enhancing apple farm profitability. Int J Appl Res 9(1):486–488
Pavek P, Granatstein D (2014) The potential for legume cover crops in Washington Apple Orchards a discussion and literature review. Plant Materials Technical Note No. 22. USDA Natural Resources Conservation Service, Spokane, Washington.
Halli SB, Swamy KR, Sankanur M, Yewale A (2016) Fruit-based agroforestry systems for food security and higher profitability. Int J Farm Sci 6(2):302–311
Hou X, Fahad S, Zhao P, Yan B, Liu T (2022) The trilogy of the Chinese apple futures market: Price discovery, risk-hedging and cointegration. Sustainability 14(19):12864
Hurd CL, Law CS, Bach LT, Britton D, Hovenden M, Paine ER, Raven JA, Tamsitt V, Boyd PW (2022) Forensic carbon accounting: assessing the role of seaweeds for carbon sequestration. J Phycol 58(3):347–363. https://doi.org/10.1111/jpy.13249
Jamar L, Rondia A, Lateur M, Minet L, Froncoux A, Stilmant D (2015) Co-design and establishment of innovative fruit-based agroforestry cropping systems in Belgium. In: International symposium on innovation in integrated and organic horticulture (INNOHORT) 1137, pp 347–350
Joshi VK, Sharma S, Kumar V (2021) Cider: the production technology. In: Winemaking. CRC Press, pp 581–607
Kansiime MK, Alawy A, Allen C, Subharwal M, Jadhav A, Parr M (2019) Effectiveness of mobile agri-advisory service extension model: evidence from Direct2Farm program in India. World Dev Perspect 13:25–33. https://doi.org/10.1016/j.wdp.2019.02.007
Kashyap SD, Dagar JC, Pant KS, Yewale AG (2014) Soil conservation and ecosystem stability: natural resource management through agroforestry in Northwestern Himalayan region. In: Agroforestry systems in India: livelihood security & ecosystem services, pp 21–55
Keerthika A, Parthiban KT (2022) Quantification and economic valuation of carbon sequestration from smallholder multifunctional agroforestry: a study from the foothills of the Nilgiris, India. Curr Sci 122(1):61. https://doi.org/10.18520/cs/v122/i1/61-69
Keprate A, Bhardwaj DR, Sharma P, Kumar D, Rana RK (2024) Biomass partitioning, carbon storage, and pea (Pisum sativum L.) crop production under a Grewia optiva-based agroforestry system in the Mid-Hills of the Northwestern Himalayas. Sustainability 16(17):7438
Kranz AJ, Wolz KJ, Miller JR (2019) Effects of shrub crop interplanting on apple pest ecology in a temperate agroforestry system. Agrofor Syst 93:1179–1189. https://doi.org/10.1007/s10457-018-0224-8
Kumar J (2024) Component interactions and their impact on production and economic viability of fruit-tree based agroforestry system in dry temperate zone of Himachal himalayas (Doctoral dissertation, UHF, NAUNI)
Kumar BM, Jose S (2018) Phenotypic plasticity of roots in mixed tree species agroforestry systems: review with examples from peninsular India. Agrofor Syst 92:59–69. https://doi.org/10.1007/s10457-016-0012-2
Kumar J, Bhardwaj DR, Thakur CL, Sharma P, Sharm P (2024) Soil organic carbon dynamics under different land use systems in dry temperate ecosystem of North-Western Himalayas. Catena 240:108022. https://doi.org/10.1016/j.catena.2024.108022
Kumari B, Pant KS, Bishist R, Prakash P, Bhatia AK (2025) Optimizing growth and yield: nutrient source effects on Brassica campestris var. brown sarson under high density Malus × domestica Borkh. based agri-horticultural system in the Western Himalayas. Agrofor Syst 99:149. https://doi.org/10.1007/s10457-025-01256-y
Lauri PÉ, Gautier A (2024) Apple trees in agroforestry–investigating the plasticity of vegetative and reproductive traits. Sci Hortic 336:113448
Lauri PÉ, Pitchers B, Dufour L, Simon S (2020) Apple farming systems - current initiatives and some prospective views on how to improve sustainability. Acta Hortic 1281:307–322. https://doi.org/10.17660/ActaHortic.2020.1281.42
Lemaire G, Garnier J, da Silveira Pontes L, de Faccio Carvalho PC, Billen G, Simioni Assmann T (2023) Domestic herbivores, the crucial trophic level for sustainable agriculture: avenues for reconnecting livestock to cropping systems. Agronomy 13(4):982

Liu Z, Jia G, Yu X (2020) Water uptake and WUE of apple tree-corn agroforestry in the loess hilly region of China. Agric Water Manag 234:106138

Liu Y, Xu M, Liu M, Cheng C, Qiao L, Li Y, Li Z (2025) Responses of soil available nutrients and microbial performance in a newly established apple orchard after five-year fertilization with different sources of livestock manure. J Agric Food Res 19:101635. https://doi.org/10.1016/j.jafr.2025.101635

Lovell ST, Dupraz C, Gold M et al (2018) Temperate agroforestry research: considering multifunctional woody polycultures and the design of long-term field trials. Agrofor Syst 92:1397–1415. https://doi.org/10.1007/s10457-017-0087-4

Malézieux E (2012) Designing cropping systems from nature. Agron Sustain Dev 32:15–29

Mehta H, Kumar P, Sharma NC et al (2024) Apple-based farming systems improve production and rhizosphere microbiome in the North-West Himalayan region's dry temperate environment. J Soil Sci Plant Nutr 24:6814–6837. https://doi.org/10.1007/s42729-024-02006-2

Meng YU, Li WJ, Shan JI, Jing CH, Chang Q, Glyn JO et al (2023) Farmers' precision pesticide technology adoption and its influencing factors: evidence from apple production areas in China. J Integr Agric 22(1):292–305

Mia MJ, Furmanczyk EM, Golian J, Kwiatkowska J, Malusá E, Neri D (2021) Living mulch with selected herbs for soil management in organic apple orchards. Horticulturae 7(3):59

Mihrete TB, Mihretu FB (2025) Crop diversification for ensuring sustainable agriculture, risk management and food security. Global Challenges 9:2400267

Mole SS (2018) Food processing and value addition: pathway to agricultural sustainability. SCMS J Indian Manag 15(1)

Mukhlis I, Rizaludin MS, Hidayah I (2022) Understanding socio-economic and environmental impacts of agroforestry on rural communities. Forests 13(4):556

Mungai NW, Motavalli PP, Kremer RJ, Nelson KA (2005) Spatial variation of soil enzyme activities and microbial functional diversity in temperate alley cropping systems. Biol Fertil Soils 42:129–136. https://doi.org/10.1007/s00374-005-0005-1

Nath AJ, Sileshi GW, Laskar SY, Pathak K, Reang D, Nath A, Das AK (2021) Quantifying carbon stocks and sequestration potential in agroforestry systems under divergent management scenarios relevant to India's nationally determined contribution. J Clean Prod 281:124831. https://doi.org/10.1016/j.jclepro.2020.124831

Nigussie Z, Fisseha G, Alemayehu G, Abele S (2019) Smallholders' apple-based agroforestry systems in the north-western highlands of Ethiopia. Agrofor Syst 93:1045–1056. https://doi.org/10.1007/s10457-018-0199-5

Ntawuruhunga D, Ngowi EE, Mangi HO, Salanga RJ, Shikuku KM (2023) Climate-smart agroforestry systems and practices: a systematic review of what works, what doesn't work, and why. Forest Policy Econ 150:102937. https://doi.org/10.1016/j.forpol.2023.102937

Nyong AP, Ngankam TM, Felicite TL (2020) Enhancement of resilience to climate variability and change through agroforestry practices in smallholder farming systems in Cameroon. Agrofor Syst 94:687–705

O'Rourke D (2021) Economic importance of the world apple industry. In: Korban SS (ed) The apple genome. Compendium of Plant Genomes. Springer, Cham, pp 1–18. https://doi.org/10.1007/978-3-030-74682-7_1

Pantera A, Mosquera-Losada MR, Herzog F, Den Herder M (2021) Agroforestry and the environment. Agrofor Syst 95(5):767–774

Pitchers B, Do FC, Pradal C, Dufour L, Lauri PÉ (2021) Apple tree adaptation to shade in agroforestry: an architectural approach. Am J Bot 108(5):732–743

Racsko J, Schrader LE (2012) Sunburn of apple fruit: historical back-ground, recent advances and future perspectives. Crit Rev Plant Sci 31:455–504

Rai R, Joshi S, Roy S, Singh O, Samir M, Chandra A (2015) Implications of changing climate on productivity of temperate fruit crops with special reference to apple. J Hortic 2(135):2376–0354

Raj SV, Chinnadurai M, Divya M, Narmadha N (2014) Markets and marketing of agroforestry products in India. Agroforestry: theory and practices. Scientific Publishers, India, p 926

Rajput BS (2010) Bio-economic appraisal and carbon sequestration potential of different land use systems in temperate North-Western Himalayas (Doctoral dissertation)

Ramananjatovo T, Guénon R, Peugeot J et al (2024) Positive influence of apple trees on soil chemical and biological activities in agroecological garden orchard system. Agrofor Syst 98:3233–3246. https://doi.org/10.1007/s10457-024-01088-2

Rinady MVP, Nuraini Y, Prayogo C, Arfarita N (2023) The effect of land management and organic matter inputs on bacterial population and soil nutrients across different types of agroforestry system. Biodiversitas 24:1333–1345. https://doi.org/10.13057/biodiv/d240302

Robinson TL, Francescatto P, Lordan J (2019) Precision pruning of gala apples. Fruit Quarterly 27(1):5–8

Salve A (2018) Structure and function of agroforestry systems in dry temperate regions of north western Himalayas (Doctoral dissertation, UHF, NAUNI)

Shahane AA (2023) Diversification: a potential options for enhancing economic viability and ecological sustainability in agricultural production systems

Sharma P, Singh MK, Tiwari P (2017a) Agroforestry: a land degradation control and mitigation approach. Bull Environ Pharmacol Life Sci 6(5):312–317

Sharma P, Singh MK, Tiwari P, Verma K (2017b) Agroforestry systems: opportunities and challenges in India. J Pharmacogn Phytochem 6(6S):953–957

Sharma P, Bhardwaj DR, Singh MK, Verma K, Vishwakarma SP, Kumar D, Thakur P (2021) Carbon trading in agroforestry: potential, prospects and constraints. In: Yadav L, Yadav SK (eds) Climate change: adaptation, loss and damage. Neel Kamal Prakashan, Delhi, pp 185–197

Sharma S, Singh P, Angmo P, Satpute S (2022) Total and labile pools of organic carbon in relation to soil biological properties under contrasting land-use systems in a dry mountainous region. Carbon Manag 13(1):352–371. https://doi.org/10.1080/17583004.2022.2089236

Sharma P, Bhardwaj DR, Singh MK, Nigam R, Pala NA, Kumar A et al (2023) Geospatial technology in agroforestry: status, prospects, and constraints. Environ Sci Pollut Res:1–29

Shen L, Wang X, Liu T, Wei W, Zhang S, Zhu Y, Tuerti T, Li L, Zhang W (2023) Apple–soybean mixed stand increased fine root distribution and soil water content with reduced soil nitrate nitrogen. Agronomy 13(2):548

Simon S, Lesueur-Jannoyer M, Plénet D, Lauri PÉ, Le Bellec F (2017) Methodology to design agroecological orchards: learnings from on-stationand on-farm experiences. Eur J Agron 82:320–330

Singh N, Sharma DP, Chand H (2016) Impact of climate change on apple production in India: a review. Curr World Environ 11(1):251

Sinha S (2023) Response of radish (Raphanus sativus L.) to natural farming practices under apple based agroforestry system (Doctoral dissertation, UHF, NAUNI)

Smith LG, Westaway S, Mullender S, Ghaley BB, Xu Y, Lehmann LM, Pisanelli A, Russo G, Borek R, Wawer R, Borzęcka M (2022) Assessing the multidimensional elements of sustainability in European agroforestry systems. Agric Syst 197:103357

Stocker BD, Zscheischler J, Keenan TF, Prentice IC, Peñuelas J, Seneviratne SI (2018) Quantifying soil moisture impacts on light use efficiency across biomes. New Phytol 218(4):1430–1449

Sun Y, Bi H, Xu H, Duan H, Peng R, Wang J (2018) Below-ground interspecific competition of apple (Malus pumila M.)–soybean (Glycine max L. Merr.) intercropping systems based on niche overlap on the Loess Plateau of China. Sustainability 10(9):3022

Tagliavini M, Tonon G, Scandellari F, Quinones A, Palmieri S, Menarbin G, Gioacchini P, Masia A (2007) Nutrient recycling during the decomposition of apple leaves (Malus domestica) and mowed grasses in an orchard. Agric Ecosyst Environ 118(1–4):191–200

Tajebe L, Gelan A (2018) Income contribution and adoption potential of apple based agroforestry on homestead farms in West and North Shoa zones of Ethiopia. J Dev Agric Econ 10(6):176–185

Tajebe L, Anjulo A, Ewnetu Z (2013) Apple based agroforestry in Dendi Woreda, Oromiya Region: income contribution and determinants for adoption. Ethiop J Agric Sci 23(1–2):61–73

Tamirat G, Muluken P (2018) Analysis of apple fruit value chain in Southern Ethiopia; the Case of Chencha. J Plant Breed Crop Sci 6(3):26–34. https://doi.org/10.15580/GJPBCS.2018.3.100218043

Ullah N, Ullah S, Khan A, Ullah I, Badshah S (2018) Preparation and evaluation of carrot and apple blended jam. J Food Process Technol 9(4):725

United Nations Sustainable Development Goals (2024). https://sdgs.un.org/goals/goal2 (accessed 7 Apr 2025)

Vallejo VE, Roldan F, Dick RP (2010) Soil enzymatic activities and microbial biomass in an integrated agroforestry chronosequence compared to monoculture and a native forest of Colombia. Biol Fertil Soils 46:577–587

Vasylieva N, Harvey J (2021) Production and trade patterns in the world apple market. Innov Mark 17(1):16

Verma K, Sharma P, Kumar D, Vishwakarma SP, Meena NK (2021) Strategies sustainable management of agroforestry in climate change mitigation and adaptation. Int J Curr Microbiol App Sci 10(01):2439–2449

Verma K, Sharma P, Bhardwaj DR, Kumar R, Kumar NM, Singh AK (2023) Land and environmental management through agriculture, forestry and other land use (AFOLU) system. In: Raj A, Jhariya MK, Banerjee A, Nema S, Bargali K (eds) Land and environmental management through forestry. https://doi.org/10.1002/9781119910527.ch10

Verma P, Sharma NC, Sharma DP, Kumar P, Chand K, Thakur H (2024) Dwarfism mechanism in Malus clonal rootstocks. Planta 260(6):133

Volk GM (2017) Ensuring the genetic diversity of apples. In: Evans K (ed) Achieving sustainable cultivation of apples. Burleigh Dodds Science Publishing, Cambridge, UK, pp 27–46

Wang S, An J, Zhao X, Gao X, Wu P, Huo G, Robinson BH (2020) Age- and climate- related water use patterns of apple trees on China's Loess Plateau. J Hydrol 582:0022–1694. https://doi.org/10.1016/j.jhydrol.2019.124462

Wang L, Zheng J, Wang G, Dang Q-L, Tissue D (2023) Combined effects of elevated CO2 and war-mer temperature on limitations to photosynthesisand carbon sequestration in yellow birch. Tree Physiol 43(3):379–389

Wei W, Liu T, Zhang S, Shen L, Wang X, Li L, Zhu Y, Zhang W (2024) Root spatial distribution and belowground competition in an apple/ryegrass agroforestry system. Agric Syst 215:103869

Wilson MH, Lovell ST (2016) Agroforestry—the next step in sustainable and resilient agriculture. Sustainability 8(6):574

Wu T, Wang Y, Yu C, Chiarawipa R, Zhang X, Han Z, Wu L (2012) Carbon sequestration by fruit trees-Chinese apple orchards as an example. PLoS One 7(6):e38883

Yadav A, Gendley MK, Sahu J, Patel PK, Chandraker K, Dubey A (2019) Silvopastoral system: a prototype of livestock agroforestry. Pharma Innov J 8(2):76–82

Yang M, Wang S, Zhao X, Gao X, Liu S (2020) Soil properties of apple orchards on China's loess plateau. Sci Total Environ 723:138041

Yinglong Q, Haibin S, Ruiping L, Ju Z, Bin L, Min L (2019) Effects of film mulching on maize growth and soil water, fertilizer and heat under fertigation of drip irrigation. Trans Chin Soc Agric Eng 35:99–110

Zahoor S, Dutt V, Mughal AH et al (2021a) Apple-based agroforestry systems for biomass production and carbon sequestration: implication for food security and climate change contemplates in temperate region of Northern Himalaya, India. Agroforest Syst 95:367–382. https://doi.org/10.1007/s10457-021-00593-y

Zahoor S, Dutt V, Mughal A, Pala NA, Rashid M, Qaisar K, Khan P (2021b) Economics and productivity of apple-based agroforestry system for livelihood generation in North Western Himalaya, India. Indian J Agrofor 23(1). Retrieved https://epubs.icar.org.in/index.php/IJA/article/view/114627

Zamora DS, Jose S, Nair PKR (2007) Morphological plasticity of cotton roots in response to interspecific competition with pecan in an alley cropping system in the southern United States. Agrofor Syst 69:107–116

Zhang F, Sarkar A, Wang H (2021) Does internet and information technology help farmers to maximize profit: a cross-sectional study of apple farmers in Shandong, China. Land 10(4):390

Zhang Y, Liu Y, Cheng C, Li Y, Li Z (2024) Effects of combined application proportion of cow manure and chemical fertilizer on soil organic carbon pool and enzyme activity in apple orchard. Sci Agric Sin 57(20):4107–4118. https://doi.org/10.3864/j.issn.0578-1752.2024.20.015

Zhao L, Gao X, He N, Zhao X (2022) Ecohydrological advantage of young apple tree-based agroforestry and its response to extreme droughts on the semiarid Loess Plateau. Agric For Meteorol 321:0168–1923. https://doi.org/10.1016/j.agrformet.2022.108969

Zheng C, Wang R, Zhou X, Li C, Dou X (2021) Effects of mulch and irrigation regimes on water distribution and root competition in an apple–soybean intercropping system in Loess Plateau, China. Agric Water Manag 246:106656

Zhou X, Wang R, Gao F, Xiao H, Xu H, Wang D (2019) Apple and maize physiological characteristics and water-use efficiency in an alley cropping system under water and fertilizer coupling in Loess Plateau, China. Agric Water Manag 221:1–12

Chapter 7
Modern Techniques in Agroforestry Practices in North–West Himalaya

Pankaj Kumar, Bhupendra Singh, Taufiq Ahmad, Atul Kumar, Indra Singh, and Naresh Singh

Abstract Traditional crop cultivation in the North–West Himalayan region often proceeds without sufficient consideration of local climate, soil, and land conditions. Diverse agroforestry systems (AFS) are practiced across different agroclimatic zones, varying in both scientific design and traditional application. These systems aim to manage allelopathic effects, crop acclimatization, soil fertility and texture, environmental suitability, tree–crop interactions, and climate change impacts. Although many stakeholders are applying appropriate AFS with satisfactory results, failures persist in tree–crop interactions due to unsuitable genotypes and allelopathic effects. AFS design decisions must also be based on productivity and income benefits. The selection of suitable genotypes of trees and crops is a primary component for establishing effective AFS. Emerging models such as Miyawaki and Milpa agroforestry show promise when adopted in different regions of India and worldwide. Therefore, such systems may be revolutionary in agroforestry, and relevant research findings must be incorporated regularly to improve AFS, as many valuable insights remain underutilized.

P. Kumar (✉)
Department of Genetics and Plant Breeding, College of Hill Agriculture, Chirbatiya, at Ranichauri, Veer Chandra Singh Garhwali Uttarakhand University of Horticulture and Forestry, Ranichauri, Tehri Garhwal, Uttarakhand, India

B. Singh · I. Singh
Department of Silviculture and Agroforestry, College of Forestry, Veer Chandra Singh Garhwali Uttarakhand University of Horticulture and Forestry, Tehri Garhwal, Uttarakhand, India

T. Ahmad
Department of Extension Education, College of Forestry, Veer Chandra Singh Garhwali Uttarakhand University of Horticulture and Forestry, Ranichauri, Tehri Garhwal, Uttarakhand, India

A. Kumar
Department of Forest Products and Utilization, College of Forestry, Veer Chandra Singh Garhwali Uttarakhand University of Horticulture and Forestry, Ranichauri, Tehri Garhwal, Uttarakhand, India

N. Singh
Department of Forest Biology and Tree Improvement, College of Forestry, Veer Chandra Singh Garhwali Uttarakhand University of Horticulture and Forestry, Tehri Garhwal, Uttarakhand, India

B. Singh et al. (eds.), *Himalayan Agroforestry Systems*, Advances in Agroforestry 15,
https://doi.org/10.1007/978-3-032-25096-4_7

Keywords AFS · Miyawaki and Milpa · North–West Himalaya · Allelopathy

1 Introduction

Indian agroforestry systems exhibit considerable structural and functional diversity, shaped by variations in temperature, elevation, soil structure, and rainfall patterns (Combe 1982; Nair and Dagar 1991). One of several definitions describes agroforestry as "a collective name for agroforestry land-use system in which woody perennials (trees, shrubs, etc.) are grown in association with herbaceous plants (crops, pastures) or livestock, in a spatial arrangement, rotation, or both; there are usually ecological and economic interactions between the trees and other components of the system" (Lundgren 1982).

Verma et al. (2023) evaluated various tree–crop combinations for biomass production, accumulation, and soil nutrient enrichment in eight traditional and commercially developed agroforestry systems (AFSs) in the mid-hill zone of the North–Western Himalaya. These include agrisilvihorticulture, agrisilviculture, silvopasture, fruit-based, fodder-based, bamboo-based, *Melia*-based, and poplar-based systems, compared with sole cropping systems.

Silvopasture (SP) functions as an important fodder resource for herders and farmers who depend on livestock for income. Essential timber, fuel, and fodder tree species in these systems include *Acacia propinqua*, *Eucalyptus tereticornis*, *Olea europaea*, *Myrica esculenta*, *Dalbergia sissoo*, *Albizia julibrissin*, *Grewia subinaequalis*, *Morus alba*, *Bauhinia variegata*, *Toona ciliata*, *Ficus carica*, *Robinia pseudoacacia*, *Populus deltoides*, *Bambusa vulgaris*, *Ulmus procera*, *Pinus roxburghii*, *Quercus leucotrichophora*, *Cedrus deodara*, *Salix alba*, and *Juniperus arizonica*.

Dominant grasses such as *Setaria italica*, *Chrysopogon fulvus*, *Cynodon dactylon*, *Arundinella nepalensis*, *Themeda triandra*, *Heteropogon contortus*, *Apluda mutica*, *Ischaemum rugosum*, *Dichanthium annulatum*, *Panicum miliaceum*, *Eragrostis pilosa*, *Chloris barbata*, and *Imperata cylindrica,* along with other natural grasses, form integral components of silvipastoral systems.

Additional natural grass species, including *Arundinella nepalensis*, *Chrysopogon aciculatus*, *Themeda triandra*, *Heteropogon contortus*, *Apluda mutica*, *Dichanthium annulatum*, *Panicum miliaceum*, *Bothriochloa ischaemum*, *Chloris barbata*, *Imperata arundinacea*, and *Cymbopogon citratus*, are also common within these grassland ecosystems. These areas often also include scattered fuelwood and timber trees such as *Pinus roxburghii*, *Acacia propinqua*, and *Pyrus malus*.

Forbs and legumes, including *Artemisia indica*, *Bidens pilosa*, *Euphorbia zeyherioides*, *Micromeria fruticosa*, *Dicliptera paniculata*, *Cassia fistula*, and *Lespedeza cuneata*, together with shrubs such as *Berberis gracilipes*, *Lantana camara*, *Leptodermis scabrida*, *Rabdosia repens*, *Meriandra bengalensis*, *Adhatoda vasica*, and *Murraya paniculata*, enhance the grassland productivity and provide valuable fodder for grazing animals, locally known as *ghasani* (Singh 2014; Singh and Kumar 2025).

Table 7.1 Carbon sequestration potential of agroforestry systems in different regions

Agroforestry system	Species	Region	Carbon stock (Mg C ha^{-1})
Agrisilviculture system	Poplar (*Populus deltoids*)-wheat (*Triticum aestivum)*	North India	34.61
Silvopastoral system	Acacia *propinqua* + *Cenchrus ciliaris*	North–Western India	6.82
Agrisilviculture	*Dendrocalamus hamiltonii*	Uttarakhand	15.91
Agrisilviculture	*Populus deltoides*	Uttarakhand	12.02
Silviculture	*Eucalyptus tereticornis*	Dehradun, Uttarakhand	4.40
			5.90
Silviculture	*Tectona grandis*	Tarai Central Division, Uttarakhand	3.74
			2.25
			2.87
Agrihorticulture	Fruit trees	Himachal Pradesh	12.15

Source: Verma et al. (2016)

2 Carbon Sequestration and Agroforestry

Since the industrial era, atmospheric concentrations of greenhouse gases have increased significantly, at rates of 0.25% per year for nitrogen oxides to 0.4% per year for CO_2 and CH_4. Carbon sequestration is one of the key strategies for capturing CO_2 from the atmosphere and storing it in long-term reservoirs. It is the provision of long-term carbon storage in the terrestrial biosphere, underground, or the ocean, and helps reduce the buildup of atmospheric CO_2, thereby improving environmental conditions and slowing environmental degradation. The reduction of atmospheric CO_2 concentration can be achieved by reducing the demand for energy and by increasing the rate of CO_2 removal through carbon sequestration (Verma et al. 2016; see Table 7.1).

3 Agroforestry Arrangements and Preferable Tree Species

Agricultural practices integrate major fruit trees and annual crops. Fruit trees such as mango (*Mangifera indica*), citrus (*Citrus limon*), guava (*Psidium guajava*), apple (*Pyrus malus*), and kiwi (*Actinidia deliciosa*), are grown in specific patterns, while principal crops are cultivated alongside them. Rabi crops include wheat (*Triticum aestivum*), barley (*Hordeum vulgare*), millets, pulses, and vegetables, whereas kharif crops such as maize (*Zea mays*), sorghum (*Sorghum bicolor*), soybean (*Glycine max*), blackgram (*Vigna mungo*), and various vegetables predominate. This integrated system ensures efficient land use and supports seasonal crop production.

The agrisilvihorticulture system (ASH) incorporates several major fodder and timber tree species, including *Grewia optiva*, *Toona ciliata*, *Terminalia bellirica*, *Terminalia chebula*, *Ficus carica*, *Sapindus mukorosii*, *Cedrus deodara*, *Juniperus communis*, *Populus spp.*, *Morus alba*, *Melia dubia*, *Bauhinia variegata*, *Albizia julibrissin*, *Anogeissus latifolia*, *Salix alba*, *Robinia pseudoacacia*, *Quercus spp.*, and *Celtis australis*. Fruit trees include *Mangifera indica*, *Emblica officinalis*, *Citrus reticulata*, *Psidium guajava*, *Carica papaya*, *Litchi chinensis*, *Prunus armeniaca*, *Pyrus communis*, *Prunus persica*, *Prunus domestica*, *Diospyros kaki*, *Vitis vinifera*, and *Pyrus malus*. Local farmer preference tends to lean towards fodder and fuelwood trees over fruit trees in this system (Singh and Kumar 2025).

In the North–Western Himalayas, agroforestry systems play a crucial role in sustainable land management and socio-economic development. Various practices such as agrisilviculture, agrihorticulture, and silvopastoral systems are prevalent, with their composition varying according to farmers' needs and local ecological conditions (Sharma et al. 2022). These systems offer multiple benefits, including soil conservation, enhanced biodiversity, and improved ecosystem stability (Kashyap et al. 2014). The production potential ranges from 5.13 to 198.20 t ha^{-1}, with benefit-cost ratios between 1.23 and 5.77 (Sharma et al. 2022). Agroforestry also contributes to carbon sequestration, with storage values ranging from 29.72 to 109.93 t ha^{-1} (Sharma et al. 2022). Overall, agroforestry systems in this region offer promising solutions for efficient land use, food security, and climate change mitigation.

According to Kumar et al. (2018), the agroforestry systems of the Indian Himalayan region have been overviewed, where four types of agroforestry systems exist in the Garhwal Himalayan region (Nautiyal et al. 1998).

1. *Arable agroforestry system*
   This system integrates various annual crops with multipurpose trees. Common crops include various plant species such as Indian barnyard millet (*Echinochloa frumentacea*), maize (*Zea mays*), soybean (*Glycine max*), finger millet (*Eleusine coracana*), gingelly (*Sesamum indicum*), and beans (*Phaseolus radiatus*). Associated tree species include haldu (*Haldina cordifolia*), sehra (*Bauhinia retusa*), kachnar (*Bauhinia variegata*), githa (*Boehmeria rugulosa*), lote tree/honeyberry (*Celtis australis*), dahman (*Cordia macleodii*), bihul (*Grewia optiva*), kamala dye tree (*Mallotus philippensis*), and Indian laurel (*Terminalia tomentosa*).
2. *Agroforestry under village community lands and shifting cultivation*
   During cropping seasons, the agroforestry sequence without trees includes mung beans (*Vigna mungo*) and soybeans (*Glycine max*) as the predominant arable crops. Under shifting cultivation, the fallow period (2–4 years) is dominated by small trees such as samakdan (*Rhus parviflora*).
3. *Homestead multi-tier fodder system*
   In this system, the upper story consists of fodder and timber trees such as khadik (*Celtis australis*), kachnar (*Bauhinia variegata*), bihul (*Grewia optiva*), and timber trees such as poplar (*Populus spp.*), willow (*Salix spp.*), Hamilton's bamboo

(*Dendrocalamus hamiltonii*), and solid bamboo (*Dendrocalamus strictus*). The middle story includes medicinal plants such as adusa (*Adhatoda vasica*) and chaste tree (*Vitex negundo*), along with fruit trees such as pear (*Pyrus* spp.), plum (*Prunus* spp.), and *Citrus* spp. The lower story is occupied by cucurbitaceous species, taro (*Colocasia esculenta*), yams (*Dioscorea alata*), and turmeric (*Curcuma longa*).

4. *Village community forests*
   For fuelwood, species such as cutch tree (*Acacia catechu*), haldu (*Haldina cordifolia*), and samakdan (*Rhus parviflora*) are used. For fodder, species such as bihul (*Grewia optiva*) and githa (*Boehmeria rugulosa*) are preferred. For fiber, fruit, and timber, species such as *Grevia optiva*, *Celtis australis*, *Terminalia spp. Bauhinia variegata, Phyllanthus embilica* and *Toona ciliata Ficus sp.*) are collected.
5. *Agrisilviculture systems*
   According to Rawat and Vishvakarma (2011), prominent tree species in these systems include bihul (*Grawia optiva*), khadik (*Celtis australis*), Persian lilac (*Syringa persica*), white mulberry (*Morus alba*), and Indian gooseberry (*Emblica officinalis*). Vegetables such as brinjal (*Solanum melongena*), chillies (*Capsicum* spp.), peas (*Pisum sativum*). and brassicas—such as cauliflower (*Brassica oleracea* var. *botrytis*), cabbage (*B. oleracea* var. *capitata*), and radish (*Raphanus sativus*)—are grown for soil erosion control and moisture conservation. Multipurpose woody hedgerows are recommended for land management in many parts of the region.

It is evident that farmers preferentially manage and introduce fodder species. In areas with poor soil, limited water, and nutrient constraints where cropping is difficult, forage and firewood species are favorable for production. The development of silvopastural systems with a 10-year rotation has dramatically increased productivity from 0.5–1.5 t ha^{-1} $year^{-1}$ to more than 10 t ha^{-1} $year^{-1}$ (Fig. 7.1).

4 Allelopathic Effects During Tree-Crop Interaction

Allelopathic substances (allelochemicals) were first identified by Davis in black walnut (*Juglans regia*). Foliar leachates of walnut containing juglone were found to inhibit germination and seedling growth of crops beneath the tree. Crop species differ markedly in their performance within agroforestry systems due to allelopathic effects.

Aqueous extracts of mature leaves from *Azadirachta indica*, *Terminalia arjuna*, *Dalbergia sissoo*, *Albizia lebbeck*, *Sesbania grandiflora*, *Acacia auriculiformis*, and *Leucaena leucocephala* significantly reduced germination of wheat (*Triticum aestivum*), paddy (*Oryza sativa*), and gram (*Cicer arietinum*) in petri-dish bioassays. The maximum inhibition of germination was caused by *Sesbania grandiflora*, followed

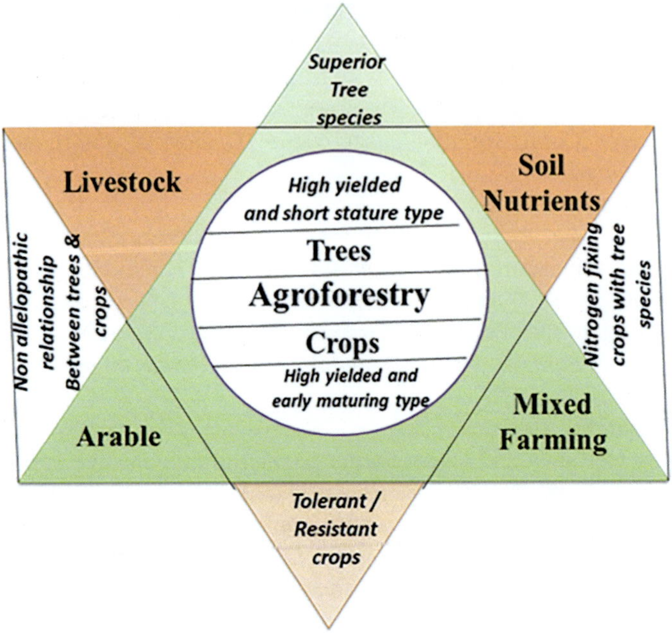

Fig. 7.1 General schematic diagram of agroforestry system

by *Acacia auriculiformis*. The magnitude of reduction was least in wheat (*Triticum aestivum*) and greatest in gram (*Cicer arietinum*). The decrease in germination was directly proportional to the increase in concentration of leaf extract (Rao et al. 1994).

Pine (*Pinus*) needles also exhibit strong allelopathic potential affecting seed germination and early seedling growth of wheat and mustard. Phenolic compounds present in the pine needles are primarily responsible for the observed reductions in crop production (Kiran 2016).

The extract of *Pinus* needles exhibits allelopathic activity against wheat and mustard crops, which are predominant in the hills of the North–Western Himalayas. Every season, pine needles fall on the forest floor, and during the rainy season, allelochemicals from decomposing needles dissolve in water and leach into the soil. This process is considered a major cause of reduced crop yields in the mountain regions of India and elsewhere (Baroniya and Baroniya 2014). To address this issue, there is a need to develop well-designed agroforestry systems that combine suitable tree species with well-performing crops.

Rizvi et al. (1999) listed over 80 tree species grown in agroforestry programs that exhibit allelopathic activity. These include *Acacia* spp., *Albizia lebbeck*, *Eucalyptus* spp., *Grewia optiva*, *Gliricidia sepium*, *Leucaena leucocephala*, *Moringa oleifera*, and *Populus deltoides*, all of which can negatively affect crop performance due to allelopathy.

According to Kohli et al. (2006), several cover crops produce allelochemicals that contribute to weed suppression. Examples include *Avena sativa* L. (oat), which produces scopoletin and phenolic acids; *Brassica* spp. (turnip, mustard, radish, rape), which produce glucosinolates and their breakdown products such as isothiocyanates, nitriles, epithionitriles, and thiocyanates; *Fagopyrum esculentum* (buckwheat), which produces fatty acids (palmitic, stearic, arachidic, behenic), fagomine, 4-piperidone, and 2-piperidinemethanol; *Hordeum vulgare* (barley), which produces gramine and hordenine; *Ipomoea tricolor* (morning glory), which produces tricolorin A; *Medicago sativa* (alfalfa), which produces medicarpin, 4-methoxy medicarpin, sativan, 5-methoxy sativan, canavanine, saponins, and phenolic acids such as ferulic, chlorogenic, and salicylic acid; *Mucuna pruriens* (velvet bean), which produces L-DOPA; *Secale cereale* (rye), which produces DIBOA, BOA, β-hydroxybutyric acid, β-phenyllactic acid, and phenolic acids; *Sorghum* spp., which produce sorgoleone, dhurrin, and breakdown products such as p-hydroxybenzoic acid and p-hydroxybenzaldehyde; *Trifolium* spp. (clover), which produce phenolic acids; *Triticum aestivum* (wheat), which produces DIMBOA and phenolic acids (syringic, vanillic, trans-ferulic, cis-ferulic, trans-p-coumaric, cis-p-coumaric, and p-hydroxybenzoic); and *Vicia villosa* (hairy vetch), which produces cyanamide.

Such allelochemicals effectively suppress weed performance and growth. Additional allelochemicals of trees and horticultural plants are listed in Table 7.2.

Factors Affecting Tree–Crop Interaction

1. *Species*
   Selection of properly compatible tree–crop combinations is essential.
2. *Sunlight*
   Management of shade-tolerant species and pruning of tree crowns are required to reduce shading on agricultural crops.

Table 7.2 Allelochemicals in Tree plants and horticultural plants

S. N.	Plant species	Allelochemicals	
1.	Wild tamarind (*Leucaena leucochephala*)	Mimosine	Anjali et al. (2022)
2.	Walnut (*Juglans regia*)	Juglone	
3.	Apple (*Malus domestica*)	Phlorizin, quercetin	
4.	Neem (*Azadirachta indica*)	Azadirachtin	
5.	Eucalyptus (*Eucalyptus* spp.)	1.4-cineole, 1.8-cineole	
6.	Guava (*Psidium guajava*)	Phenolics	
7.	Peach (*Prunus persica*)	Amygdalin	

3. *Density*
   Tree density must be optimized to reduce competition; trees should be planted at appropriate numbers per hectare.
4. *Age*
   Competition is reduced when tree–crop combinations are established during the early growth stages of trees.
5. *Site*
   Site quality determines the carrying capacity of the agroforestry system.
6. *Management*
   Appropriate management practices enhance agricultural crop performance and overall system productivity in tree–crop combinations.

Tree-Crop Combination Advantages

- *Species choice*
  Both tree and agricultural crop species must be selected carefully for compatibility.
- *Agroforestry system design*
  System design may include parallel tree and crop rows or concentric crop arrangements around trees.
- *Agroforestry management system*
  Timely and appropriate levels of management are required for effective agroforestry system performance (see Table 7.3).

5 Components of Agro-Forestry

Agroforestry systems generally consist of four to five basic sets of components that humans structurally manage. These systems are categorized as: (a) agrisilviculture, (b) agrihorticulture, (c) silvopastoral, and (d) agrisilvipastoral. Based on the nature of components, the following common agroforestry systems are found in different agro-ecological regions of India (Fanish and Priya 2013):

1. Agrisilviculture (trees + crops)
2. Boundary plantation (tree on boundary + crops)
3. Block plantation (block of trees + block of crops)
4. Energy plantation (trees + crops during initial years)
5. Alley cropping (hedges + crops)
6. Agrihorticulture (fruit trees + crops)
7. Agrisilvihorticulture (trees + fruit trees + crops)
8. Agrisilvipasture (trees + crops + pasture/animals)
9. Silviolericulture (tree + vegetables)
10. Hortipasture (fruit trees + pasture/animals)
11. Hortiolericulture (fruit tree + vegetables)
12. Silvopasture (trees + pasture/animals)
13. Forage forestry (forage trees + pasture)

Table 7.3 Agroforestry species and their allelopathic activity

S. N.	Agroforestry species	Target species	Plant parts/ allelochemicals	Uses of agroforestry species	Ref
1.	*Azadirachta indica*	Rice, peanut, wheat, maize, various microorganisms	Leaf, wood, and leaf litter leachates; litter, mature leaf extracts	Timber, manure, oil, fuel, food, pest control	Venkatesh et al. (2024)
2.	*Leucaena leucocephala*	Lettuce, rice, sorghum	Aqueous leachate and extracts of leaves, litter, dry leaf mulch (mimmosine)	Fuelwood, poles, timber, food, soil conservation	
3.	*Melia azedarach*	Cabbage, cress (*Lepidium sativum*)	Leaf leachate	Crop shade, fuel wood, timber, lumber	
4.	*Populus deltoides*	Sugarcane, wheat	Soil, leachate	Timber	
5.	*Tamarindus indica*	*Amaranthus spinosus*	Ethanolic extracts of leaves and seeds	Beverage, fruit, fuelwood, shade tools, rituals	
6.	*Eucalyptus globulus*	Cucumber, blackgram, lettuce	Leaf extract and leachate, soil percolate, canopy effect	Lumber, essential oil, poles	Anjali et al. (2022)
7.	*Acacia tortilis*	Leaf, stem, and soil extract	Pearl millet (*Pennisetum glaucum*), sesame (*Sesamum indicum*), cluster bean (*Cyamopsis tetragonoloba*), wheat (*Triticum aestivum*)	Germination, growth, yield	
8.	*Bamboo*	Leaf extract	Groundnut (*Arachis hypogaea*)	Oil, growth, chlorophyll, protein content	
9.	*Eucalyptus citridora*	Leaf, stem, and root extract	Okra (*Abelmoschus esculentus*) wheat (*Triticum aestivum*). cowpea (*Vigna unguiculata*), maize (*Zea mays*)	Growth	
10.	*Eucalyptus tereticornis*	Leaf, stem, and root extract	Sorghum (*Sorghum bicolor*), cowpea (*Vigna unguiculata*), sunflower (*Helianthus annuus*)	Germination, growth	
11.	*Leucaena leucocephala*	Mimosine, leaf extract, soil extract	Green gram (*Vigna radiata*), rice (*Oryza sativa*), wheat (*Triticum aestivum*), maize (*Zea mays*)	Germination, growth, yield	

14. Shelterbelts (trees + crops)
15. Windbreaks (trees + crops)
16. Live fence (shrubs and understory trees on boundary)
17. Silvi-horti-sericulture (trees/fruit trees + sericulture)
18. Horti-apiculture (fruit trees + honeybees)
19. Aquaforestry (trees + fish)
20. Homestead system (trees, fruit trees, vegetables, and other multiple combinations)

6 Agro Climatic Zones

Several agroclimatic conditions exist in Uttarakhand due to wide variations in slope and elevation. The state has several physiographic zones and farming situations, which are presented in Table 7.4 (Official Website of State Horticulture Mission, Government of Uttarakhand) (Table 7.5).

7 Criteria for Species Selection

Several criteria are used for selecting suitable species in advanced agroforestry systems. These include tolerance to low temperatures, deep root systems capable of penetrating rocky soils, moderate or controlled growth to minimize competition with crops, ability to withstand strong winds, multipurpose uses (such as fodder or fruit production), and good adaptation to the region's specific climate and soil conditions. The primary criterion for selecting a tree species is determined by the farmers themselves. It is generally preferable to choose well-known and locally accepted species rather than unfamiliar ones. When introducing a new species, detailed studies should be conducted on the unknown tree species. Promoting the extension of a tree species that is disliked by farmers for any reason is often very difficult (www.worldagroforestry.org; see Tables 7.6, 7.7, and 7.8).

8 Miyawaki Method of Afforestation

Natural forest development is a slow process that can take a long time. Over the past two decades, scientists worldwide have gained fresh insights into both theoretical and practical approaches for restoring and reconstructing natural ecosystems (Clewell and Aronson 2007). The Miyawaki method, developed by Japanese botanist Akira Miyawaki, is a technique designed to accelerate the growth of dense and natural forests. The process involves the careful selection and planting of native tree species along with heavy mulching to increase soil fertility. Mulching is

Table 7.4 Different agro-climatic conditions in Uttarakhand

S.N.	Zone type	Farming condition	Soil situation	Rainfall condition	Area	Principal farm and livestock produces
1.	Zone A range: <1000 m	Tarai irrigated region	Alluvial soil	1400 mm/year	Udham Singh Nagar and Haridwar districts	*Crops (cereal, pulses, oilseeds):* Rice (*Oryza Sativa*), wheat (*Triticum aestivum*), sugarcane (*Saccharum officanarum*), lentil (*Lens culinaris*), chickpea (*Cicer arietinum*), rapeseed mustard (*Brassica* spp.) *Horticultural crops:* Mango (*Mangifera indica*), litchi (*Litchi chinensis*)), guava (*Psidium guajava*), peach (*Prunus persica*), plum (*Prunus domestica)* *Livestock:* Buffalo, cattle
		Bhabar irrigated region	Alluvial mixed soil with boulders and shingle	1400 mm/year	Nainital, Dehradun, and Pauri Garhwal districts	*Crops (cereal, pulses, vegetables):* Rice (*Oryza Sativa*), wheat (*Triticum aestivum*), sugarcane (*Saccharum officanarum*), rapeseed mustard (*Brassica* spp.), potato (*Solanum tubersosum*), lentil (*Lens culinaris)* *Horticultural crops:* Mango (*Mangifera indica*), litchi (*Litchi chinensis*), guava (*Psidium guajava*) *Livestock:* Buffalo, cattle
		Irrigated lower hills region (600–1000 m)	Alluvial sandy soil	2000–2400 mm/ year	Champawat, Pauri Garhwal, Dehradun, Nainital, and Tehri Garhwal districts	*Crops:* Rice (*Oryza sativa*), wheat (*Triticum aestivum*) *Vegetable crops:* Onion (*Allium cepa),* chilli (*Capsicum annuum)*, pea (*Pisum sativum*), potato (*Solanum tuberosum*), radish (*Raphanus sativus*), cauliflower (*Brassica oleracea)*, pulses, oilseeds, soybean (*Glycine max)* *Horticultural crops:* Mango (*Mangifera indica*), guava (*Psidium guajava*), peach (*Prunus persica*), plums (*Prunus domestica)* *Livestock:* Buffalo, cattle

(continued)

Table 7.4 (continued)

S.N.	Zone type	Farming condition	Soil situation	Rainfall condition	Area	Principal farm and livestock produces
		Rainfed lower hills region (600–1000 m)	Residual sandy loam	2000–2400 mm/year	Champawat, Nainital, Pauri Garhwal, Dehradun, Tehri Garhwal, and Bageshwar districts	*Crops (cereals, millets):* Finger millet (*Eleusine coracana*), maize (*Zea mays*), rice (*Oryza sativa*), wheat (*Triticum aestivum*), pulses *Horticultural crops:* Mango (*Mangifera indica*), guava (*Psidium guajava*), peach (*Prunus persica*), and plum (*Prunus domestica)* *Livestock:* Buffalo, cattle, goats
2.	Zone B range: 1000–1500 m	Mid-hills region of south aspect (1000–1500 m)	Sandy loam soil	1200–1300 mm/year	Champawat, Nainital, Almora, Dehradun, Tehri Garhwal, and Bageshwar districts	*Crops (cereals, millets):* Rice (*Oryza sativa*), finger millet (*Eleusine coracana*), wheat (*Triticum aestivum*), pulses *Vegetable crops:* Potato (*Solanum tubersosum*), tomato (*Solanum lycopersicum*), pea (*Pisum sativum*), cole crops (*Brassica* spp.) *Horticultural crops:* Peach (*Prunus persica*), plum (*Prunus domestica)* *Livestock:* Cattle, sheep, goats
3.	Zone C Range: 1500–2400 m	High hills region (1500–2400 m)	Red to dark soil	1200–2500 mm/year	Pithoragarh, Almora, Chamoli, and Bageshwar districts	*Crops (cereals, millets, pulses):* Amaranth (*Amaranthus hypochondriacus)*, finger millet (*Eleusine coracana)*, French bean (*Phaseolus vulgaris)* *Vegetable crops:* Cole crops (*Brassica* spp.*)*, potato (*Solanum tubersosum)*, pea (*Pisum sativum*) *Horticultural crops:* Peach (*Prunus persica*), plum (*Prunus domestica),* pear (*Pyrus communis*), apple (*Pyrus malus*), other stone fruits *Livestock:* Cattle, sheep, goats

4.	Zone D Range: >2400 m	Very high hills region (>2400 m)	Red to dark black clay soil	1300 mm/year	Pithoragarh, Chamoli, and Uttarkashi districts	*Crops (potential crops):* Amaranth (*Amaranthus hypochondriacus*), buckwheat (*Fagopyrum esculentum*) *Vegetable crops:* Pea (*Pisum sativum*), cole crops (*Brassica* spp.), potato (*Solanum tuberosum)* *Horticultural crops:* Apple (*Pyrus malus)* *Livestock:* Sheep and goats (mainly)

Source: https://shm.uk.gov.in/pages/display/6-state-profile. Accessed 21/02/2025

Table 7.5 Different Ecological Zones in Himachal Pradesh, India

S. N.	Type of Zone	Ranges of Elevation (meters)	Area covered (Km^2)	Per cent (%)	Zone Under Districts
1.	Zone I: Sub-montane and low hills sub-tropical zone	240–1000 m	10,260	18.44	Eight districts, namely Kangra, Una, Hamirpur, Bilaspur, Solan, Chamba, Mandi, and Sirmaur
2.	Zone II: Mid-hills sub-humid zone	1000–1500 m	4664	8.38	Ten districts, namely Chamba, Kangra, Mandi, Shimla, Solan Sirmaur, Kullu, Kinnaur, Hamirpur, and Bilaspur
3.	Zone III: High hills wet temperate zone	1500–2500 m	9217	16.56	Ten districts, namely Shimla, Mandi, Chamba, Kangra, Kullu, Solan, Sirmaur, Kinnaur, Lahaul, and Spiti
4.	Zone IV: High hills dry temperate zone	>2500 m	31,509	56.62	Eight districts, including Kangra, Lahaul and Spiti, Kinnaur, Chamba, Mandi, Sirmaur, and Shimla

Source: Sharma et al. (2022), DoA (2009)

Table 7.6 Multipurpose tree species in agroforestry

S. N.	Tree species	Uses
1.	*Albizia lebbeck*	Fuel, fodder, timber, nitrogen fixing
2.	*Alnus nepalensis*	Fodder, fuel, timber, nitrogen fixing
3.	*Bauhinia variegata*	Fodder, fuel, edible pods, timber, medicine, nitrogen fixing, bee forage
4.	*Bauhinia semla*	Fodder, fuel, timber, medicine, nitrogen fixing, bee forage
5.	*Boehmeria regulosa*	Fodder, fuel, medicine, timber
6.	*Celtis australis*	Fodder, fuel, edible fruit, timber, medicine
7.	*Emblica officinalis*	Medicine, fodder, fuel, edible fruit
8.	*Ficus* spp.	Fodder, fuel, edible fruit, timber, medicine
9.	*Grewia optiva*	Fodder, fuel, edible fruit, fiber, fertilizer
10.	*Ougeinia oojeinensis*	Fodder, bee forage, fuel, medicine, timber
11.	*Prunus cerasoides*	Bee forage, fuel, timber, fodder, medicine, religious
12.	*Pyrus pashia*	Fuel, edible fruit, timber, medicine, bee forage
13.	*Quercus leucotrichophora*	Timber, fuelwood
14.	*Terminalia* spp.	Fodder, fuel, timber, medicine
15.	*Toona ciliata*	Fuel, timber

Source: Bhatt et al. (2010), Yadav and Bisht (2013), Singh (2004)

Table 7.7 Different varieties/genotypes tolerant to abiotic stresses

S. N.	Drought-tolerant varieties	
1.	Pigeonpea (*Cajanus cajan)*	Palnadu, Co 5, ICPL-87, Maruti, Pragati, Abhay, Jagrati (ICPL 151), Asha, Paras, MAL 13, BRG-1, BRG-2, PRG-158, VL Arhar-1, Rajeev, BDN-711, BDN-716, GT-102, LRG-52, GRG 811, BDN-708, BRG 5, PRG 176, Prakash
2.	Blackgram (*Vigna mungo*)	Shekhar-2, Azad-3, Pratap Urd-1, (KPU 07-08), BDU-1, Pant Urd-35
3.	Greengram (*Vigna radiata*)	RMG-268, Pratap, PDM-139, RMG-344, Pusa Vishal, GM-4, CO. (Gg) 8, Yadadri, VBN 4 VGG 10-008, BM-2003-2
4.	Chickpea (*Cicer arietinum*)	Pant Kabuli gram-2, Pant gram-4, Indira Chana, Nandyal Gram 49, JG 16, Pusa Chickpea 10216
5.	Cowpea (*Vigna unguiculata*)	UPC 618, UPC 622, C 519, IT-38956-1, UPC 628, Hidrudaya
6.	Horsegram (*Macrotyloma uniflorum*)	PHG-9, CRIDA-18R, Indira Kulthi-1, CRIDALATHA, VL Gahat-19 (VLG-19), GHG-5, Bilasa Kulthi (BSP 15-1), Dapoli Kulthi-1
7.	Lentil (*Lens culinaris*)	Kota Masoor 3 (RKL 605-03), L 4729
8.	Groundnut (*Arachis hypogaea)*	Abhaya, Narayani, Prasuna, AK-265, SG-99, VRI (Gn)-7, Girnar-2 &3; Greeshma Mallika, Kadiri Haritandhra, GJG-HPS-1, Raj Durga, Girnar-3, R-2001-2, RARS-T-1,2; Divya, Raj Mongfali-1, Dharani, Birsa Groundnut 4, Dheeraj, Kadri lepakshi
9.	Soybean (*Glycine max*)	Pant Soybean 24, VL SOYA-2, JS-80-21, JS 71-05, Ahilya-3, JS-335, Indira Soya- 9, PS-1092, MACS-450, Parbhani Sona, Ahilya-4, MAU-71, JS-9305, JS-95-60, PS-1225, Pratap Soya-2, RKS-24, Pratap Soya 45 and NRC 136
10.	Linseed (*Linum usitatissimum)*	Jawahar-552, 165, KL-31, NL-97, Shekhar, Sharda, JLS-27 Suyog, Binwa, Indira Alsi-32, Deepika, RLC 92, Himani, JLS-67, JLS-73, Mau Azad Alsi-2
11.	Sesame (*Sesamum indicum*)	GT-10, PB NO.1, TYPE-13, JT-21, TKG-22, JTS-8, Nirmala (OR-Sel-164), Pragati (MT-75), Jawahar Til PKDS-11
12.	Rape seed and mustard (*Brassica* spp.)	Geeta, Pusa Bold, Puas Tarak, Rohini, PT-507, KBS-3, Agrani, Aravali, Karan Tara, Pusa Mustard-21, Divya-33, Pant Rai 20, RH 761, DRMR 150-35
13.	Citrus (*Citrus* spp.)	Nagpur mandarin, Nagpur Seedless, PDKV lime, Katol Gold
14.	Pomegranate (*Punica granatum*)	Goma Khatta, Ganesh, P-23, P-26, Mridula, Phule Arakta, Bhagwa
15.	Brinjal (*Solanum melongena)*	PKM-1, KashiSandesh, KashiTaru, KashiHimani
16.	Chilli (*Capsicum annuum*)	Samrudhi, Kashi Anmol
17.	Cluster bean (*Cyamopsis tetragonoloba*)	RGC-936, 1017, 1003, 1066, Thar Bhadavi
18.	Onion (*Allium cepa*)	Agrifound Dark Red, Arka Kalyan, Raseedpura local

(continued)

Table 7.7 (continued)

S. N.	Drought-tolerant varieties	
19.	Tomato (*Solanum lycopersicum*)	Arka Meghali, Arka Vikas, Thar Annant
20.	Black pepper (*Piper nigrum*)	Panniyur-6, Panniyur-7, Panniyur-8, Panniyur-9 Sigandhini, Pournami
21.	Cardamom (*Elettaria cardamomum*)	ICRI-5, ICRI-6, ICRI-4
22.	Ginger (*Zingiber officinale*)	CDPK1, Suruchi, Suravi, Himgiri
23.	Turmeric (*Curcuma longa*)	IISR-Pragathi, CO-2, BSR-1
24.	Coriander (*Coriandrum sativum*)	CO-1, CO-2, RCr-20
25.	Fennel (*Foeniculum vulgare*)	Co-1, GF-1, S-7-9
26.	Ajwain (*Trachyspermum ammi*)	Ajmer Ajwain-1, Ajmer Ajwain-2, Ajmer Ajwain-93
Waterlogging-tolerant varieties		
27.	Rice (*Oryza sativa)*	Sarala, Samba sub-1, Swarna Sub1, CSR43, Reeta, CR Dhan 500, Jalamani, Jayantidhan, CR Dhan505, 801, 802, Sabour Shree, DRR Dhan 50, BahadurSub-1, jalmagna, Jaladhi, Jalpriya, DRR Dhan 50, Tripura Jala
28.	Wheat (*Triticum aestivum)*	KRL 19, KRL 210 & KRL 283
29.	Maize (*Zea mays*)	Jawahar Maize 218, Pusa Jawahar Hybrid Maize-1, Pragati
30.	Pigeonpea (*Cajanus cajan)*	Maruti, ICPL 84023, Asha, Mal 13
31.	Soybean (*Glycine max*)	NRC-37, JS 97-52
Heat-tolerant varieties		
32.	Rice (*Oryza sativa)*	DRR Dhan 47 & 52
33.	Wheat (*Triticum aestivum)*	HW 2045, Pusa gold, DWR17, DBW 173, HI 1634 (Pusa Ahilya)
34.	Maize (*Zea mays*)	Suwan, PMH-7, RCRMH2
35.	Pearlmillet (*Pennisetum glaucum)*	GHB-558,538, 732
36.	Pigeonpea (*Cajanus cajan)*	Bahar, UPAS-120
37.	Chickpea (*Cicer arietinum*)	Indira Chana, Pant G 186, JG -6, JAKI 9218, JG-14

(continued)

Table 7.7 (continued)

S. N.	Drought-tolerant varieties	
38.	Lentil (*Lens culinaris*)	Kota Masoor 3
39.	Groundnut (*Arachis hypogea*)	Phulepragati (JL-24), Ambar (CSMG-84-1), Vemana (K-134), Kadiri-6 (K-1240)
40.	Soybean (*Glycine max*))	JS-335
41.	Linseed (*Linum usitatissimum*)	RLC 92 (IC 555926)
42.	Sesame (*Sesamum indicum*)	CUMS-17 (Suprava)
43.	Rape seed and mustard (*Brassica* spp.)	Pusa Mustard 22 & Vijay, NRCDR-02, Pusa Tarak, Pant Rai-19 (PR-2006-1), Vasundhara
44.	Bottle gourd (*Lagenaria siceraria*)	Thar Samridhi, Pusa Santushti
45.	Brinjal (*Solanum melongena*)	Kashi Sandesh, Kashi Taru, Thar Rachit
46.	Chilli (*Capsicum annuum*)	Kashi Abha
47.	Cucumber (*Cucumis sativus*)	Pusa Barkha
48.	Okra (*Abelmoschus esculentus*)	Kashi Pragati, Kashi Kranti
49.	Potato (*Solanum tuberosum*)	Kufri Surya Surya, Kufri Lauvkar
50.	Radish (*Raphanus sativus*)	Pusa Chetaki, Kashi Mooli-40
51.	Ridge gourd (*Luffa acutangula*)	Thar Karni, Pusa Sneha, Thar Tapish
52.	Tomato (*Solanum lycopersicon*)	Pusa Sadabahar, Arka Meghali, Varkha Bahar-1 & 2, Arka Vikas
Cold- and frost-tolerant crop varieties		
53.	Rice (*Oryza sativa*)	Kalinga 1, Tellahamsa, Pant Dhan 11, Varun Dhan
54.	Wheat (*Triticum aestivum*)	Buland, Mansarovar, Shalimar Wheat-1, RSP 561
55.	Pigeonpea (*Cajanus cajan*)	Bahar
56.	Pearl millet (*Pennisetum glaucum*)	GHB-538
57.	Chickpea (*Cicer arietnum*)	PDG 4

(continued)

Table 7.7 (continued)

S. N.	Drought-tolerant varieties	
58.	Rape seed and mustard (*Brassica* spp.)	RH-9801 (SWARNA), RGN-48, 73
59.	Setaria grass (Foxtail millet) (*Setaria italica*)	PSS-1, Nandi, Setaria-92, S-18, Him Palam Setaria - 2 (S -25)
60.	Bottle gourd (*Lagenaria siceraria*)	Pusa Santushti
61.	Tomato (*Solanum lycopersicon*)	Pusa Sadabahar
Salt-tolerant crop varieties		
62.	Rice (*Oryza sativa*)	Luna Suvarna, CR Dhan 402, 403, 405, 406, Vikas DRR Dhan 9, 33, 39, CSR10, CSR13, CSR23, CSR27, Panvel 3, CARI Dhan 5
63.	Wheat (*Triticum aestivum*)	KRL 1-4, KRL 19, KRL 210, KRL 283, KRL-213
64.	Pigeonpea (*Cajanus cajan*)	UPAS 120, Jagriti (ICPL 151), C 11
65.	Chickpea (*Cicer arietinum*)	Karnal chana-1
66.	Cowpea (*Vigna unguiculata*)	Hissar Cowpea-46, Bidhan Rice bean-3
67.	Rice bean (*Vigna umbellata*)	Vikrant (VH-82-1)
68.	Lentil (*Lens culinaris*)	PSL-9 and PDL-1
69.	Rape seed and Mustard (*Brassica* spp.)	Pusa Vijay, Pusa mustard 25, 22, 29 & 30, CS 52, CS 54, CS 56, Narendra Rai, NRCDR 601 & NRCDR-02, Giriraj (DRMRIJ 31), RGN 229, RGN-145 & RGN 236

Source: Boraiah et al. (2023)

recommended to avoid soil dryness, prevent erosion on steep slopes, suppress weeds, shield seedlings from cold, and serve as manure when materials degrade (Miyawaki 2004).

This strategy uses a dense planting approach that combines intermediate and late successional species, along with several companion species (Miyawaki 1998). The high planting density creates a state of dynamic balance and cooperation among the many species (Padilla and Pugnaire 2006). It is anticipated that the technique will result in a plantation that is 30 times denser than usual and grows 10 times faster than traditional plantations. The Miyawaki technique has been applied in South American countries, Malaysia, Japan, and the Far East in both urban and ecologically damaged areas. These rapidly established urban woodlands provide multiple benefits, such as lowering local temperatures, improving air quality, sequestering CO_2, enhancing human wellbeing, and increasing real-estate values. Consequently,

Table 7.8 Horticultural crop varieties/genotypes tolerant to abiotic stresses

Crop for specific abiotic stress		Varieties/genotypes
Drought-tolerant varieties		
70.	Apple (*Mallous domestica*)	M4,7,26,104,106, 111 *Malus baccata*, *M. toringoides*, *M. sieversii*
71.	Pear (*Pyrus communis*)	Oregon 211 and 249, Oregon 260, 261, 264, Quince A, Provence Quince
72.	Peach (*Prunus persica*)	GF677, 577, Myrobalan, Bright Hybrid (Drought and Salinity tolerant varieties)
73.	Plum (*Prunus domestica*)	Mariana 4001, Peach–almond hybrid, Myrobalan 27, GF 667, GF 577
Cold-tolerant varieties		
74.	Apple (*Mallous domestica*)	G 11, 30, 41, 935, B 9, P 2, M7, 26, 106
75.	Apricot (*Prunus armeniaca*)	*Prunus armeniaca* 'Suka' (wild type)
Heat-tolerant varieties		
76.	Apricot (*Prunus armeniaca*)	*P. armeniaca* 'Sahara' (wild type)
Waterlogging-tolerant varieties		
77.	Apricot (*Prunus armeniaca*)	Line FA-3-6, Pollizo prune
78.	Kiwi Fruit (*Actinidia deliciosa*)	*A. valvata* (KR5 & KR3)
Salinity-tolerant varieties		
79.	Avocado (*Persea americana*)	West Indian race
80.	Citrus (*Citrus* spp.)	Rangpur lime, Marmalade orange, Rough Lemon, Cleopatra mandarin, Sour Orange, Karna Khatta, Nasnaran, Attani 2, Gou Tou Cheng (drought- and salinity-tolerant varieties)
81.	Guava (*Psidium guajava*)	'Crioula'
82.	Grapes (*Vitis vinifera*)	*V. champini*, *V. vinifera*, Dogridge, Salt Creek, *V. berlandieri* × *V. rupestris*, 1613, 1103 Paulsen
83.	Mango (*Mangifera indica*)	Nekkare, Terpentine, Olour, Bappakai, Gomera-1, *M. zeylanica*, ML-2, GPL-1
84.	Pear (*Pyrus communis*)	*Pyrus betulifolia*
85.	Pomegranate (*Punica granatum*)	Tab-o-Larz

Source: Boraiah et al. (2023)

many urban real-estate developers are now incorporating green Miyawaki-based projects in metropolitan areas to attract more clients (Poddar 2021). According to Poddar 2021 the features and six steps of miyawaki methods are mentioned below:

9 Features of the Miyawaki Method

(a) Plantations are approximately 30 times denser than those produced by traditional methods.
(b) At least 50–100 distinct natural species are planted in the same area.
(c) Noise and dust levels are reduced by up to 30 times.
(d) Carbon dioxide absorption is up to 30 times higher than in monoculture plantations.
(e) Annual growth of at least 1 m is guaranteed.
(f) The natural forest becomes totally maintenance-free and self-sustaining after the first 3 years.
(g) The system requires no chemical fertilizers and supports local wildlife.

10 Six Steps of the Miyawaki Method

1. Determine which nutrients the soil lacks.
2. Identify native plant species suited to the local soil type and climate.
3. Provide the required soil nutrients using locally available biomass, such as agricultural or industrial by-products, ideally sourced within 50 km of the site.
4. Plant saplings up to 80 cm tall at a density of 3–5 saplings per square meter in soil modified to a depth of one meter.
5. The forest patch should be at least 100 m^2. Within approximately 8 months, the vegetation becomes dense enough that sunlight barely reaches the ground. Rainwater is fully absorbed, and fallen leaves decompose into humus. As the forest develops, it produces more of its own nutrients, accelerating growth. The high planting density causes the trees to compete for sunlight, which drives rapid vertical growth.
6. During the first 2–3 years, the forest requires regular watering and weeding. After this initial period, it becomes self-sustaining and should be left undisturbed to allow natural ecological processes, including wildlife colonization, to occur.

11 Miyawaki Method in India

This strategy is steadily gaining traction in India. Afforest, a social company, has collaborated with various businesses and communities to establish these forests. They recently collaborated with the Department of Biotechnology (DBT), Government of India, to transform a sewage-infested plot of land near the Barapullah drain. The drain was cleaned as part of a larger initiative called Local Treatment of

Urban Sewage Streams for Healthy Reuse (LOTUS HR). DBT, the Ministry of Science and Technology, and the Government of India collaborated to build a demonstration facility to clean up the Barapullah drain. The foundation of the forest was set in July 2018 after early site inspections in January 2018.

12 Milpa Agroforestry System

According to Fonteyne et al. (2023), the milpa agroforestry system is a traditional agricultural practice originating in Mesoamerica. It is a production system based on the cultivation of the *Mesoamerican triad*—maize (*Zea mays* L.), bean (*Phaseolus* spp.), and squash (*Cucurbita* spp.)—but is also associated with a great diversity of additional crops and wild species (Linares and Bye 2011). The term *milpa* is a Spanish and Mexican word that comes from the Nahuatl language (*milli* meaning sown field, and *pan* meaning on top of). The term is known by several names in other regions; for example, in the United States, it is known as *the three sisters*, while in Mesoamerica it is simply called milpa.

Milpa helps to safeguard enormous biodiversity and serves as a cornerstone of food security for many traditional rural communities. Milpa has been cultivated under a wide range of climatic, soil, and cultural conditions. Depending on the local environment and traditions, milpa systems vary significantly in the plant species cultivated, land preparation methods, use of fire, water regime, and length of the fallow cycle. Broadly, milpa systems can be classified into three types: extensive rainfed systems with long fallow cycles, intensive rainfed systems with short fallow periods, and intensive rainfed or irrigated systems without fallow (Aguilar et al. 2003).

The area of milpa is decreasing rapidly due to the system's low productivity and profitability. Additional contributing factors include cultural rejection, migration of youth from rural areas, and soil degradation caused by shortened fallow periods (Isakson 2009; Novotny et al. 2021; Orozco Ramırez and Astier 2022).

Milpa crops are highly complementary in that:

1. Plants receive light at different canopy levels: maize is the tallest, squash grows at ground level, and beans climb maize stalks.
2. Beans fix nitrogen, which can be used by the companion crops.
3. Squash leaves shade the soil, reducing weeds and conserving moisture.
4. Flowers of beans, squash, and other plants attract beneficial insects, reducing pest pressure.
5. Greater root volume and improved soil structure enhance water and nutrient availability compared to monocultures.

13 Conclusion

Agroforestry, as an effective multicrop system, provides an excellent option for integrating different crops and trees suited to match specific locations and desired traits. Strategic land-use planning that incorporates high-performing crops and tree species while avoiding allelopathic conflicts can enhance productivity and stakeholder satisfaction. Several agroforestry systems have traditionally been practiced in the North Himalayan region, and these systems should be applied scientifically using high-yielding and climate-resilient crops and tree species to achieve greater economic value. Given that most hilly areas depend primarily on rainfall, stakeholders commonly adopt rainfed agroforestry systems. Therefore, the selection of rainfed, cold-tolerant, and resilient traditional crops, along with high-yielding, good-quality tree species, should be carefully matched to local altitudinal zones. The Milpa and Miyawaki systems are now well-accepted approaches in agroforestry, and such systems may represent a breakthrough in the development of well-established agroforestry models in the North Himalayan region.

References

https://apps.worldagroforestry.org/Units/Library/Books/Book%2006/html/7.2_criter_f_a_gd_agrfr_tre.htm?n=79. Accessed 04/03/2025

Aguilar J, Illsley C, Marielle C (2003) Los Sistemas agrıcolas de maız y sus procesos técnicos. In: Esteva G, Marielle C (eds) Sin maız no hay pa´ ıś. Consejo Nacional para la Cultura y las Artes, Mexico City, pp 83–122

Anjali KS, Krishnamoorthi S, Madiwalar AF (2022) Allelopathy in agroforestry. In: Agroforestry prospective, strategies and future aspects. Taran Publication, pp 121–126

Baroniya SS, Baroniya MB (2014) Germination and early seedling growth of mustard and wheat as affected by allelopathic activity of Pinus needle extracts. Int Res J Environ Sci 3(5):27–34

Bhatt V, Purohit V, Negi V (2010) Multipurpose tree species of Western Himalaya with an agroforestry perspective for rural needs. J Am Sci 6(1):73–80

Boraiah K.M., Basavaraj P.S., Vijaysinha D. Kakade, Harisha C.B., Khapte, P.S., Halagundegowda G.R., Krishnamurthy D., Kulshreshtha, N., Vijayakumar H.P., Naik, B., Sammi, J.R., Reddy, K. and Pathak, H. (2023). Abiotic stress-tolerant crop varieties in India: status and a way forward. In: Mohamed A. El-Esawi, Recent trends in plant breeding and genetic improvement. DOI: https://doi.org/10.5772/intechopen.1001916

Clewell AF, Aronson J (2007) Ecological restoration: principles, values, and structure of an emerging profession. Island Press, Washington DC

Combe J (1982) Agroforestry techniques in tropical countries: potential and limitation. Agrofor Syst 1:13–28. https://doi.org/10.1007/BF00044326

DoA (Department OF Agriculture) (2009) The study on diversified agriculture for enhanced farm income in the state of Himanchal Pradesh final report volume –II (annexure part I). JICA, Nippon, Koei, Co. Ltd, p 188. Available online at: https://openjicareport.jica.go.jp./pdf/11925534_01.pdf

Fanish SA, Priya RS (2013) Review on benefits of agro forestry system. Int J Educ Res 1:1–12

Fonteyne S, Caamal JBC, Lopez-Ridaura S, Loon JV, Balbuena JE, Alcalá LO, Hernández FM, Odjo S, Verhulst N (2023) Review of agronomic research on the milpa, the traditional polyculture system of Mesoamerica. Front Agron 5:1115490. https://doi.org/10.3389/fagro.2023.1115490

Isakson SR (2009) No hay ganancia en la milpa: the agrarian question, food sovereignty, and the on-farm conservation of agrobiodiversity in the Guatemalan highlands. J Peasant Stud 36:725–759. https://doi.org/10.1080/03066150903353876

Kiran R (2016) Tree-crop interactions in agroforestry systems. Indian J Plant Soil 3(2):93–97. https://doi.org/10.21088/ijps.2348.9677.3216.6

Kohli RK, Batish DR, Singh HP (2006) Allelopathic interactions in agroecosystems. In: Reigosa MJ, Pedrol N, González L (eds) Allelopathy: a physiological process with ecological implications, pp 465–493

Kumar BM, Handa AK, Dhyani SK, Arunachalam A (2018) Agroforestry in the Indian Himalayan Region: an overview, Chapter 6. CAB International, pp 153–172. https://doi.org/10.1079/9781780644851.0153. file:///C:/Users/DELL/Downloads/Kumaretal2018Chapter6TemperateAFsystems.pdf. Accessed 22/12/2025

Linares E, Bye R (2011) ¡La milpa no es solo maiz! In: Alvarez-Buylla Roces E, Carreon Garcia A, San Vicente Tello XXXA (eds) Haciendo milpa. UNAM, Coyoacan, Mexico, pp 9–12

Lundgren B (1982) Introduction (editorial). Agrofor Syst 1:1–12

Miyawaki A (1998) Restoration of urban green environments based on the theories of vegetation ecology. Ecol Eng 11:157–165

Miyawaki A (2004) Restoration of living environment based on vegetation ecology: theory and practice. Ecol Res 19(1):83–90

Nair PKR, Dagar JC (1991) An approach to developing methodologies for evaluating agroforestry system in India. Agrofor Syst 16:55–81. https://doi.org/10.1007/BF00053197

Nautiyal S, Maikhuri RK, Semwal RL, Rao KS, Saxena KG (1998) Agroforestry systems in the rural landscape—a case study in Garhwal Himalaya, India. Agrofor Syst 41:151–165

Novotny IP, Tittonell P, Fuentes-Ponce MH, López-Ridaura S, Rossing WAH (2021) The importance of the traditional milpa in food security and nutritional selfsufficiency in the highlands of Oaxaca, Mexico. PLoS One 16:1–21. https://doi.org/10.1371/journal.pone.0246281

Orozco Ramırez Q, Astier M (2022) La agricultura familiar y la milpa ante la polıtica neoliberal en mexico. In: Borrego A, Ramırez M, Astier M (eds) Integración global: el nuevo entorno de los territorios locales. Centro de Investigaciones en′ Geografıa Ambiental, Morelia, pp 1–17

Padilla FM, Pugnaire FI (2006) The role of nurse plants in the restoration of degraded environments. Front Ecol Environ 4(4):196–202

Poddar S (2021) Miyawaki technique of afforestation. Krishi Sci eMagazine Agric Sci 2(9):1–4

Rao OP, Saxena AK, Singh BP (1994) Allelopathic effects of certain agroforestry tree species on the germination of wheat, paddy and gram. Ann For 2:60–64

Rawat YS, Vishvakarma SCR (2011) Pattern of fodder utilization in relation to sustainability under indigenous agroforestry systems, North-Western Himalaya, India. Environ We: Int J Sci Technol 6:1–13

Rizvi SJH, Tahir M, Rizvi V, Kohli RK, Ansari A (1999) Allelopathic interaction agroforestry systems. Crit Rev Plant Sci 18:773–796

Sharma H, Pant KS, Bishist R, Sharma PP, Gautam KL (2022) Agroforestry system in North-Western Himalayas, India: an overview. Agrobiodivers Agroecol 02(01):1–20. https://doi.org/10.33002/aa020101

Singh B (2004) Altitudinal variation in relation to seed, seedling and fodder quality of Celtis australis L. In: Garhwal & Kumaon (eds) A promising agroforestry tree-crop of Central Himalaya, India. D.Phil thesis, HNB Garhwal University, Srinagar Garhwal, Uttaranchal, p 122

Singh M (2014) Pattern, composition and vegetation dynamics of agroforestry systems in Giri catchment, HP. Ph.D. Thesis. Dr. Y. S. Parmar University of Horticulture and Forestry, Nauni, Solan (HP), India, 173 p

Singh G, Kumar D (2025) Potential of agroforestry in livelihood security in the north-west himalya with special reference to Himachal Pradesh, India. Krishi Sci eMagazine Agric Sci 6(3):50–54. https://krishiscience.co.in

Venkatesh L, Maheshnaik BL, Shwetha VR, Ali S (2024) Agroforestry: tree crop interaction and its management. In: Modern horizons in agricultural sciences. K D Publications, pp 239–249

Verma A, Tewari JC, Kaushal R, Saresh NV (2016) Agroforestry for carbon sequestration in North Western Himalaya. In: Dagar JC, Tewari JC (eds) Agroforestry research developments. Nova Science Publishers, Inc., New York, pp 307–325

Verma T, Bhardwaj DR, Sharma U, Sharma P, Kumar D, Kumar A, Kumar A (2023) Agroforestry systems in the mid-hills of the north-western Himalaya: a sustainable pathway to improved soil health and climate resilience. J Environ Manag 348:119264. https://doi.org/10.1016/j.jenvman.2023.119264

Yadav RP, Bisht JK (2013) Agroforestry: a way to conserve MPTs in North Western Himalaya. Res J Agric Forestry Sci 2320:6063

Chapter 8
Tree Biomass and Carbon Sequestration in Traditional Agroforestry Systems of Western Himalaya

Sandeep Kumar, Vinod Prasad Khanduri, Sarswati Prakash Sati, and Manoj Kumar Riyal

Abstract In the Western Himalayas, local inhabitants have long practiced agroforestry by intentionally retaining multipurpose trees on agricultural land to obtain food, fuel, fodder and other essential benefits. In addition, agroforestry systems play a significant role in carbon sequestration and climate change mitigation amid the current era of global warming and climate change. Traditional agroforestry systems are widespread in the Western Himalayan region. These systems sequester substantial amounts of atmospheric carbon in the various carbon pools, including trees and crop biomass, soil, and leaf litter. The amount of vegetation carbon stocks depends on several factors, such as species, age, structure, and spatial distribution. The traditional agroforestry systems of the Western Himalayas include agrisilviculture, agrihorticulture, hortiagriculture, agrohortisilviculture, silvopastoral, and hortipastoral practices, which support a wide variety of plant species. Within these systems, carbon stock in tree biomass varies considerably between elevational zones and agroforestry types, with the highest carbon stock in tree biomass of 51.14 Mg ha^{-1} observed in the silvopastoral system. Soil carbon stocks are also an important carbon pool in the traditional agroforestry systems of the Western Himalayas. These stocks generally decrease with increasing soil depth and respond differently depending on elevation. Overall, silvopastoral and agrisilviculture systems exhibit higher carbon sequestration rates, highlighting their potential for climate change mitigation and

---

S. Kumar (✉)
Department of Silviculture and Agroforestry, Veer Chandra Singh Garhwali Uttarakhand University of Horticulture and Forestry, Ranichauri, Tehri Garhwal, Uttarakhand, India

Department of Basic and Social Sciences, Veer Chandra Singh Garhwali Uttarakhand University of Horticulture and Forestry, Ranichauri, Tehri Garhwal, Uttarakhand, India

V. P. Khanduri
Department of Silviculture and Agroforestry, Veer Chandra Singh Garhwali Uttarakhand University of Horticulture and Forestry, Ranichauri, Tehri Garhwal, Uttarakhand, India

S. P. Sati · M. K. Riyal
Department of Basic and Social Sciences, Veer Chandra Singh Garhwali Uttarakhand University of Horticulture and Forestry, Ranichauri, Tehri Garhwal, Uttarakhand, India

B. Singh et al. (eds.), *Himalayan Agroforestry Systems*, Advances in Agroforestry 15,
https://doi.org/10.1007/978-3-032-25096-4_8

sustainable land management. Despite the significant benefits of agroforestry systems in the Western Himalayas, several challenges persist, including limited awareness of their overall benefits, restricted market access, and the adverse impacts of climate change on agricultural productivity. However, several opportunities exist to improve the adoption and effectiveness of agroforestry systems in the Western Himalayas. These include promoting climate-resilient tree and crop species, developing market linkages for agroforestry products, providing training and technical support to farmers, and leveraging traditional agroforestry systems for agroecotourism activities.

Keywords Agrisilviculture · Agrihorticulture · Climate change · Soil carbon stock · Silvopastoral

1 Introduction

The Himalayas are not only a geographical marvel but also a region of immense cultural, ecological, and economic importance, stretching across multiple countries, including India, Nepal, Bhutan, Pakistan, and China. The Western Himalayas encompass the Indian union territories and states of Jammu and Kashmir, Ladakh, Himachal Pradesh, and Uttarakhand, characterized by rugged terrain, high-elevation peaks, and diverse climatic zones. The geographical diversity of the Western Himalayas is further accentuated by its major river systems, including the Indus, Ganga, Yamuna, Kali, Chenab, and Jhelum, among others. These rivers not only support agriculture and hydropower generation but also play a crucial role in sustaining the region's ecosystems. The Western Himalayas are renowned for their rich biodiversity and ecosystem services. The region is home to a wide variety of flora and fauna. Additionally, most of the human population in the Western Himalayas lives in villages or rural areas and mostly depends on agriculture or livestock to sustain their livelihood. The forests of the Western Himalayas, which include coniferous, broadleaf, and alpine ecosystems, provide critical ecosystem services such as carbon sequestration, soil erosion prevention, and water regulation (Singh 2007; Joshi and Negi 2011). Agroforestry practices in this region provide ecosystem services similar to those of natural forests, while also offering significant economic benefits to farmers.

Climate change is one of the most pressing global challenges, driven by rising atmospheric carbon dioxide (CO_2) levels, which have increased dramatically from a pre-industrial level of approximately 278 ppm to over 421 ppm by early 2023 (Venegas et al. 2023) and 427.09 ppm by February 2025 (www.co2.earth). Carbon, primarily in the form of carbon dioxide (CO_2), plays a crucial role in climate change by acting as a primary greenhouse gas, trapping heat in the atmosphere and causing global warming. This increase is primarily due to human activities, including fossil fuel consumption and land-use changes, which release more CO_2 than natural sinks can absorb (Hertzberg and Schreuder 2016). In this era of global warming, the World Meteorological Organization asserts that the CO_2 concentration in the earth's atmosphere is about 413.20 ± 0.2 ppm, which has increased the world's average temperature by 1.11 °C (WMO 2021). Climate change poses a significant threat to global as well as Himalayan ecosystems, with the Western Himalayas facing challenges such as reduced agricultural productivity, water scarcity, and loss of biodiversity. Forests and other tree-based land-use systems are among the most efficient natural mechanisms for carbon sequestration. They absorb atmospheric CO_2 through photosynthesis and store it in plant biomass. Additionally, they transfer sequestered carbon to the soil through litterfall and root decomposition, further enhancing long-term carbon storage, acting as vital carbon sinks that mitigate the impacts of climate change. The carbon sequestration potential of trees varies depending on factors such as species, growth rate, size, soil, and environmental conditions (Vacek et al. 2023). Rising temperatures, increased frequency of droughts, and extreme weather events can impair the carbon sequestration capacity of trees by altering their structure and function, often resulting in reduced growth, higher mortality rates, and accelerated decomposition.

The forest ecosystem is characterized by perennial vegetation, which provides additional carbon storage capacity. In addition to natural forests, agroforestry systems contribute significantly to carbon sequestration and offer a promising solution to the climate crisis, and reduce the load on natural forests (Ghale et al. 2022). Agroforestry has been practiced for centuries, as evidenced by the intentional retention of multipurpose trees (e.g., for fodder, fruit, fuelwood, and timber) on farm bunds (Semere 2019; Chavan et al. 2020), and continues to play a crucial role in sustaining farming systems. According to the World Agroforestry Centre, agroforestry is a land-use system that intentionally integrates woody perennials—such as trees, shrubs, and palms—with crops and/or animals on the same land-management unit to promote sustainable land use and enhance farm productivity. As a combination of agricultural and forestry systems, agroforestry seeks to balance multiple objectives, producing trees for timber and commercial purposes, ensuring a diverse and sufficient supply of nutritious food to meet both global demand and local needs, and protecting the natural environment to sustain essential resources and ecosystem services for present and future generations. In the wake of climate change and declining factor productivity, agroforestry systems have the capacity to mitigate the effects of climate change and increase overall productivity through their contributions to soil health improvement, nutrient cycling, plants and soil carbon sequestration, and better economic returns relative to sole cropping systems. Agroforestry systems have been recognized as a climate change adaptation strategy by 29 countries, including India, and as a mitigation strategy by 23 countries in their Intended Nationally Determined Contributions (INDCs) submitted to the UNFCCC (Richards et al. 2016). Agroforestry has significant potential to contribute to 9 out of the 17 Sustainable Development Goals (SDGs) (Van Noordwijk 2020), including reducing poverty (SDG 1: No Poverty), eliminating hunger (SDG 2: Zero

Hunger), combating climate change (SDG 13: Climate Action), and conserving biodiversity while promoting sustainable land management (SDG 15: Life on Land). Furthermore, agroforestry can help to achieve other goals by improving gender equality (SDG 5: Gender Equality), health (SDG 3: Good Health and Well-Being), access to clean water (SDG 6: Clean Water and Sanitation), sustainable energy solutions (SDG 7: Affordable and Clean Energy), and responsible agricultural production (SDG 12: Responsible Consumption and Production). In the Himalayas, both traditional and modern agroforestry are widely distributed. Traditional agroforestry involves indigenous knowledge-based practices of intercropping crops with multipurpose tree species. In these systems, trees are not planted in a specific sequence, but farmers are protected, and naturally regenerated tree species on farm bunds to meet household needs. In contrast, modern agroforestry includes scientifically designed systems where tree species are intentionally planted in specific patterns to minimize their impact on crops and provide an additional source of income. Agroforestry systems include agrisilviculture, agrihorticulture, hortiagriculture, agrohortisilviculture, silvopastoral, and hortipastoral practices. The names indicate various land-use combinations, with *agri* referring to agriculture, *silvi* denoting silviculture or cultivation of forest trees, *horti* indicating horticulture, and *pastoral* referring to cattle grazing. The combination of these terms defines the nature and focus of each agroforestry system. For instance, *agrisilviculture* involves cultivating agricultural crops alongside forest tree species, with agriculture being the primary focus. Similarly, in *hortiagriculture*, agricultural crops are grown under fruit trees, but the emphasis is primarily on horticulture, with agriculture playing a secondary role. Generally, the term mentioned first in each system name indicates the main focus or dominant component of that particular agroforestry practice. In the tough Himalayan terrain, these agroforestry systems optimize land usage, promote sustainability, and support local livelihoods by integrating trees, crops, and animals. Despite their critical role in carbon sequestration, these agroforestry systems face numerous challenges and are sometimes not managed effectively.

2 Scope and Importance of Agroforestry in the Western Himalayas

Agroforestry has considerable potential to address economic, social, and environmental challenges (Mbow et al. 2014). By incorporating multipurpose tree species, agroforestry systems can help to restore degraded lands while generating employment, increasing farmers' income, and reducing poverty (Jinger et al. 2023; Akter et al. 2022). Additionally, agroforestry minimizes surface runoff, nutrient leaching, and soil erosion via the stabilizing effect of tree roots, while also improving nutrient recycling, maintaining soil organic matter, reducing pressure on natural forests, and enhancing ecological stability (Sileshi et al. 2020; Teo et al. 2025). Initially recognized for its ecological contributions, agroforestry is now also acknowledged for its significant social impact, earning the description of "a very good social science" (Atangana et al. 2014). It supports food and nutritional security by providing tree-based foods, sustaining staple crop production, and generating income through tree product sales (Jamnadass et al. 2013). Agroforestry also has the potential to influence human migration patterns, a pressing issue in the Western Himalayan region. Various studies in the Western Himalayas and other regions have demonstrated that agroforestry-based farming is more profitable than conventional agriculture. Fast-growing and multipurpose trees provide quick and diverse sources of revenue, making agroforestry a viable and attractive alternative to conventional farming. Since animals form a vital component of agroforestry systems, farmers can further diversify their income by raising dairy animals such as cows and buffaloes, as well as meat-producing animals and birds, including goats, sheep, and chickens. In the Western Himalayas, intercropping within agrisilviculture and agrihorticulture systems has proven more profitable than traditional agriculture or monocropping, as evidenced by their higher net present values. Agroforestry promotes eco-intensification through efficient resource use (Pantera et al. 2021), carbon sequestration, biodiversity conservation, and soil enrichment (Udawatta et al. 2019; Kumar et al. 2021). Research indicates higher microbial diversity in agroforestry systems compared to monocropping, with increased activity of beneficial bacteria and fungi (Udawatta et al. 2019; Beule et al. 2020). These systems enhance soil organic carbon and mitigate greenhouse gas emissions (Mayer et al. 2022). In addition, the presence of trees in agroforestry systems improves air quality by absorbing atmospheric pollutants such as ozone, sulfur dioxide, and nitrogen dioxide through leaf stomata (Nowak et al. 2014).

Agroforestry practices in the Western Himalayas play a pivotal role in addressing the region's ecological, economic, and social challenges. These practices, which integrate trees with agricultural crops and/or livestock, have been traditionally practiced in the region and are increasingly recognized for their potential to enhance biodiversity, ecosystem services, and climate resilience (Pancholi et al. 2023). Traditional agroforestry systems in the Himalayas support diverse indigenous and exotic tree and crop species. For example, agrisilviculture systems support greater tree species diversity than sequential agroforestry systems (Kumar et al. 2018). Similarly, homegarden systems, common in the region, often include a mix of fruit trees, multipurpose tree species, and crops, to further enhance biodiversity (Vibhuti et al. 2019). Agroforestry systems also provide important habitats for herbaceous plants, shrubs, and grasses, which are essential for maintaining ecosystem balance and supporting livestock grazing.

Agroforestry systems in the Western Himalayas provide a wide range of ecosystem services that benefit both local communities and the global environment. These services include provisioning services, such as food, fodder, and fuelwood, as well as regulating services, such as carbon sequestration and soil conservation. Among these, provisioning services offer the most direct benefits to local livelihoods (Hemel et al. 2024). For example, agrisilviculture systems in the Western Himalayas produce a variety of crops, including cereals, pulses, and vegetables, which are vital for local food security. Similarly, silvopastoral systems provide high-quality fodder for livestock, which is critical for animal husbandry in the region. In addition to food and fodder, agroforestry systems also provide fuelwood, which is a critical energy source for many households in the Western Himalayas. The production of fuelwood in agroforestry systems helps to reduce the pressure on natural forests, thereby contributing to forest conservation. Regulating services provided by agroforestry systems include carbon sequestration, soil conservation, and water regulation (Kuyah et al. 2017). These services are essential for mitigating climate change and ensuring the long-term sustainability of agricultural systems in the region. Carbon sequestration is one of the most significant regulating services provided by agroforestry systems. Studies have shown that agroforestry systems in the Western Himalayas have the potential to sequester significant amounts of carbon, thereby contributing towards global efforts to mitigate climate change (Kumar et al. 2021; Tariyal et al. 2022; Singh et al. 2024; Sharma et al. 2023). Soil conservation is another critical regulating service provided by agroforestry systems. The integration of trees in agricultural landscapes helps to reduce soil erosion, improve soil fertility, and enhance water infiltration. Agroforestry systems also help to regulate water cycles by enhancing water infiltration and reducing runoff (Awazi et al. 2024). This is particularly important in the Western Himalayas, where water scarcity is a significant challenge during the dry season. Supportive and cultural ecosystem services are also provided by various agroforestry practices. These systems contribute to nutrient cycling and offer important ethical, recreational, and ecotourism values.

Agroforestry systems in the Western Himalayas offer significant economic benefits (Yadav et al. 2018). These systems produce a variety of marketable products, including fruits, nuts, and timber. For instance, the production of apples, walnuts, and other fruits in agroforestry systems generates substantial income for many households in the region (Zahoor et al. 2021; Bhushan et al. 2023). In addition to these direct economic benefits, agroforestry enhances overall economic resilience by diversifying agricultural production, reducing the risk of crop failure, and stabilizing household incomes. The integration of multiple crops and tree species enables farmers to better adapt to changing market conditions and climate variability. Moreover, agroforestry systems contribute to rural employment by creating job opportunities in tree planting, crop management, and product processing. Fuelwood and fodder production, in particular, provide important income sources during the off-farming season. Overall, the Western Himalayas benefit significantly from agroforestry, which helps address ecological challenges, deliver essential ecosystem services, and strengthen economic resilience in the region.

3 Agroforestry for Tree Biomass and Carbon Sequestration

The biomass of an individual tree represents the total living organic matter it contains and is divided into above-ground and below-ground components. The below-ground tree biomass consists of core and fine roots, while the above-ground tree biomass includes the stem, bark, branches, and foliage (leaves). In agroforestry systems, tree biomass can be estimated using either destructive or non-destructive methods. The non-destructive approach calculates above-ground biomass based on tree height, diameter, and wood density. Below-ground biomass is estimated by multiplying the above-ground biomass by allometric equations or root-shoot ratios (a standard factor of 0.26) (Cairns et al. 1997). The amount of carbon stored in tree biomass is directly proportional to the total biomass, as trees typically store approximately half of their biomass as carbon. Therefore, tree carbon is calculated by multiplying the biomass by the standard conversion factor of 0.5 (IPCC 2003).

The concept of carbon sequestration is fundamentally consistent across all land-use systems. While the underlying mechanisms are similar, they may manifest differently in different systems depending on the unique characteristics of each system (Nair et al. 2010). Several factors significantly influence the carbon storage potential of plant biomass. Estimating carbon stocks in agroforestry is critical to addressing the global climate emergency. In recent decades, carbon sequestration has become a major research priority as scientists seek the most effective strategies for reducing atmospheric CO_2 levels. Agroforestry is increasingly recognized as one of the most promising approaches to achieving the goal of net-zero emissions (Tirkey et al. 2024). Although trees are the primary contributors to carbon storage in agroforestry systems, crops, shrubs, herbs, and soil also play essential roles in absorbing and storing carbon. Agroforestry systems comprise multiple components that sequester and store carbon in various pools, including tree biomass, crop biomass, shrub and herb biomass, leaf litter, and soil (see Fig. 8.1). Compared to single-species crop or pasture systems, agroforestry systems offer greater potential for carbon sequestration because they can more efficiently capture and utilize growth resources such as light, nutrients, and water (Nair et al. 2010).

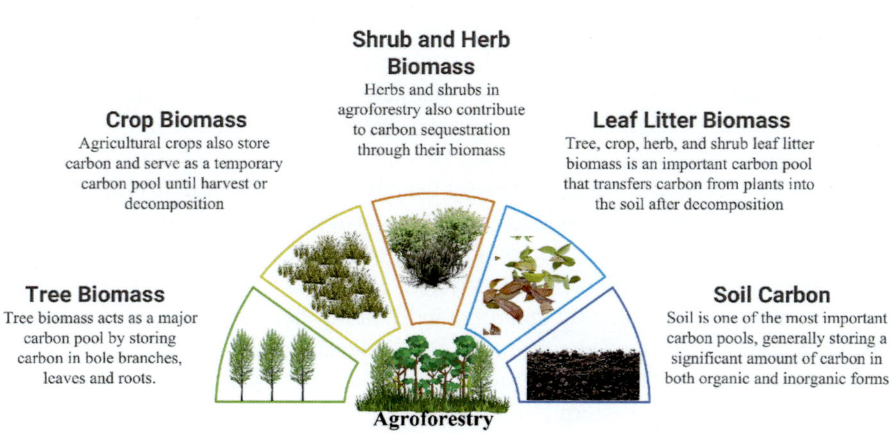

Fig. 8.1 Different carbon pools in agroforestry systems

4 Traditional Agroforestry Systems of Western Himalaya

Mountain communities in the Western Himalayas are highly dependent on agricultural ecosystems due to their socio-cultural practices and religious beliefs (Pandey 2009). The retention of trees on agricultural land for fodder, firewood, and housing needs exemplifies the various traditional agroforestry practices in the region. Traditional agroforestry systems are ancient land-use practices found worldwide that differ in structure, function, socio-economic characteristics, and ecological services (Viswanath and Lubina 2017). These traditional agroforestry systems represent one of the finest examples of human–nature coexistence and co-evolution, as local communities have developed these systems over generations to meet their needs while fully respecting the surrounding environment. The Himalayas are among the world's most diverse, resource-rich, and fragile ecosystems where indigenous rural communities practice a variety of land-use strategies. Traditional agroforestry systems have a long history dating back to antiquity. They are characterized by the natural regeneration and protection of native tree species on farmers' fields, with diverse structures and functions that have been refined rather than intensified over countless generations (Verma et al. 2024). The traditional agroforestry systems in the Western Himalayan region are well adapted to local agro-climatic conditions and have evolved gradually to fulfill community needs. These systems play a vital role in biodiversity conservation by providing habitats for a wide variety of plants and animals. They deliver critical ecosystem services, including soil fertility maintenance, water retention, soil erosion control, and increasing the tolerance of agricultural systems to climate variability, thereby maintaining ecological balance. In addition, they supply a range of goods such as food, fuel, fodder, and timber, which can help alleviate poverty and support stable livelihoods. Singh et al. (2024) estimated that the traditional agroforestry systems in the Garhwal Himalayas generate carbon credit values ranging from ₹11,709.64 to ₹15,659.82 per hectare, can be provide additional income to farmers beyond what their agricultural and tree-based enterprises produce. Traditional agroforestry systems in the Himalayas are also important for carbon storage, which can help combat climate change. Carbon credits generated through these systems may have the potential to provide additional revenue sources for local communities in the near future. These systems are not only ecologically and environmentally beneficial but also culturally significant. They reflect the traditional knowledge and customs of particular groups, which have been passed down through centuries. Overall, traditional agroforestry systems in the Himalayan region, particularly the Western Himalayas, are critical for environmental sustainability, economic stability, and cultural heritage. They demonstrate a holistic approach to land management that balances environmental health and human requirements. This section reviews some of the common traditional agroforestry systems of the Western Himalayas (see Fig. 8.2). The name of the system typically indicates the dominant tree species component of that system.

Fig. 8.2 Overview of Western Himalayan traditional agroforestry systems (**a**) with dominance of *Grewia optiva* (**b**), *Boehmeria rugulosa* (**c**), *Celtis australis* (**d**), and *Quercus leucotrichophora* (**e**) in Western Himalaya

4.1 Grewia optiva-*Based Traditional Agroforestry Systems*

Grewia optiva (syn. *Grewia oppositifolia*) commonly known as Bhimal, is a valuable multipurpose agroforestry tree species native to the Western and Central Himalayas. It is widely used for nutritious green fodder during winter, fuelwood, and bast fiber. Its tender twigs are traditionally used to make baskets and other domestic products. Primarily found in agroforestry systems and rarely reported in nearby forests, *G. optiva* thrives at elevations between 500 and 2000 m above sea level, with frequent occurrences up to 1500 m. This species plays a crucial role in enhancing soil health and conservation, particularly when integrated with crops such as millet (Mehta et al. 2024). In the Western Himalayas, *G. optiva*-based agroforestry systems contribute significantly to the livelihood security of local communities by supplying essential resources, including fuelwood, fodder, and crops. Moreover, *G. optiva* enhances food security by improving crop yields when intercropped with peas and millets, with reported yields ranging from 5.72 to 6.04 t ha^{-1} in the Western Himalayas (Keprate et al. 2024). Beyond agriculture, *G. optiva* plays a vital role in carbon sequestration, helping mitigate climate change. *Grewia optiva*-based traditional agroforestry systems incorporating this species exhibit high carbon stocks, such as in the Garhwal Himalayas, where tree carbon stock ranges from 15.15 Mg ha^{-1} to 2.32 Mg ha^{-1}under different diameter classes (Tariyal et al. 2022). Similarly, in other regions of the Western Himalayas, *G. optiva*-based systems contribute to significant carbon storage (Bhardwaj et al. 2023). Overall, *G. optiva*-based agroforestry systems enhance soil fertility, support biodiversity, improve agricultural productivity, and play a crucial role in climate change mitigation, making them an essential component of sustainable agroecosystems in the Western Himalayas.

4.2 **Boehmeria rugulosa**-*Based Traditional Agroforestry Systems*

Although *Boehmeria rugulosa*-based traditional agroforestry systems are not explicitly detailed in publications, useful insights can be drawn from the general characteristics and benefits of traditional agroforestry practices in the regions. These systems are widely distributed in the foothills of the Western Himalayas and are commonly found on farmland at elevations of 1200–1400 m (Maikhuri et al. 2000). These traditional agroforestry systems also integrate *B. rugulosa* as a dominant tree species alongside crops and livestock. The resulting systems provide a range of ecological, economic, and social benefits. They enhance biodiversity, improve soil fertility, and supply essential resources such as food, fuelwood, and fodder. Their adaptability and sustainability play a crucial role in community welfare and environmental conservation.

4.3 **Celtis australis**-*Based Traditional Agroforestry Systems*

Celtis australis is a significant tree species in traditional agroforestry systems of the Western Himalayas, thriving at elevations of 500–2500 m. This multipurpose tree species supports rural livelihoods by providing fodder, fuel, and timber, while also enhancing soil fertility through its leaf litter. The integration of *C. australis* into agroforestry systems plays a crucial role in sustaining livestock during periods of fodder scarcity and contributes to the socio-economic stability of rural communities (Yadav and Bisht 2015; Singh et al. 2022). Its foliage is nutrient-rich, with crude protein content ranging from 9.17% to 16.97%, depending on seasonal and elevational variations (Singh et al. 2022). The tree also produces a substantial amount of leaf litter, with an annual leaf fall of 81.80 g m^{-2}, which returns essential nutrients to the soil and improves fertility (Panwar and Gupta 2015). *Celtis australis* is often found in association with other tree species such as *Ficus, Bauhinia,* and *Albizia*, forming diverse agroforestry systems that maximize land-use efficiency. Its adaptability to various soil types and elevations makes it a resilient choice for agroforestry, particularly in rainfed agricultural systems (Yadav and Bisht 2015). Megha et al. (2025) reported that *C. australis* stores 47.47 Mg ha^{-1} of carbon in its bole biomass, 18.13 Mg ha^{-1} in branch biomass, and 16.51 Mg ha^{-1} in root biomass, resulting in a total tree carbon stock of 82.11 Mg ha^{-1}. Additionally, the species produces 5.43 Mg ha^{-1} of oxygen annually in the Doon Valley of the Western Himalayan region.

4.4 Quercus leucotrichophora-*Based Traditional Agroforestry Systems*

In the Western Himalayas, traditional agroforestry systems dominated by *Quercus leucotrichophora* (commonly known as ban oak or banj) are widely distributed across elevations ranging from 1400 to 2200 m. These systems not only contribute to carbon sequestration but also serve as valuable sources of fuelwood, fodder, and medicinal resources (Joshi and Juyal 2017), thereby sustaining livelihoods. Beyond their ecological and economic benefits, these systems embody traditional knowledge and practices that help preserve biodiversity and cultural heritage, a feature observed in similar agroforestry systems worldwide. Agroforestry practices dominated by *Q. leucotrichophora*, such as agrisilviculture and silvopastoral systems, play a crucial role in carbon sequestration. Significant amounts of carbon are stored at different elevations, contributing to climate change mitigation. Specifically, *Q. leucotrichophora*-based traditional agroforestry systems sequester substantial carbon in tree biomass. Agrisilviculture systems store between 36.94 and 38.83 Mg ha^{-1}, while silvopastural systems store between 42.34 and 51.14 Mg ha^{-1}, depending on elevation (Kumar et al. 2021). These systems also enhance soil quality, as evidenced by the increased soil organic carbon and other soil properties (Kumar et al. 2021).

4.5 *Homegarden*

Traditional homegardens in the Western Himalayas constitute a vital agroforestry system that supports plant biodiversity, cultural practices, and sustainable livelihoods. These homegardens are characterized by a diverse range of plant species, including fruit trees, cucurbitaceous vegetables, field crops, and ethnomedicinal plants. They are often integrated with domestic animals, which play an important role in the health and cultural traditions of local communities (Bargali and Bargali 2020). Beyond their ecological importance, homegardens serve as repositories of traditional knowledge and contribute significantly to rural livelihoods and biodiversity conservation. They are typically managed with minimal capital investment and simple technology, relying heavily on family labor, particularly women, to maintain their sustainability and productivity. Homegardens also have considerable potential for carbon sequestration while simultaneously producing agricultural yields and tree-based products. Research findings of Bargali and Bargali (2020) indicate that soil carbon sequestration in these systems ranges between 22.56 and 81.51 Mg ha^{-1}, with values increasing with elevation in smaller homegardens. However, carbon sequestration in tree biomass remains largely unexplored in the Western Himalayan region.

5 Carbon Sequestration in Traditional Agroforestry Systems of Western Himalaya

Agroforestry systems in the Western Himalaya region play a crucial role in carbon sequestration and contribute significantly to global climate change mitigation efforts. Elevation is a key factor influencing carbon sequestration potential of agroforestry systems in this region. Studies indicate that lower elevation generally exhibit higher carbon stocks due to favorable climatic conditions and greater tree densities. For example, in the Tehri District of Uttarakhand, agrihortisilviculture systems showed higher carbon stocks at lower elevation compared to middle and upper elevation (Vikrant et al. 2021). Similarly, in the Garhwal Himalayas, tree biomass carbon stocks were higher at lower elevations, with *Grewia optiva*-based agrisilviculture systems sequestering 23.29 Mg ha^{-1} at lower elevations versus 18.09 Mg ha^{-1} at upper elevations (Tariyal et al. 2022). However, some studies suggest that certain systems may store more carbon at higher elevations. In the Kumaun Himalayas, soil carbon stocks increased with elevation, particularly in homegarden systems (Bargali and Bargali 2020). Additionally, Kumar et al. (2025) reported higher tree biomass carbon stocks at higher elevations, emphasizing the complexity of elevational effects on carbon sequestration. Tree species composition also significantly influences the carbon sequestration potential of traditional agroforestry systems in the Western Himalayas. Species such as *G. optiva*, *C. australis*, and *Q. leucotrichophora* are commonly found in these traditional systems and are known for their high carbon storage capacity. These multipurpose species not only sequester carbon but also provide fodder, fuel, and fiber, making them valuable components of agroforestry systems (Kumar et al. 2021; Tariyal et al. 2022). In the Tehri Garhwal district, *Q. leucotrichophora*-based systems demonstrated significant carbon sequestration potential, with higher carbon stocks at elevations between 1800 and 2200 m amsl (above mean sea level) compared to lower elevations (Kumar et al. 2021). This underscores the importance of species selection in optimizing carbon sequestration.

Soil organic carbon (SOC) is another critical component of total carbon stocks in agroforestry systems. Research consistently shows that SOC contributes more to total carbon stocks than plant biomass (see Table 8.1). For instance, in the Garhwal Himalayas, soil carbon stocks in traditional agroforestry systems reached 56.74 Mg ha^{-1}, surpassing the carbon stored in tree biomass (Kumar et al. 2012). Likewise, in the Kumaun Himalayas, SOC stocks were highest in homegarden systems, emphasizing the importance of soil management practices for carbon sequestration (Bargali and Bargali 2020). Soil depth and elevation also influence SOC stocks; as SOC decreases with soil depth and elevation, while it is also reported that SOC increases with elevation (Kumar et al. 2021).

Among different agroforestry systems, agrihortisilviculture systems, which integrate trees, crops, and fruits, have shown high carbon sequestration potential. In the Tehri District of Uttarakhand, agrihortisilviculture systems exhibited the highest carbon stock (32.56 t ha^{-1}) among all studied agroforestry systems (Vikrant et al.

Table 8.1 Reported tree species and total carbon stock in tree biomass and soil across various traditional agroforestry systems in the Western Himalayas

Location/Region	Type of traditional agroforestry system	Reported species in traditional agroforestry	Carbon Stock (Mg ha^{-1})		References
			In tree biomass	Soil (depth in cm)	
Uttarkashi district, Uttarakhand Himalaya	Agrisilviculture (800–1200 m elevation)	*C. australis, G. optiva, Melia azedarach, Prunus cerasoides, Pyrus pashia,*	30.49	41.48 (0–15 cm)	Kumar et al. (2025)
				40.72 (15–30 cm)	
				40.90 (30–45 cm)	
	Agrisilviculture (1200–1600 m elevation)	*C. australis, G. optiva, M. azedarach, P. cerasoides, P. pashia, Q. leucotrichophora*	33.80	54.60 (0–15 cm)	
				45.53 (15–30 cm)	
				42.46 (30–45 cm)	
	Agrisilviculture (1600–2000 m elevation)	*C. australis, G. optiva, M. azedarach, P. cerasoides, P. pashia, Q. leucotrichophora*	34.43	55.60 (0–15 cm)	
				56.18 (15–30 cm)	
				54.81 (30–45 cm)	
Tehri Garhwal, Uttarakhand Himalaya	Agrisilviculture (700–1200 m elevation)	*Bauhinia variegata, Boehmeria rugulosa, C. australis, Cinnamomum tamala, Ficus auriculata, Ficus benghalensis, Ficus religiosa, Ficus semicordata, G. optiva, M. azedarach, Ougeinia oojeinensis, Toona ciliata*	18.94	46.55 (0–15 cm)	Singh et al. (2024)
				44.10 (15–30 cm)	
	Agrisilviculture (1200–1700 m elevation)	*Azadirachta indica, B. rugulosa, Bauhinia variegata, Bauhinia semla, C. australis, Ficus religiosa, F. roxburghii, F. palmata, G. optiva, Lannea coromandelica, Mangifera indica, Morus serrata, T. ciliata*	23.76	59.20 (0–15 cm)	
				58.64 (15–30 cm)	
	Agrisilviculture (1700–2200 m elevation)	*Bombex ceiba, C. australis, Ficus palmata, Ficus roxburghii, G. optiva, Juglans regia, Lannea coromandelica, Morus serrata, P. cerasoides, Phyllanthus emblica, Q. leucotrichophora, T. ciliata*	27.01	51.46 (0–15 cm)	
				50.80 (15–30 cm)	

(continued)

Table 8.1 (continued)

Location/Region	Type of traditional agroforestry system	Reported species in traditional agroforestry	Carbon Stock (Mg ha^{-1})		References
			In tree biomass	Soil (depth in cm)	
Tehri Garhwal, Uttarakhand Himalaya	Agrisilviculture (800–1200 m elevation)	*Bauhinia variegata, C. australis, F. auriculata, F. palmata, G. optiva, P. cerasoides, P. pashia, Q. leucotrichophora, T. ciliata*	28.49	–	Tariyal et al. (2022)
	Agrisilviculture (800–1200 m elevation)	*C. australis, F. palmata, G. optiva, J. regia, Malus domestica, Myrica esculenta, P. pashia, Pinus roxburghii, Prunus armeniaca, Q. leucotrichophora, Rhododendron arboreum, T. ciliata*	25.93	–	
Tehri Garhwal, Uttarakhand Himalaya	Agrisilviculture (1400–1800 m elevation)	*C. australis, F. auriculata, G. optiva, J. regia, M. domestica, P. armeniaca, P. cerasoides, P. roxburghii, Prunus domestica, P. pashia, Q. leucotrichophora, R. arboreum*	36.94	40.88 (0–15 cm)	Kumar et al. (2021)
				43.24 (15–30 cm)	
				44.25 (30–45 cm)	
	Silvopasture (1400–1800 m elevation)	*C. australis, Cornus capitata, F. auriculata, F. palmata, G. optiva, J. regia, P. armeniaca, P. cerasoides, P. roxburghii, Q. leucotrichophora, R. arboreum, T. ciliata*	51.14	40.69 (0–15 cm)	
				40.53 (15–30 cm)	
				38.18 (30–45 cm)	
	Agrisilviculture (1800–2200 m elevation)	*C. australis, Cornus capitata, F. auriculata, G. optiva, J. regia, M. domestica, Myrica esculenta, P. armeniaca, P. cerasoides, P. domestica, P. pashia, P. roxburghii, Pyrus pyrifolia, Q. leucotrichophora, R. arboreum*	38.84	49.52 (0–15 cm)	
				39.22 (15–30 cm)	
				30.45 (30–45 cm)	
	Silvopasture (1800–2200 m elevation)	*C. australis, C. capitata, F. auriculata, F. palmata, G. optiva, J. regia, M. domestica, M. esculenta, P. armeniaca, P. cerasoides, P. pashia, P. roxburghii, Prunus persica, Q. leucotrichophora, R. arboreum, T. ciliata,*	42.34	49.75 (0–15 cm)	
				42.38 (15–30 cm)	
				36.36 (30–45 cm)	

Kirtinagar block of Tehri Garhwal, Uttarakhand	Agrisilviculture (550–600 m elevation)	*G. optiva*-dominated traditional agroforestry systems	20.47	16.38 (0–15 cm)	Rana et al. (2020)
				12.72 (15–30 cm)	
	Agrisilviculture (600–900 m elevation)		20.83	16.41 (0–15 cm)	
				13.23 (15–30 cm)	
	Agrisilviculture (1200–1500 m elevation)		19.17	21.45 (0–15 cm)	
				15.37 (15–30 cm)	
Chamba district, Himachal Pradesh	Agrisilviculture	*Abies pindrow, Albizia chinensis, B. ceiba, B. variegata, Broussonetia papyrifera, C. australis, Cedrus deodara, F. palmata, G. optiva, Melia composita, Morus alba, P. pashia, Picea smithiana, Pinus wallichiana, Populus ciliata, Q. leucotrichophora, Quercus semecarpifolia, Robinia pseudoacacia, T. ciliata*	24.51	25.37 (0–30 cm)	Sharma et al. (2023)
	Agrihorticulture	*Citrus* sp., *J. regia, Litchi chinensis, M. indica, M. domestica, P. armeniaca, Prunus amygdalus, P. domestica, P. persica, Psidium guajava, Pyrus communis*	17.46	22.73 (0–30 cm)	
	Agrisilvihorticulture	*Aesculus indica, Albizia chinensis, B. variegata, C. australis, C. deodara, F. palmata, G. optiva, Leucaena leucocephala, Melia composita, Morus alba, P. roxburghii, P. wallichiana, P. ciliata, Q. leucotrichophora, Q. semecarpifolia, Quercus dilatata, Salix tetrasperma, T. ciliata, Vitex negundo*	22.35	31.66 (0–30 cm)	

(continued)

Table 8.1 (continued)

Location/Region	Type of traditional agroforestry system	Reported species in traditional agroforestry	Carbon Stock (Mg ha^{-1})		References
			In tree biomass	Soil (depth in cm)	
Chamba district, Himachal Pradesh	Agrihortisilviculture	*A. chinensis, A. indica, B. variegata, C. australis, C. deodara, Citrus sp, F. palmata, F. palmata, G. optiva, J. regia, L. leucocephala, M. alba, M. composita, M. domestica, M. indica, P. armeniaca, P. domestica, P. guajava, P. persica, P. roxburghii, P. wallichiana, P. ciliata, P. granatum, Pyrus communis, Q. leucotrichophora, Q. semecarpifolia, Quercus dilatata, S. tetrasperma, T. ciliata, V. negundo*	29.83	25.92 (0–30 cm)	Sharma et al. (2023)
	Hortiagriculture	*Corylus jacquemontii, J. regia, M. domestica, P. persica, Pyrus communis*	22.73	22.25 (0–30 cm)	
	Silvopastoral	*Albizia chinensis, B. variegata, Broussonetia papyrifera, Butea monosperma, C. australis, F. palmata, F. roxburghii, L. leucocephala, Melia composita, Olea cuspidata, P. cerasoides, P. pashia, P. roxburghii, Pistacia integerrima, T. ciliata, V. negundo,*	24.00	35.77 (0–30 cm)	
	Pastoral silviculture	*A. chinensis, A. indica, B. monosperma, B. variegata, C. australis, C. deodara, Crataegus songarica, Erythrina suberosa, F. palmata, Olea cuspidata, P. ciliata, P. pashia, P. roxburghii, Pinus gerardiana, Pistacia integerrima, S. tetrasperma, T. ciliata, Ulmus wallichiana, V. negundo*	17.36	28.15 (0–30 cm)	
	Hortipastoral	*J. regia, Malus domestica, Pyrus communis*	19.94	23.96 (0–30 cm)	

Kwalkhad Watershed, Himachal Pradesh	Agrisilvihorticulture	*Acacia catechu, C. australis, Dalbergia sissoo, Ficus* sp., *G. optiva, P. armeniaca, P. granatum, Q. leucotrichophora, T. ciliata*	14.78	52.14 (0–20 cm)	Goswami et al. (2014)
				95.46 (0–40 cm)	
	Agrihortisilviculture	*A. chinensis, C. australis, Citrus* sp., *G. optiva, P. granatum, Prunus* sp., *Pyrus communis, T. ciliata*	14.45	46.56 (0–20 cm)	
				84.14 (0–40 cm)	
	Agrisilviculture	*C. australis, Ficus* sp., *G. optiva, Melia azedarach, P. pashia, T. ciliata, Ulmus villosa*	10.64	50.58 (0–20 cm)	
				92.65 (0–40 cm)	
	Agrihorticulture	*Citrus* sp., *Emblica officinalis, P. armeniaca, P. granatum*	12.15	47.73 (0–20 cm)	
				90.07 (0–40 cm)	
	Silvopastoral	*A. catechu, B variegata, C australis, P roxburghii, Q leucotrichophora*	3.36	63.54(0–20 cm)	
				115.45 (0–40 cm)	
Ladakh and Karakoram ranges, Himalayas	Agrisilviculture	–	11.57	–	Namgial et al. (2022)
	Agrihortisilviculture	–	13.58	–	
	Silvopasture	–	13.21	–	
	Hortisilvipasture	–	15.44	–	
	Agrihorticulture	–	11.56	–	

(continued)

Table 8.1 (continued)

Location/Region	Type of traditional agroforestry system	Reported species in traditional agroforestry	Carbon Stock (Mg ha^{-1})		References
			In tree biomass	Soil (depth in cm)	
Karsog Tehsil of Mandi district, Himachal Pradesh	Agrisilviculture	*C. australis, Ficus* sp., *Ficus* sp., *G. optiwa, Morus alba,* *P. roxburghii, Q. leucotrichophora, Robiniapseudoacacia, T. ciliata*	*10.17*	–	Sharma (2024)
	Agrihorticulture	*J. regia, M. domestica, P. armeniaca, P. domestica, P. granatum, Prunus dulcis*	7.92	–	
	Agrihortisilviculture	*C. australis, Ficus* sp., *M. alba, M. domestica, P. armeniaca,* *P. domestica, P. granatum, P. roxburghii, Q. leucotrichophora, R. arboretum, R. pseudoacacia, T. ciliata*	6.99	–	
	Agrisilvihorticulture	*C. australis, Ficus* sp., *M. domestica, P. armeniaca, P. domestica, P. granatum, P. roxburghii, Q. leucotrichophora,* *R. arboretum, R. pseudoacacia*	9.74	–	
	Hortipastoral	*M. domestica, P. dulcis, P. armeniaca, P. domestica, P. granatum*	9.16	–	
	Silvopastoral	*C. australis, Ficus sp., P.roxburghii, Q. eucotrichophora, R. arboretum, R. pseudoacacia, T. ciliate*	6.67	–	
	Pastoral silviculture	*C. australis, C. deodara, P. roxburghii, P. wallichiana, Q. leucotrichophora, R. arboreum, T. ciliata*	4.55	–	
Kupwara district, Kashmir valley	Agrisilviculture	*Juglans* sp., *Malus* sp., *Populus* sp., *Robinia* sp., *Salix* sp., etc.	25.99	–	Ajit et al. (2017)

2021). Similarly, silvopastoral systems, which combine trees with pasture, have demonstrated notable carbon sequestration potential. In the Garhwal Himalayas, silvopastoral systems sequestered 51.14 Mg ha^{-1} of carbon at lower elevations, significantly surpassing agrisilviculture systems (Kumar et al. 2021). The high tree density in silvopastoral systems enhances carbon storage in both tree biomass and soil. Agrisilvihorticulture systems, which integrate crops, pasture, and horticultural species, have also been studied for their carbon sequestration potential. In the Kwalkhad Watershed of Himachal Pradesh, agrisilvihorticulture systems sequestered 14.78 Mg ha^{-1} of carbon, comparable to agrihortisilviculture systems (Goswami et al. 2014). These systems help in maintaining soil fertility and biodiversity while enhancing carbon sequestration. In traditional agroforestry systems, the average total carbon stock in tree biomass was 32.56 t ha^{-1}, while the average soil carbon stock was 56.74 t ha^{-1} (Kumar et al. 2012). Panwar et al. (2022) reported that carbon stocks in tree biomass of agrisilviculture, agrihorticulture, and silvopasture ranged from 15.94–202.59, 18.19–171.95, and 43.51–136.42 Mg ha^{-1}, respectively, in the Western Himalayan states of India. The variation in tree density directly influences carbon stocks, as higher tree densities typically result in more significant biomass accumulation and carbon sequestration potential (Goswami et al. 2014). The total carbon pool in the soil was assessed in two layers: 0–20 cm and 20–40 cm. In the 0–20 cm layer, abandoned land exhibited the highest total carbon pool at 86.75 Mg ha^{-1}, followed by silvopasture at 63.54 Mg ha^{-1}. Different agroforestry systems showed statistically similar carbon pools, reinforcing their effectiveness in maintaining soil carbon stocks. By contrast, pure agriculture exhibited the lowest carbon pool at 35.93 Mg ha^{-1}, suggesting that agroforestry systems are more effective in carbon storage than conventional agricultural practices.

Table 8.1 shows the traditional agroforestry systems from several regions of the Western Himalayas, emphasizing tree species diversity and carbon stored in tree biomass and soil. Various agroforestry types, such as agrisilviculture, silvopasture, and agrihorticulture, support a diverse range of tree species, with *C. australis*, *G. optiva*, *Q. leucotrichophora*, *Prunus cerasoides*, *Ficus* sp., and *Toona ciliata* being commonly reported forest tree species, and *Malus domestica*, *Prunus armeniaca*, *Prunus domestica,* and *Punica granatum* being commonly reported horticulture trees. The carbon stock in tree biomass varies significantly across elevation gradients and agroforestry types, with the highest recorded tree biomass carbon stock of 51.14 Mg ha^{-1} was observed in the silvopastoral system at 1400–1800 m in Tehri Garhwal, and the lowest (3.36 Mg ha^{-1}) found in the silvopastoral system of the Kwalkhad Watershed, Himachal Pradesh. Soil carbon stock also varies significantly, with the maximum value of 115.45 Mg ha^{-1} observed in the silvipastoral system at 0–40 cm soil depth of Kwalkhad Watershed and the lowest (12.72 Mg ha^{-1} at 15–30 cm depth) in the lower elevation agrisilviculture systems of Kirtinagar, Uttarakhand. Soil carbon stocks tend to rise with elevation in Uttarkashi and Tehri Garhwal, whereas in Himachal Pradesh, agroforestry systems such as agrohortisilviculture and silvopasture have relatively high soil carbon stocks. In the Kashmir Valley, tree biomass carbon stock is considerable (25.99 Mg ha^{-1}), but in Ladakh and the Karakoram ranges, it stays relatively low (11.57–15.44 Mg ha^{-1}), possibly due to difficult climatic circumstances. Kumar et al. (2021) also reported that the *Q. leucotrichophora*-based traditional agroforestry systems of Garhwal Himalaya have the potential to store 30–45% of carbon stock compared to natural forestry areas dominated by the same species. Overall, traditional agroforestry systems in the Western Himalayas play an important role in carbon sequestration, with variations depending on tree species composition, elevation, and land-use patterns. Silvopastoral and agrisilviculture systems have higher carbon storage rates, underlining their potential for climate change mitigation and sustainable land management.

6 Challenges and Opportunities for Traditional Agroforestry Systems in the Western Himalaya

Agroforestry research and development aim to improve agricultural economies, generate employment, alleviate poverty, achieve zero hunger, ensure food security, mitigate climate change, reduce greenhouse gas emissions, and promote sustainable land use, paving the way with increased greenery toward a evergreen revolution. Despite the significant benefits of agroforestry systems in the Western Himalayas, several challenges need to be addressed to realize their full potential. These challenges include limited awareness of the advantages of agroforestry, restricted market access to agroforestry products, and the adverse effects of climate change on agricultural productivity. Many farmers in the Western Himalayas are unaware of agroforestry's potential to enhance biodiversity, improve soil health, and increase income. This lack of awareness often leads to the continued use of monoculture practices, which are less sustainable and more vulnerable to the effects of climate change, while agroforestry practices are more climate-resilient (Vijayanathan et al. 2022). Another challenge is limited access to markets for agroforestry products. Many farmers in the Western Himalayas face difficulties in accessing markets for their agroforestry products—particularly timber and other wood products—which limits their ability to generate income from these systems. This challenge is particularly significant for smallholder farmers, who often lack the resources and knowledge to navigate complex market systems. Climate change presents another major obstacle facing agroforestry systems in the Western Himalayas. Crop and tree growth conditions are changing due to rising temperatures, shifting precipitation patterns, and an increase in the frequency of extreme weather events. This makes it difficult for farmers to maintain the productivity of their agroforestry systems and requires the adoption of climate-resilient practices. Agroforestry systems themselves are vulnerable to the impacts of climate change (Quandt et al. 2023). Despite these challenges, several opportunities exist to enhance the adoption and effectiveness of agroforestry systems in the Western Himalayas. Key strategies include promoting climate-resilient tree and crop species, developing market linkages for agroforestry products, and providing training and technical support to farmers. Agroecotourism also offers considerable potential for generating income through diverse agroforestry systems such as agrihorticulture and agrisilviculture. These systems not only create economic opportunities but also help in rural development. Agroforestry also contributes significantly to women's empowerment by providing long-term income sources and greater independence. It also helps to address the high demand for fodder and contributes to milk and meat production through animal husbandry, thereby strengthening the rural economy. Farmers benefit economically from agroforestry by diversifying income streams with timber, fuelwood, fodder, medicinal plants, and high-value crops such as apples and walnuts. Furthermore, agroforestry promotes biodiversity, protects wildlife habitats, preserves local knowledge, enhances resilience to extreme weather, regulates microclimates, and mitigates risks from droughts and floods, making it a sustainable and environmentally beneficial practice.

7 Conclusion

The Western Himalayas form part of one of the world's most diverse, resource-rich, and fragile ecosystems. Indigenous communities in this region have long practiced a variety of land-use systems, among which traditional agroforestry holds a prominent significance. These ancient practices are characterized by the natural regeneration and protection of indigenous tree species on farmlands. Over centuries, these systems have been refined into diverse structures and functions that support both ecological integrity and rural livelihoods. Traditional agroforestry systems in the Himalayas include agrisilviculture, agrihorticulture, hortiagriculture, agrohortisilviculture, silvopastoral, hortipastoral, and other integrated practices. While primarily developed to sustain Himalayan rural livelihoods, these systems have emerged in the context of modern climate change as sustainable solutions for mitigating global warming and playing an imperative role in carbon sequestration by capturing and long-term storage of excess atmospheric carbon in tree biomass. While trees are central to carbon storage in agroforestry systems, other components, such as crops, shrubs, herbs, and soil, also contribute significantly to carbon sequestration and retention. Studies in the Western Himalayas report tree biomass carbon stocks ranging from 3.36 to 51.14 Mg ha^{-1} and soil carbon storage ranging from 12.72 to 115.54 Mg ha^{-1} at different soil depths. Despite the numerous benefits of agroforestry, several challenges continue to limit its full potential. Overall, traditional agroforestry has long served as a lifeline for rural livelihoods and economic prosperity in the Himalayas, delivering numerous direct and indirect advantages.

Declarations

Financial Interests The authors declare they have no financial interests.

Competing Interests The authors declare that they have no known competing financial interests or personal relationships that could have appeared to influence the work reported in this paper.

Ethics Approval Not applicable.

References

Ajit HAK, Dhyani SK, Bhat GM, Malik AR, Dutt V, Masoodi TH, Uma JA (2017) Quantification of carbon stocks and sequestration potential through existing agroforestry systems in the hilly Kupwara district of Kashmir valley in India. Curr Sci 113(4):782–785

Akter R, Hasan MK, Kabir KH, Darr D, Roshni NA (2022) Agroforestry systems and their impact on livelihood improvement of tribal farmers in a tropical moist deciduous forest in Bangladesh. Trees Forests People 9:100315. https://doi.org/10.1016/j.tfp.2022.100315

Atangana A, Khasa D, Chang S, Degrande A (2014) Sociocultural aspects of agroforestry and adoption. In: Tropical agroforestry. Springer, Netherland, Dordrecht, pp 323–332. https://doi.org/10.1007/978-94-007-7723-1_17

Awazi NP, Njamnjubo NA, Ambebe TF (2024) Agroforestry for ecosystem services: assessing the role of agroforestry for water regulation. Forestist, Published online July 22. https://doi.org/10.5152/forestist.2024.24032

Bargali K, Bargali SS (2020) Effect of size and altitude on soil organic carbon stock in homegarden agroforestry system in Central Himalaya, India. Acta Ecologica Sinica 40(6):483–491. https://doi.org/10.1016/j.chnaes.2020.10.002

Beule L, Lehtsaar E, Corre MD, Schmidt M, Veldkamp E, Karlovsky P (2020) Poplar rows in temperate agroforestry croplands promote bacteria, fungi and denitrification genes in soils. Front Microbiol 10:3108. https://doi.org/10.3389/fmicb.2019.03108

Bhardwaj DR, Sharma P, Kumar D, Panwar P, Kumar A, Pala NA, Rajput BS, Kumar R, Negi V, Rajput P, Salve A (2023) Carbon stock inventory and biomass production in different land use systems of NorthWestern Himalaya. In: Climate change in the Himalayas. Academic Press, pp 217–233

Bhushan P, Bhardwaj DR, Sharma P, Kumar D, Midhun N, Kumar AKS (2023) Fruit based agroforestry systems in Himachal Pradesh. Van Sangyan 10(2):16–23

Cairns MA, Brown S, Helmer EH, Baumgardner GA (1997) Root biomass allocation in the world's upland forests. Oecologia 111:1–11. https://doi.org/10.1007/s004420050201

Chavan SB, Newaj R, Rizvi RH, Ajit PR, Alam B, Handa AK, Dhyani SK, Jain A, Tripathi D (2020) Reduction of global warming potential vis-a-vis greenhouse gases through traditional agroforestry systems in Rajasthan, India. Environ Dev Sustain 23:4573–4593. https://doi.org/10.1007/s10668-020-00788

Ghale B, Mitra E, Sodhi HS, Verma AK, Kumar S (2022) Carbon sequestration potential of agroforestry systems and its potential in climate change mitigation. Water Air Soil Pollut 233(7):228. https://doi.org/10.1007/s11270-022-05689-4

Goswami S, Verma KS, Kaushal R (2014) Biomass and carbon sequestration in different agroforestry systems of a Western Himalayan watershed. Biol Agric Hortic 30(2):88–96. https://doi.org/10.1080/01448765.2013.855990

Hemel SAK, Hasan MK, Akter R, Roshni NA, Sayem A, Rasul S, Islam MT (2024) Delving into Piper chaba-based agroforestry system in northern Bangladesh: ecosystem services, environmental benefits and potential conservation initiatives. Trees For People 16:100523. https://doi.org/10.1016/j.tfp.2024.100523

Hertzberg M, Schreuder H (2016) Role of atmospheric carbon dioxide in climate change. Energy Environ 27(6–7):785–797. https://doi.org/10.1177/0958305X16674637

IPCC (2003) In: Penman J, Gytarsky M, Hiraishi T, Krug T, Kruger D, Pipatti R, Buendia L, Miwa K, Ngara T, Tanabe K, Wagner F (eds) Good practice guidance for land use, land-use change and forestry. IPCC National Greenhouse Gas Inventories Programme, Institute of Global Environmental Strategies, Hayama, Kanagawa, Japan, 590p

Jamnadass R, Place F, Torquebiau E, Malezieux E, Iiyama M, Sileshi GW, Kehlenbeck K, Masters E, McMullin S, Dawson I (2013) Agroforestry for food and nutritional security. Unasylva 64(2):23–29

Jinger D, Kaushal R, Kumar R, Paramesh V, Verma A, Shukla M, Chavan SB, Kakade V, Dobhal S, Uthappa AR, Roy T (2023) Degraded land rehabilitation through agroforestry in India:

achievements, current understanding and future prospectives. Front Ecol Evol 11:1088796. https://doi.org/10.3389/fevo.2023.1088796

Joshi AK, Juyal D (2017) Traditional and ethnobotanical uses of *Quercus leucotrichophora* a. Camus (*Quercus oblongata* D. Don) in Kumaun and Garhwal regions of Uttarakhand, India: a review. Int J Herb Med 5(5):06–08

Joshi G, Negi GC (2011) Quantification and valuation of forest ecosystem services in the Western Himalayan region of India. Int J Biodivers Sci Ecosyst Serv Manag 7(1):2–11. https://doi.org/10.1080/21513732.2011.598134

Keprate A, Bhardwaj DR, Sharma P, Kumar D, Rana RK (2024) Biomass partitioning, carbon storage and pea (*Pisum sativum* L.) crop production under a *Grewia optiva*-based agroforestry system in the Mid-Hills of the NorthWestern Himalayas. Sustainability 16(17):7438. https://doi.org/10.3390/su16177438

Kumar M, Anemsh K, Sheikh MA, Raj AJ (2012) Structure and carbon stock potential in traditional agroforestry system of Garhwal Himalaya. J Agric Technol 8(7):2187–2200

Kumar BM, Handa AK, Dhyani SK, Arunachalam A (2018) Agroforestry in the Indian Himalayan region: an overview. In: Gordon AM, Newman SM, Coleman BRW (eds) Temperate agroforestry systems, 2nd edn. CAB International, Wallingford, pp 153–172

Kumar S, Bijalwan A, Singh B, Rawat D, Yewale AG, Riyal MK, Thakur TK (2021) Comparison of carbon sequestration potential of *Quercus leucotrichophora*–based agroforestry systems and Natural Forest in Central Himalaya, India. Water Air Soil Pollution 232:350. https://doi.org/10.1007/s11270-021-05294-x

Kumar S, Kumar S, Khanduri VP, Singh B, Joshi R, Riyal MK, Rawat D, Kumar KS (2025) Tree diversity, carbon sequestration and production potential of Oryza sativa L. in traditional agroforestry systems of Garhwal Himalaya, India. Carbon Res 4(1):6. https://doi.org/10.1007/s44246-024-00158-5

Kuyah S, Öborn I, Jonsson M (2017) Regulating ecosystem services delivered in agroforestry systems. In: Dagar J, Tewari V (eds) Agroforestry. Springer, Singapore. https://doi.org/10.1007/978-981-10-7650-3_33

Maikhuri RK, Semwal RL, Rao KS, Singh K, Saxena KG (2000) Growth and ecological impacts of traditional agroforestry tree species in Central Himalaya, India. Agrofor Syst 48:257–271

Mayer S, Wiesmeier M, Sakamoto E, Hubner R, Cardinael R, Kuhnel A, Kogel-Knabner I (2022) Soil organic carbon sequestration in temperate agroforestry systems—a meta-analysis. Agric Ecosyst Environ 323:107689. https://doi.org/10.1016/j.agee.2021.107689

Mbow C, Van Noordwijk M, Luedeling E, Neufeldt H, Minang PA, Kowero G (2014) Agroforestry solutions to address food security and climate change challenges in Africa. Curr Opin Environ Sustain 6:61–67. https://doi.org/10.1016/j.cosust.2013.10.014

Megha VR, Kaushal R, Islam S, Mandal U, Mehta H, Rajkumar, Tomar JMS, Gupta AK, Barh A, VM P, Singhal V, Pandey R, Khatri D, Madhu M (2025) Allometric scaling and carbon sequestration in agroforestry species of the Western Himalayas: a model-based approach. Environ Monit Assess 197(3):311. https://doi.org/10.1007/s10661-025-13735-x

Mehta H, Rathore AC, Tomar JMS, Mandal D, Kumar P, Kumar S, Sharma SK, Kaushal R, Singh C, Chaturvedi OP, Madhu M (2024) Minor millets based agroforestry of multipurpose tree species of Bhimal (*Grewia optiva* Drummond JR ex Burret) and Mulberry (*Morus alba* L.) for resource conservation and production in North Western Himalayas–10-year study. Agric Ecosyst Environ 359:108761. https://doi.org/10.1016/j.agee.2023.108761

Nair PKR, Nair VD, Mohan Kumar B, Showalter JM (2010) Carbon sequestration in agroforestry systems. In: Advances in Agronomy, pp 237–307. https://doi.org/10.1016/s0065-2113(10)08005-3

Namgial J, Prabhakar M, Gupta SK, Gautam KL (2022) Productivity and carbon storage potential of different land use systems in Leh Region of Himalayan Cold Desert. Indian J Ecol 49(6):2343–2346

Nowak DJ, Hirabayashi S, Bodine A, Greenfield E (2014) Tree and forest effects on air quality and human health in the United States. Environ Pollut 193:119–129. https://doi.org/10.1016/j.envpol.2014.05.028

Pancholi R, Yadav R, Gupta H, Vasure N, Choudhary S, Singh MN, Rastogi M (2023) The role of agroforestry systems in enhancing climate resilience and sustainability-a review. Int J Environ Clim Change 13(11):4342–4353. https://www.sdiarticle5.com/review-history/109110

Pandey R (2009) Forest resource utilization by tribal community of Jaunsar, Uttarakhand. Indian For 135(5):655

Pantera A, Mosquera-Losada MR, Herzog F, den Herder M (2021) Agroforestry and the Environment. Agrofor Syst 95(5):767–774. https://doi.org/10.1007/s10457-021-00640-8

Panwar VP, Gupta MK (2015) Litter fall, nutrient return and soil fertility under *Celtis australis*: an indigenous agroforestry tree species in Himachal Pradesh, India. Int J Environ Sci 6(3):318–329

Panwar P, Mahalingappa DG, Kaushal R, Bhardwaj DR, Chakravarty S, Shukla G, Thakur NS, Chavan SB, Pal S, Nayak BG et al (2022) Biomass production and carbon sequestration potential of different agroforestry systems in India: a critical review. Forests 13:1274. https://doi.org/10.3390/f13081274

Quandt A, Neufeldt H, Gorman K (2023) Climate change adaptation through agroforestry: opportunities and gaps. Curr Opin Environ Sustain 60:101244. https://doi.org/10.1016/j.cosust.2022.101244

Rana K, Kumar M, Kumar A (2020) Assessment of annual shoot biomass and carbon storage potential of Grewia optiva: an approach to combat climate change in Garhwal Himalaya. Water Air Soil Pollut 231:1–13. https://doi.org/10.1007/s11270-020-04825-2

Richards M, Bruun TB, Campbell B, Gregersen LE, Huyer S, Kuntze V, Madsen STN, Oldvig MB, Vasileiou I (2016) How countries plan to address agricultural adaptation and mitigation: an analysis of intended nationally determined contributions. CCAFS dataset version 1.3. Copenhagen, Denmark: CGIAR Research Program on Climate Change, Agriculture and Food Security (CCAFS)

Semere M (2019) Biomass and soil carbon stocks assessment of agroforestry systems and adjacent cultivated land, in Chehawereda, Gurage zone, Ethiopia. Int J Res Agric For 6(7):17–24

Sharma P (2024) Economics and carbon sequestration potential of existing agroforestry systems in Karsog Tehsil of Himachal Pradesh. Master's thesis, Agroforestry and Silviculture, Dr. Yashwant Singh Parmar University of Horticulture and Forestry, Solan, Himachal Pradesh, India

Sharma H, Pant KS, Bishist R, Gautam KL, Dogra R, Kumar M, Kumar A (2023) Estimation of biomass and carbon storage potential in agroforestry systems of North Western Himalayas, India. Catena 225:107009. https://doi.org/10.1016/j.catena.2023.107009

Sileshi GW, Mafongoya PL, Nath AJ (2020) Agroforestry systems for improving nutrient recycling and soil fertility on degraded lands. In: Agroforestry for degraded landscapes: recent advances and emerging challenges, vol 1, pp 225–253. https://doi.org/10.1007/978-981-15-4136-0_8

Singh SP (2007) Himalayan forests ecosystem services: incorporating in national accounting, central himalayan environment association (CHEA). Nainital, Uttarakhand

Singh B, Kumar M, Cabral-Pinto MM, Bhatt BP (2022) Seasonal and altitudinal variation in chemical composition of *Celtis australis* L. tree foliage. Land 11(12):2271. https://doi.org/10.3390/land11122271

Singh N, Riyal MK, Singh B, Khanduri VP, Rawat D, Singh C, Pinto MMC, Kumar M (2024) Carbon sequestration potential of agroforestry versus adjoining forests at different altitudes in the Garhwal Himalayas. Atmosphere 15(3):313. https://doi.org/10.3390/atmos15030313

Tariyal N, Bijalwan A, Chaudhary S, Singh B, Dhanai CS, Tewari S, Kumar M, Kumar S, Cabral Pinto MM, Thakur TK (2022) Crop production and carbon sequestration potential of Grewia *optiva*-based traditional agroforestry systems in Indian Himalayan region. Land 11(6):839. https://doi.org/10.3390/land11060839

Teo HC, Lamba A, Ng SJ, Nguyen AT, Dwiputra A, Lim AJY, Nguyen MN, Tor-ngern P, Zeng Y, Dewi S, Koh LP (2025) Reduction of deforestation by agroforestry in high carbon stock forests of Southeast Asia. Nat Sustainability:1–5. https://doi.org/10.1038/s41893-025-01532-w

Tirkey AIP, Kumar R, Prasad RS, Malakar A (2024) Agroforestry for carbon neutrality: an effective pathway to net zero. J Adv Biol Biotechnol 27(11):591–602. https://doi.org/10.9734/jabb/2024/v27i111643

Udawatta PR, Rankoth L, Jose S (2019) Agroforestry and Biodiversity. Sustainability 11(10):2879. https://doi.org/10.3390/su11102879

Vacek Z, Vacek S, Cukor J (2023) European forests under global climate change: review of tree growth processes, crises and management strategies. J Environ Manag 332:117353. https://doi.org/10.1016/j.jenvman.2023.117353

Van Noordwijk M (2020) Agroforestry as nexus of sustainable development goals. IOP Conf Ser: Earth Environ Sci 449(1):1–9

Venegas RM, Acevedo J, Treml EA (2023) Three decades of ocean warming impacts on marine ecosystems: a review and perspective. Deep-Sea Res II Top Stud Oceanogr 212:105318

Verma K, Sharma P, Bhardwaj DR, Singh MK, Verma T, Thakur P (2024) Traditional agroforestry practices in the Indian Himalayas. In: Borthakur A, Singh P (eds) Learning 'from' and 'with' the locals. Advances in Asian human-environmental research. Springer, Cham. https://doi.org/10.1007/978-3-031-51696-2_7

Vibhuti, Bargali K, Bargali SS (2019) Species composition, diversity and traditional uses of homegarden in Kumaun Himalaya, India. Indian J Agric Sci 89(9):1415–1418

Vijayanathan J, Rawichandran D, Abdullah MZ, Koter R, Ahmad R, Anuar MAH (2022) Agroforestry practices to achieve sustainable and climate resilient forests. J Trop Plant Physiol 14(2):1–12. https://doi.org/10.56999/jtpp.2022.14.2.22

Vikrant KK, Chauhan DS, Rizvi RH (2021) Carbon storage potential in traditional agroforestry system-case study of Tehri district in Uttarakhand, North-Western Himalaya, India. Indian For 147(3):240–249

Viswanath S, Lubina PA (2017) Traditional agroforestry systems. In: Dagar J, Tewari V (eds) Agroforestry. Springer, Singapore. https://doi.org/10.1007/978-981-10-7650-3_3

WMO (2021) State of the global climate 2021- WMO provisional report. World Meteorological Organization, Geneva, 47p

Yadav RP, Bisht JK (2015) *Celtis australis* Linn: a multipurpose tree species in North West Himalaya. Int J Life-Sci Sci Res 1:66–70

Yadav RP, Gupta B, Bhutia PL, Bisht JK (2018) Sustainable agroforestry systems for livelihood security and their economic appraisal in Indian Himalayas. Econ Aff 63(3):633–639

Zahoor S, Dutt V, Mughal AH, Pala NA, Qaisar KN, Khan PA (2021) Apple-based agroforestry systems for biomass production and carbon sequestration: implication for food security and climate change contemplates in temperate region of Northern Himalaya, India. Agrofor Syst 95:367–382. https://doi.org/10.1007/s10457-021-00593-y

Chapter 9
Variation in Nutritive Value of Fodder in the Western Himalayas

Naresh Singh, Aman Dabral, Bhupendra Singh, and Munesh Kumar

Abstract The Western Himalayas represent a unique ecological region where livestock rearing is an integral component of rural livelihoods, yet productivity is constrained by limited and variable fodder resources. This review synthesizes current knowledge on the availability, diversity, and nutritive composition of fodder species across altitudinal gradients and seasons. Fodder availability in the region is dominated by a few tree and shrub species such as *Quercus leucotrichophora*, *Celtis australis,* and *Grewia optiva*, which are widely preferred by local communities due to their higher crude protein (6–15% DM) and mineral content compared to grasses. Seasonal variation strongly influences nutrient quality, with protein levels generally peaking during monsoon and declining in winter, while crude fiber and lignin increase with plant maturity. Altitude further shapes fodder quality, as high-elevation species often exhibit reduced digestibility due to secondary metabolites. Despite their importance, the nutritive value of many indigenous fodder species remains under-researched, leading to feed imbalances and mineral deficiencies in livestock. Sustainable management strategies, including agroforestry integration, balanced supplementation, and conservation of high-quality native species, are therefore essential to address fodder scarcity and enhance livestock productivity in the Western Himalayas.

Keywords Livestock productivity · Fodder quality · Hill agroforestry system · Sustainable management · Conservation

N. Singh · A. Dabral
Department of Forest Biology and Tree Improvement, College of Forestry, Ranichauri, Veer Chandra Singh Garhwali Uttarakhand University of Horticulture and Forestry, Pauri, Uttarakhand, India

B. Singh (✉)
Department of Silviculture and Agroforestry, College of Forestry, Veer Chandra Singh Garhwali Uttarakhand University of Horticulture and Forestry, Pauri, Uttarakhand, India

M. Kumar
Department of Forestry and Natural Resources, Hemvati Nandan Bahuguna Garhwal University (A Central University), Srinagar Garhwal, Uttarakhand, India

B. Singh et al. (eds.), *Himalayan Agroforestry Systems*, Advances in Agroforestry 15,
https://doi.org/10.1007/978-3-032-25096-4_9

1 Introduction

Livestock farming plays a crucial role in the livelihoods of people living in the Western Himalayan region. Here, the mid-hills, subalpine, and alpine pastures provide essential feed, along with crop leftovers and grazing areas found in forests and wastelands (Islam et al. 2023). Despite the availability of these resources, still a scarcity of abundant biomass to support the current livestock population, leading to serious feed shortages. Notably, in Himachal Pradesh, there is a significant deficit of about 26% in green fodder and a staggering 54% in dry fodder. To cope with these challenges, farmers often turn to fodder trees, especially during tough times, but they typically rely on only around 20 species out of the 84 fodder tree species that thrive in the Himalayas (Inder et al. 2006).

1.1 Importance of Fodder Quality in Sustaining Rural Livelihoods and Livestock Productivity

The nutritional makeup of fodder plays a crucial role in determining how productive livestock can be, which in turn affects the economic sustainability of smallholder farming systems that support more than two-thirds of rural communities in semi-arid areas (Navale et al. 2022). In India, where crop residues make up 54% of the feed for ruminants, the low crude protein content (just 3–5%) and high lignin levels limit feed digestion. This situation calls for the addition of quality green fodder to meet the animals' metabolic needs. Leguminous forages such as *Medica gosativa*, which contains 14.71% crude protein and *Trifolium pratense,* are rich in essential amino acids and minerals [for instance, calcium at 132.05 milligram/kilogram dry matter (mg/kg DM) and zinc at 2.17 mg/kg DM], both of which are vital for improving lactation efficiency and boosting disease resistance in livestock. Research shows that incorporating high-protein fodder crops into animal diets can increase milk production by 18–23% compared to feeding them only crop residues, allowing animals that typically produce 7–8 kg/day to maintain their output without needing concentrated feeds. However, these nutritional advantages are often overlooked due to regional disparities—importantly, farmers in Punjab and Haryana tend to burn excess residues, while areas such as Himachal Pradesh struggle with a 26–54% shortfall in the availability of green and dry fodder. Agronomic practices can enhance these nutritional benefits: applying nitrogen (120 kg/ha) to cereals can raise crude protein levels by 28%, while adding phosphorus (60 kg/ha) to legumes can boost dry matter yield by 34%. Additionally, irrigation timing strategically and using microbial seed inoculants (like Azotobacter and phosphate-solubilizing bacteria) can improve in vitro digestibility by 15–22%, which is directly linked to increased milk production. Still, there are challenges to adoption, especially since small landholdings (85% are less than 2 hectares) often prioritize food crops over fodder cultivation. With India's projected fodder needs for 2050 estimated at 1012

Table 9.1 Fodder productivity (Kg/ha/year ± SD) in the Himalaya region (Sachan et al. 2007)

	Altitudinal range (above mean sea level (amsl))			
Tree species	380–500	500–1000	1000–1500	1500–2000
Bauhinia variegate	114.2 ± 20.6	119.5 ± 35.7	222.8 ± 44.6	–
Boehmeria rugulosa	342.1 ± 53.6	165.2 ± 44.4	–	–
Celtis australis	8.3 ± 1.8	412.5 ± 111.0	614.2 ± 131.6	311.9 ± 72.0
Ficus palmate	75.2 ± 13.1	–	–	–
Quercus leucotricophora	–	–	303.6 ± 70.2	791.4 ± 154.4
Morus alba	–	175.8 ± 47.7	128.2 ± 44.6	–
Grewia optiva	478.3 ± 96.6	515.5 ± 130.1	265.8 ± 71.3	152.9 ± 45.9
Ficus cunia	48.2 ± 8.3	–	44.5 ± 11.3	–
Ficus hispida	38.2 ± 6.5	–	–	–
Ougeinia oojeinensis	108.0 ± 22.2	415.6 ± 113.5	348.1 ± 94.2	90.5 ± 18.1
Melia azedarach	216.1 ± 46.1	196.4 ± 54.8	74.3 ± 12.4	–

million tons of green fodder and 631 million tons of dry fodder, addressing quality gaps will require a combination of genetic improvements in forage species (like *Festuca kashmiriana*, which has 37.27 mg/100 g of carbohydrates) (Gautam et al. 2023) (Table 9.1).

2 Factors Influencing the Nutritive Value of Fodder

The nutritional value of fodder is influenced by a variety of biological, environmental, and management factors that work together in a complex way. Some of the main factors include the plant's maturity stage, the species of the plant, the climate, and the soil type, and how it is managed. Younger plants usually have more protein and less fiber, which makes them easier to digest, while mature plants tend to build up lignin and fiber, which can lower their nutritional quality (Oelberg 1956). The type of plant also matters; legumes often have more crude protein and minerals than grasses (Jakubus and Graczyk 2022). Weather conditions like temperature, humidity, rainfall, and light levels can affect photosynthesis and how nutrients are absorbed, changing the chemical makeup of the fodder (Moore et al. 2020). Soil health plays a big role in nutrient availability; for instance, adding compost can boost levels of calcium, magnesium, and phosphorus in fodder plants. Additionally, environmental stresses like drought and heat can lower protein levels and digestibility by disrupting photosynthesis and metabolic processes (Jakubus and Graczyk 2022). The time of fodder harvest also makes a difference; for example, shoots picked in the spring tend to have more crude protein and digestible nutrients than those harvested in winter (Tsombou et al. 2024). Lastly, certain antiquality factors, like tannins or lignin, can further hinder digestibility in some plant species (Moore et al. 2020).

The nutritional value of fodder is influenced by a variety of environmental, biological, and management factors that work together in a complex way. Here are some key influences highlighted in Fig. 9.1.

- *Soil texture*: Nutrient retention and root development are greatly impacted by soil texture. Because sandy soils drain more quickly, they leach more nitrogen (9.1% of applied N lost) than silt loam (3.5%), which lowers plant nutrient availability. Although clay-rich soils retain nutrients better, they run the risk of waterlogging, which interferes with root oxygenation. Aeration and water-holding capacity are balanced in the ideal soil texture to optimize the quality of fodder.
- *Tree species*: Different tree species vary markedly in nutrient profiles. For example, *Moringa stenopetala* exhibits high crude protein (22.35%) and ash content (13%), while *Sesbania sesban* has lower acid detergent fiber (11.8%), enhancing digestibility (Sebola 2015). These variations highlight the importance of species selection in fodder systems to meet specific livestock nutritional needs.
- *Microbial activity*: Soil microbes enhance nutrient cycling and organic matter decomposition. Inoculation with plant-growth-promoting rhizobacteria (PGPR) alongside organic amendments like cow urine improved pearl millet's total digestible nutrients (TDN) by 12–15%, demonstrating the role of microbial activity in boosting fodder quality (Verma et al. 2023a, b).

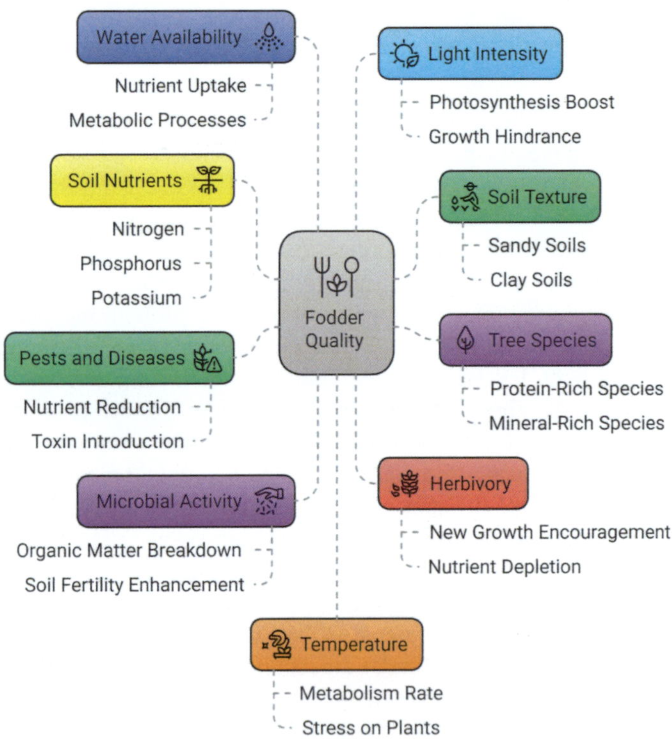

Fig. 9.1 Factors affecting fodder quality

- *Soil nutrients*: Nitrogen fertilization directly increases crude protein and energy content in fodder. Pearl millet treated with 100% recommended dose of fertilizer (RDF) + PGPR + cow urine showed 22% higher TDN yield compared to control plots. However, excessive nitrogen in sandy soils raises leaching risks, underscoring the need for balanced nutrient management (Kumar et al. 2023).
- *Water availability*: Drought stress reduces protein synthesis and increases fiber content, lowering fodder digestibility. Adequate irrigation ensures optimal nutrient uptake, with studies linking water scarcity to a 30–40% decline in metabolizable energy in fodder crops (Nandi et al. 2024).
- *Light intensity*: Photosynthetic efficiency driven by sufficient light intensity enhances carbohydrate accumulation. Suboptimal light conditions, common in dense agroforestry systems, can reduce dry matter production by up to 25%, affecting fodder yield and energy density (Subrahmanya et al. 2017).
- *Temperature*: Extreme temperatures disrupt metabolic processes, reducing protein synthesis. For instance, heat stress during flowering stages decreases leaf nitrogen content by 15–20%, impacting fodder's nutritional value (Bajwa et al. 2023).
- *Seasonality*: Harvesting stage critically influences nutrient composition. Late-harvested fodder exhibits 18–23% higher neutral detergent fiber (NDF) and 10–15% lower crude protein compared to early-cut plants, reducing digestibility. Seasonal growth patterns also affect mineral availability, with young foliage offering higher calcium and phosphorus (Mulat and Yong 2022).

3 Dominant Tree Species

Table 9.2 showcases a fascinating classification of the key fodder tree species that are essential for agroforestry and livestock systems, especially in the Himalayan and sub-Himalayan regions. For instance, Grewia optiva, commonly known as Bhimal, is a medium-sized tree from the Malvaceae family, reaching heights of

Table 9.2 Classification of dominant tree species

Dominant fodder tree species	Family	Common name	Height (m)	Season of harvesting
Grewia optiva	Malvaceae	Bhimal	05–10	October–February
Quercus semecarpifolia	Fagaceae	Kharsu	15–25	May–June
Quercus leucotrichophora	Fagaceae	Banj	15–20	March–December
Quercus floribunda	Fagaceae	Moru	15–20	January–November
Bauhinia variegata	Fabaceae	Guiyal	5–15	April–February
Morus alba	Moraceae	Sehtoot	10–15	April–June
Ficus carica	Moraceae	Fig	03–11	May–October
Salix alba	Salicaceae	White Willow	11–15	April–June
Morus serrata	Moraceae	Kemu	11–15	May–July
Celtis australis	Ulmaceae	Khareek	11–20	May–October
Ficus palmate	Moraceae	Bedu	05–11	April–June

5–10 m. It is highly valued for its protein-rich fodder, which is typically harvested between April and June. The Fagaceae family makes a strong appearance with notable species like *Quercus semecarpifolia* (Kharsu), *Quercus leucotrichophora* (Banj), and *Quercus floribunda* (Moru). These impressive trees, standing tall at 15–25 m, provide crucial fodder for livestock during the monsoon months of July to September. Another important player is *Bauhinia variegata*, or Guiyal, from the Fabaceae family, which grows between 5 and 15 m and offers nutritious leaves that are harvested in the spring (March–May). The Moraceae family also contributes significantly with species like *Morus alba* (Sehtoot), *Morus serrata* (Kemu), and *Ficus carica* (Fig), all of which provide both fodder and delicious fruits, mainly collected from June to September. *Salix alba*, or White Willow, from the Salicaceae family, thrives in damp areas and supplies fodder during the summer months (April–June). Another noteworthy species is *Celtis australis* (Khareek) from the Ulmaceae family, a resilient tree that plays a dual role in providing fodder and stabilizing soil, with harvesting taking place between July and September. Lastly, *Ficus palmata* (Bedu) adds to the mix with its edible fruits and fodder value. Collectively, these species not only meet the livestock feed needs during critical times but also contribute significantly to ecological health by aiding in soil conservation, water retention, and supporting rural livelihoods (Fig. 9.2).

Fig. 9.2 Fodder trees of the Himalayan region: (**a**) *Celtis australis*, (**b**) *Quercus leucotrichophora*, (**c**) *Grewia optiva*, (**d**) *Phanera roxburghiana* and (**e**) *Bauhinia variegata*

3.1 Traditional Uses of Fodder in Local Communities

In the Himalayan region, fodder is vital for keeping traditional livestock-based livelihoods thriving. For many rural communities, agriculture mixed with livestock plays a key role in their economic stability. These local communities gather fodder from a range of places, such as agroforestry systems, grasslands, forests, and agricultural fields. Remarkably, more than half of the feed used by ruminant animals originates from forest resources, highlighting the importance of these animals to the pastoral economy of the area. The communities make use of a wide range of plant species for fodder, with research identifying 89 different species that include trees, shrubs, herbs, grasses, and climbers. Among the multipurpose tree species, *G. optiva*, known locally as Bhimal, is the standout favorite and is commonly found throughout the Garhwal region. Other popular species include *Celtis australis*, various Bauhinia and *Ficus* species and *Terminalia* species. The seasonal availability of fodder greatly impacts how it is managed. While there is plenty of fodder during the rainy season, it tends to dwindle during dry spells, leading to nutritional shortages for milking animals. To tackle this issue, locals adapt by using fodder in both fresh and dried forms, depending on what is available.

4 Nutritional Composition of Himalayan Fodder (Table 9.3)

4.1 Macronutrients: Proteins, Carbohydrates, and Fats

The macronutrient makeup of fodder trees in the Garhwal Himalaya shows quite a bit of variation among different species, with *G. optiva*, commonly known as Bhimal, standing out as a particularly nutritious option. Its leaves boast crude protein levels between 11.52% and 17.35% dry matter (DM) (Dhavalkumar Rajeshbhai 2015), which is better than many traditional supplements like cottonseed cake (CSC), especially during those tough winter months. This impressive protein content is due to the plant's ability to maintain over 164 g/kg DM of crude protein even as it matures, ensuring that livestock get enough nitrogen for their metabolic needs and milk production. Importantly, the carbohydrates, represented as nitrogen-free extract (NFE), Bhimal provides a solid 52.99–59.60% DM, which is crucial for giving ruminants the energy they need. Although fats are present in smaller amounts (1.62–5.31% ether extract) (Khan and Habib 2012), they play a role in helping absorb lipid-soluble nutrients and improving metabolic efficiency. It was observed that Bhimal leaves can compete with CSC in promoting weight gain and wool production in sheep (Khan and Habib 2012), proving their worth as protein supplements. Other notable fodder trees are *Ficus* spp. and *Quercus* species, but limited studies carried out for specific data on their macronutrient content. The community

Table 9.3 Nutritional composition of Himalayan fodder

Species	Dry matter (DM)	Ether extract (EE)	Crude fiber (CF)	Crude protein (CP)	Total ash	Nitrogen-free extract (NFE)	References
Grewia optiva	90.0	3.2	28.4	12.5	7.8	48.1	Kumar et al. (2026)
Celtis australis	36.81	3.40	21.77	13.34	16.58	36.68	Singh et al. (2022)
Acacia catechu	62.50	6.55	21.95	8.49	9.56	53.45	Navale et al. (2022)
Albizia chinensis	43.35	2.65	39.86	19.40	4.93	33.49	Navale et al. (2022)
Bauhinia variegata	50.23	4.70	28.87	8.75	6.98	51.47	Yadav et al. (2023)
Ficus roxburghii	34.22	3.74	18.19	11.70	16.22	50.16	Navale et al. (2022)
Leucaena leucocephala	35.02	2.73	13.75	19.35	9.45	53.11	Navale et al. (2022)
Melia composite	42.18	2.32	13.43	16.04	12.52	55.58	Navale et al. (2022)
Morus serrata	39.32	4.33	13.58	14.97	15.35	51.77	Navale et al. (2022)
Quercus glauca	54.90	5.25	41.11	10.06	7.46	36.12	Navale et al. (2022)
Quercus leucotrichophora	59.69	3.39	38.02	9.48	5.66	43.88	Navale et al. (2022)
Salix tetrasperma	36.16	3.69	18.94	11.67	10.63	55.02	Navale et al. (2022)
Bauhinia retusa	55.74	1.76	25.71	10.01	7.83	54.70	Yadav et al. (2023)
Quercus semecarpifolia	69.30	2.20	33.21	8.38	4.47	29.66	Singh and Todaria (2012)

tends to favor these species because they have seen improvements in milk yield, likely due to their balanced macronutrient profiles. Interestingly, seasonal changes have little impact on Bhimal's protein stability, unlike grasses that lose nutritional value during dry spells. Additionally, the altitude also plays a role in nutrient composition, with Bhimal populations at mid-altitudes (1300–1600 m) showing the best protein (17.35%) and carbohydrate (59.60%) levels, reflecting their adaptation to the specific growing conditions at those elevations (Khatta and Katoch 1983).

Table 9.4 Fiber content and digestibility of different fodder species

Fodder species	Fiber content (% DM)	Digestibility (%)	Notes
Grewia optiva (Bhimal)	20.0–30.0	45–55	High protein and energy content; preferred by livestock
Salix spp.	25.0–35.0	40–50	Rich in minerals; used extensively in winter fodder
Medicago sativa (Alfalfa)	25.0–30.0	50–60	High digestibility; excellent protein source for livestock
Ficus spp.	30.0–40.0	35–50	Commonly used; moderate fiber digestibility
Lindelofia longiflora	15.0–20.0	60–70	Low fiber content; highly digestible, suitable for young stock
Dracocephalum heterophyllum	25.0–35.0	30–40	Used during summer; lower digestibility due to higher lignin content
Stipa jacquemontii	30.0–40.0	35–45	Grass species with moderate fiber; seasonal availability affects quality

4.2 *Fiber Content and Digestibility*

The fiber content and digestibility of fodder species in the Garhwal and Northwestern Himalayas vary significantly among different plant types. Table 9.4 shows a summary of key findings regarding fiber content and digestibility in these regions.

5 Seasonal and Altitudinal Variation in Nutritive Value

Research into the nutritional value of fodder across different altitudes uncovers some fascinating patterns that are crucial for managing livestock in mountainous areas. For instance, studies on *C. australis* in the Central Himalaya show that altitude has a significant impact on the nutritional makeup of fodder, with noticeable differences between adult and juvenile leaves (He et al. 2016). Interestingly, the dry matter content in both adult and juvenile leaves tends to decrease as you go higher up in elevation. On the flip side, the mineral content, including phosphorus, potassium, and calcium, actually increases with altitude, showing a positive correlation. Additionally, crude protein levels in adult leaves also rise with altitude, suggesting that the nutritional quality improves in populations found at higher elevations (Singh et al. 2010). In the cold Spiti valley desert, similar trends across various fodder types, such as trees, shrubs, grasses, and legumes, were observed. While the dry matter content can vary widely between species (ranging from 24.83% in *Carexatrofusca* to 68.20% in *Stipa jacquemontii*), leguminous fodders consistently have a higher protein content, averaging around 19.41%, compared to grasses, which sit at about 8.21% (Bishist et al. 2023).

5.1 Influence of Elevation on Nutrient Composition of Fodder

Elevation plays a crucial role in shaping the nutrient makeup of fodder in mountainous areas, with unique patterns emerging along the altitude changes in the Himalayas. Research on *Celtis australis* leaves shows that increasing altitude up to 1500 meters, the levels of crude protein, phosphorus, potassium, and calcium in both adult and young leaves increase, with crude protein having a strong positive link ($r = 0.91$) to elevation (Singh et al. 2010). Interestingly, dry matter content tends to decrease as altitude rises ($r = -0.64$), likely due to cooler temperatures and less evapotranspiration (Shah et al. 2019). In the high-altitude pastures of the Western Himalaya, forage species like *Medicago sativa* and *Trifolium pratense* show higher crude protein levels (up to 14.71 mg/100 g) and essential amino acids, such as histidine and leucine, when compared to grasses found in lower regions. Likewise, in Tajikistan's grasslands, adding nitrogen and phosphorus at higher elevations can double the aboveground biomass, as the interplay of precipitation and temperature gradients boosts nutrient absorption (Fan et al. 2021). At lower elevations (below 1500 m), grasses like *Setaria pumila* thrive, boasting higher dry matter (68.20%) but lower mineral density (Kumar et al. 2019).

5.2 Influence of Biotic and Abiotic Factors on Fodder Quality

Biotic and abiotic factors significantly impact fodder quality, directly affecting livestock health and dairy product quality. In a brief analysis of these influences, multiple dimensions emerge as critical determinants of overall fodder nutritional value.

5.3 Factors Influencing Fodder Quality

Both biotic and abiotic stressors play a significant role in shaping the quality of fodder, impacting plant growth, nutrient makeup, and overall biomass production through intricate mechanisms. Biotic factors, like diseases such as anthracnose, stalk rot, and graymold, can drastically lower biomass yield and change the nutritional profile of fodder crops (Kumar et al. 2022a, b). These pathogens weaken plant health, resulting in reduced crude protein levels and digestibility. On the other hand, abiotic factors such as drought, heat, cold, salinity, and waterlogging also have a major influence on fodder quality (Roberts and Mattoo 2018). For example, drought stress can impact the stay-green trait, which is vital for maintaining nutritional quality, while salinity can disrupt root structure by altering the formation of casparian strips and suberin lamellae (Hanaka et al. 2021).

6 Seasonal Variation in Nutrient Composition

Seasonal changes play a big role in the nutrient makeup of fodder, which in turn affects livestock nutrition and productivity. In the wet seasons, fodder usually has a higher content of crude protein, metabolizable energy, and minerals because the growth conditions are just right. On the flip side, during dry seasons, the nutritional quality tends to drop, with more fiber and less digestibility. This variation calls for tailored feeding strategies for each season to keep livestock healthy and performing well.

6.1 Impact of Seasonal Shifts on Leaf Nutrient Density

Seasonal changes play a big role in shaping the nutritional makeup of fodder tree leaves in the Himalayan region. This is largely due to how these plants respond to factors like temperature, moisture, and their growth stages. For instance, research on Bauhinia retusa and *B. variegata* shows that nutrient levels peak at different times of the year. During the monsoon, *B. retusa* can hit a crude protein level of 18.4%. In the summer, *B. retusa* dry matter content soars to 32.6% as it thickens its leaves to conserve water. Interestingly, the ash content, which indicates mineral concentration, tends to drop with increased rainfall. In winter, *B. variegata* can reach an ash content of 11.2% due to lower transpiration, which helps concentrate minerals (Yadav et al. 2023). Carbohydrate levels also vary seasonally, fluctuating between 42% and 58%. Notably, antinutritional factors like tannins increase from 1.8% during the monsoon to 4.3% in the dry season, as plants ramp up their chemical defenses against herbivores when resources are scarce. These biochemical changes create specific times when *G. optiva* offers the best nitrogen-free extract (54.8%) in autumn, while Morusserrata shines with higher crude protein (22.1%) during the spring growth spurt. This dynamic situation calls for careful harvesting strategies, as there is a notable 29-40% seasonal variation in the relative nutritive value indices across different species (Navale et al. 2022).

6.2 Comparative Analysis of Summer vs. Winter Foliage

The differences in temperature and water availability between summer and winter lead to distinct types of fodder. In the summer, plants are about 12–18% more digestible because they have less lignin (14.2% in summer compared to 19.8% in winter). This is especially true for *Leucaena leucocephala*, which still boasts an impressive 68.3% in vitro digestibility even during the hottest months. Winter leaves may not taste as good, but they make up for it with a 38% increase in calcium levels (1.92% compared to 1.39% in summer) in species like *Celtis australis*, which is vital for lactating animals (Navale et al. 2022). In *Bauhinia variegata*, summer leaves are rich in crude protein at 19.4%, perfect for growth, while the winter leaves provide 24.3% ether extracts (lipids) that help with temperature regulation (Yadav et al. 2023). The neutral detergent fiber (NDF) content rises from 42.1% in summer to 57.9% in winter across the species studied, which can lower how much livestock want to eat but increases the time food stays in their rumen. Farmers in Nepal's middle hills take advantage of these seasonal differences, using 83% of the grasses harvested during the monsoon for productive livestock, while only 62% of winter crop residues go to maintenance feeders.

Notably, winter-harvested *Q. leucotrichophora* has 22% more metabolizable energy (9.8 MJ/kg DM) than summer samples, even though it has less protein, making it a great option for energy-rich supplementation (Navale et al. 2022).

6.3 Influence of Monsoon and Dry Seasons on Fodder Quality

The cycle between monsoon and dry seasons brings about significant changes in fodder ecosystems. During the monsoon, the moisture content in leaves of species like *Ficus roxburghii* can soar to 72–78%. While this dilutes the nutrients, it also allows for a remarkable 17% increase in biomass yields due to rapid growth. Interestingly, the crude fiber in the monsoon foliage of *Albizia chinensis* drops to 14.3%, compared to 28.1% during the dry season, which makes it easier to digest. On the flip side, during the dry season, we see some fascinating ecological adaptations: *Olea glandulifera* boosts its phenolic compounds by 140% as a defense against drought, which means it needs urea treatment to be safely used (Navale et al. 2022). Farming communities show striking differences in how they manage feed between the seasons. For instance, Rai and Limbu groups change 94% of their diet components during the dry months, while Chetri and Brahmin farmers only adjust 31%. In terms of diet composition, monsoon meals consist of 61% grasses, but this shifts to a reliance on 73% tree fodder during the drier times, as observed in the Nepalese terai systems (Paswan et al. 2008). A critical point to note is that phosphorus availability drops significantly to 0.19% in dry-season *Acacia catechu* compared to 0.42% during the monsoon, highlighting the need for mineral supplementation. These trends emphasize the importance of harvesting high-protein foliage, like the 19.1% crude protein found in *Salix tetrasperma* during the monsoon and relying on lignin-rich species such as *Pittosporum floribundum* for rumen fill in the dry season (Navale et al. 2022).

7 Nutritive Fodder Conservation and Sustainable Management

In the Himalayan region, biotic and abiotic factors are important in determining the quality of fodder, which impacts the productivity and nutrition of livestock. Fungal diseases like anthracnose and stalk rot are examples of biotic stressors that can drastically reduce biomass yields and degrade vital nutrients like digestibility and crude protein (Katoch 2022). Additionally, antinutrients found in certain forage species can further jeopardize the safety of the fodder (Navale et al. 2022). On the abiotic side, challenges like extreme temperatures, drought, and salinity can change how plants function. For instance, drought can reduce the stay-green trait that is vital for ongoing nutrition, while salinity can interfere with root development by altering the casparian strip (Beigh et al. 2020). In high-altitude areas, winter conditions worsen feed shortages, leading to a dependence on nutrient-poor tree leaves and homemade concentrates (Arya et al. 2011).

Fodder quality changes with the seasons. During the summer, we see a boost in crude protein levels (ranging from 3.34 to 14.71 mg/100 mg) and mineral content, with calcium reaching as high as 132.05 mg/kg DM. However, winter feeding often leads to nutritional gaps (Toky et al. 1989). Agroforestry systems that mix crops like maize with nitrogen-fixing trees such as *G. optiva* and *B. variegata* play a crucial role in enhancing nutrient cycling, contributing about 48–63% of biomass from their branches and leaves (Beigh et al. 2020). Nowadays, there is a strong focus on developing stress-resistant varieties, like low-lignin sorghum, to boost digestibility and help crops adapt to seasonal changes (Katoch 2022).

7.1 *Threats to Nutritional Stability and Fodder Diversity*

Fodder diversity and nutritional stability are under serious threat from genetic erosion, environmental pollutants, and the push for uniform agriculture, putting global food security at risk. Our heavy reliance on high-yield livestock breeds has led to a significant drop in genetic diversity, with a staggering 25% of these breeds now facing extinction and 79% of poultry breeds in critical danger (Nte et al. 2023). This narrowing of genetic options weakens our defenses against diseases like avian influenza and African swine fever, as these specialized breeds often lack the adaptive traits found in rarer populations, such as the parasite-resistant Gulf Coast sheep (Untea et al. 2023). At the same time, antinutritional factors like tannins, nitrates, and mycotoxins are compromising the safety of fodder, which in turn hampers livestock productivity by interfering with nutrient absorption and causing chronic toxicity (Nicholson et al. 2021). Climate change is making these issues even worse by changing how pathogens behave and affecting the quality of silage, with Listeria-contaminated silage leading to serious infections in animals (Jakubus and Graczyk 2022). Moreover, indiscriminate cross-breeding and a preference for non-native breeds are speeding up the loss of locally adapted genetic resources. A prime example is Brazil's drought-resistant Pantaneiro cattle, whose numbers have plummeted despite their ability to thrive in tough climates (Untea et al. 2023). While global crop diversity has seen an uptick since 1961, nutritional stability has stagnated due to a decline in the "nutritional links" within crop networks, exposing vulnerabilities in nutrient redundancy (Nicholson et al. 2021). Developing countries are caught in a bind, facing challenges from both market-dependent imports and limitations in domestic production, with 70% of livestock production costs linked to increasingly contaminated feed. Tackling these threats calls for a comprehensive approach that includes conserving cryopreserved genetic material, improving anti-toxin processing methods and implementing policies that prioritize agroecological resilience over short-term productivity gains (Geng et al. 2020). In the stunning Himalayan region, the collection of sustainable fodder through traditional agroforestry systems plays a crucial role in mountain farming. These age-old practices account for about 45–50% of the total fodder needs, with nearby forests contributing 29–40% and any shortfall being covered by purchased grass. The Northwestern Himalaya is home to around 279 species of fodder trees that are essential for livestock, especially during the winter and summer months when fresh fodder is hard to come by (Yadav and Bisht 2013). Over the years, various agroforestry systems like agrisilvihorticulture, silvipasture, and those based on fodder trees have developed to meet the diverse needs of mountain communities. The way fodder is collected varies with altitude, with households gathering between 25–90 kg each day and totaling 2760–13,770 kg annually, mostly by women (60%) (Islam et al. 2023). Preferences for different species change with elevation; for instance, *Quercus* and *Ficus* are favored at higher altitudes, while *Celtis australis* and *Bauhinia* species are more popular at lower elevations (Yadav and Bisht 2013). Sustainable management practices include regulating livestock grazing, encouraging the planting of frequently collected species and making better use of agricultural by-products (Islam et al. 2023). These methods not only secure fodder availability but also boost carbon sequestration, improve soil health, and enhance the overall resilience of the ecosystem (Verma et al. 2023a, b).

8 Challenges and Opportunities

The Western Himalayan region is famous for its incredible biodiversity and traditional farming methods, but it also faces some tough challenges and exciting opportunities when it comes to fodder resources. Research in the Nainital district shows that the annual fodder consumption is quite significant, ranging from about 300 to over 3700 tons each year. This highlights just how crucial fodder is for the local communities (Pandey 2022). While the area is home to a diverse range of fodder plants, managing these resources sustainably can be quite tricky. Another challenge is the seasonal availability of fodder; 41% of the species are used in winter, 38% in summer, and only 21% are available throughout the year (Rawat and Vishvakarma 2010). This seasonal fluctuation calls for careful planning to ensure livestock get the nutrition they need consistently. When we look at the nutritional profiles of high-altitude forage species, there is a lot of potential to improve livestock health. Studies on five key species have shown impressive nutrient levels, including carbohydrates (17.44–37.27 mg/100 mg), crude protein (3.34–14.71 mg/100 mg), and beneficial crude fiber (11.16–24.16%). These plants are also rich in essential minerals, with calcium being the most abundant (13.91–132.05 mg/kg DM), followed by magnesium, iron, and zinc. In addition, they provide valuable essential amino acids like histidine, tryptophan, valine, leucine, and phenylalanine (Kumar et al. 2022a, b). This nutritional wealth opens up opportunities to boost livestock health and productivity. Many fodder species serve multiple purposes; they can be used for fuel, medicine, timber, and even agricultural tools, which adds to their overall value. By advancing agroforestry systems in the region, we can create more sustainable fodder production while also tackling ecological issues and helping with carbon sequestration (Singh et al. 2024). To protect this vital resource, conservation efforts like prioritization, mass multiplication, and reforestation are absolutely essential (Pandey et al. 2018).

9 Conclusion

Fodder resources in the Western Himalayas exhibit considerable variation in nutritive value depending on species, season, and altitude. While leguminous forages and multipurpose tree species contribute significantly to crude protein and essential minerals, reliance on a narrow range of species and seasonal scarcity often results in nutrient deficiencies for livestock. The gradual decline in protein and digestibility with advancing plant maturity highlights the importance of timely harvesting, while the influence of altitude underscores the need for region-specific fodder management practices. Current gaps include limited characterization of lesser-known indigenous species and inadequate strategies for mineral supplementation. Strengthening agroforestry systems, expanding the cultivation of nutrient-rich species, and promoting community-based fodder management can enhance both the availability and quality of forage. Overall, sustainable fodder management in the Western Himalayas is crucial not only for improving livestock health and productivity but also for ensuring the resilience of rural livelihoods in the face of climatic and ecological challenges.

References

Arya D, Bohra CPS, Tewari A (2011) Use of major fodder species in oak and pine dominant zones of Garhwal Himalaya, India – a case study. E-Int Sci Res J 3(3):187–191

Bajwa RT, Shakeel Q, Shahid M, Rizwan SA, Aqueel MA, Mubeen M et al (2023) 6 Physiological effects of thermal, mineral and water stresses on summer fodders. In: Sustainable Summer Fodder. CRC Press, pp 97–120

Beigh YA, Ganai AM, Rather MA, Athar H, Hamdani SA (2020) Livestock feed resources availability, feeding practices and nutrient balances in high Himalaya (Gurez) valley of Kashmir. Trop Anim Health Prod 52:2469–2480
Bishist R, Wangmo D, Pant KS, Gautam KL (2023) Nutritional evaluation of fodder resources utilized in animal feeding in Spiti valley of Himalyan cold desert. Indian J Agroforest 24(2)
Dhavalkumar Rajeshbhai P (2015) Fodder quality assessment and morphological variation of Grewia optiva Drumm. ex. Burr. at three different altitudes in a part of Garhwal Himalaya. Doctoral dissertation, College of Forestry, Ranichauri Campus, VCSG Uttarakhand University of Horticulture and Forestry
Fan LL, Mekrovar O, Li YM, Li KH, Ma XX, Mao JF (2021) Effect of nutrient addition on the productivity and species richness of grassland along with an elevational gradient in Tajikistan. Front Plant Sci 12:765077
Gautam P, Bhunia SR, Sahoo A, Sawal RK, Rakshit S, Yadav VK, Lal B, Ramniwas, Gograj, Bishnoi R, Rathore VS (2023) Drip irrigation and sulphur fertilization influenced fodder yield, quality and water use efficiency of groundnut in arid region. PLoS One 18(8):e0288090
Geng Y, Ranjitkar S, Yan Q, He Z, Su B, Gao S, Niu J, Bu D, Xu J (2020) Nutrient value of wild fodder species and the implications for improving the diet of mithun (*Bos frontalis*) in Dulongjiang area, Yunnan Province, China. Plant Divers 42(6):455–463
Hanaka A, Majewska M, Jaroszuk-Ściseł J (2021) Study of the influence of abiotic and biotic stress factors on horticultural plants. Horticulturae 8(1):6
He X, Hou E, Liu Y, Wen D (2016) Altitudinal patterns and controls of plant and soil nutrient concentrations and stoichiometry in subtropical China. Sci Rep 6:24261
Inder D, Misri, B, Pathania MS (2006) Forage demand and supply in western Himalaya: a balance sheet for Himachal Pradesh. Indian J Anim Sci 76(9):720–726
Islam T, Ali L, Nawchoo IA, Khuroo AA (2023) Diversity and utilization patterns of fodder resources in a Himalayan protected area. Environ Monit Assess 195(9):1117
Jakubus M, Graczyk M (2022) Quantitative changes in various nutrient ratios in fodder plants as an effect of compost and fly ash application. Int J Environ Res Public Health 19(13):8136
Katoch R (2022) Nutritional quality management of forages in the Himalayan Region. Springer, Singapore, pp 99–149
Khan NA, Habib G (2012) Assessment of Grewiaoppositifolia leaves as crude protein supplement to low-quality forage diets of sheep. Trop Anim Health Prod 44(7):1375–1381
Khatta V, Katoch BS (1983) Nutrient composition of some fodder tree leaves available in submountainous region of Himachal Pradesh
Kumar S, Suyal DC, Yadav A, Shouche Y, Goel R (2019) Microbial diversity and soil physiochemical characteristic of higher altitude. PLoS One 14(3):e0213844
Kumar R, Joshi R, Kumar R, Srivatsan V, Satyakam, Chawla A, Kumar S (2022a) Nutritional quality evaluation and proteome profile of forage species of Western Himalaya. Grassl Sci 68(3):214–225
Kumar P, Singh J, Kaur G, Adunola PM, Biswas A, Bazzer S, Tseng TM (2022b) OMICS in fodder crops: applications, challenges and prospects. Curr Issues Mol Biol 44(11):5440–5473
Kumar R, Ram H, Meena RK, Kumar S, Meena VK (2023) Nutritional value and energy balance of pearl millet fodder as influenced by different nutrient management practices. Indian J Dairy Sci 76(3):284
Kumar A, Bishist R, Bhardwaj DR, Kaushal R, Prakash P, Sharma R, Gautam KL, Tomar S (2026) Traditional Grewia optiva based agroforestry systems in the northwestern himalayas: a comprehensive study of carbon, soil health, and nutritional value. Agrofor Syst 100(1):7
Moore KJ, Lenssen AW, Fales SL (2020) Factors affecting forage quality. In: Forages: the science of grassland agriculture, vol 2, pp 701–717
Mulat B, Yong H (2022) Effect of maturity, silage and hay of various feeds on lactating cows feed intake and productivity; a meta-analysis. J Dairy Vet Anim Res 11(1):1–11
Nandi T, Menon S, Das S (2024) Impact of micro-nutrients on growth and development of fodder crops under water stress condition: a review. Int J Plant Soil Sci 36(5):989–999

Navale MR, Bhardwaj DR, Bishist R, Thakur CL, Sharma S, Sharma P, Kumar D, Probo M (2022) Seasonal variations in the nutritive value of fifteen multipurpose fodder tree species: a case study of north-western Himalayan mid-hills. PLoS One 17(10):e0276689. https://doi.org/10.1371/journal.pone.0276689
Nicholson CC, Emery BF, Niles MT (2021) Global relationships between crop diversity and nutritional stability. Nat Commun 12(1):5310
Nte IJ, Owen OJ, Owuno F (2023) Anti-nutritional factors in animal feedstuffs: a review. Int J Res Rev 10(2):226–244
Oelberg K (1956) Factors affecting the nutritive value of range forage. Rangeland Ecol Manag J Range Manage Arch 9(5):220–225
Pandey J (2022) Fuel wood and fodder consumption in Van Panchayat forests of district Nainital in western Himalaya. Int J Plant Sci 17:142
Pandey N, Joshi GC, Tewari LM, Pangtey YP (2018) Diversity of fodder plants of Betalghat Block, Nainital district, Western Himalaya. Int J Environ 6:1–27
Paswan VK, Mahapatra RK, Meena HR, Sahoo A (2008) Nutrient composition and phenolic constituents in some feed and fodder samples from temperate regions of Kumaon Himalaya. Indian J Animal Nutr 25(1):19–24
Rawat YS, Vishvakarma SC (2010) Diversity, distribution and utilization of fodder species in subtemperate, temperate and cold desert region of the Himachal Pradesh, northwestern, Himalaya. J Am Sci 6(6):73–81
Roberts DP, Mattoo AK (2018) Sustainable agriculture—enhancing environmental benefits, food nutritional quality and building crop resilience to abiotic and biotic stresses. Agriculture 8(1):8
Sachan, Bhatt BP, Todaria NP (2007) Energy efficiency of agrisilvicultural traditional agroforestry system along an altitudinal gradient in Garhwal Himalaya, India
Sebola NA (2015) A nutritional assessment of *Moringa oleifera* leaf meal for chickens commonly reared under extensive production systems: effect on growth performance, serum biochemistry and meat quality. Doctoral dissertation, North-West University (South Africa)
Shah MK, Tamang BB, Dhakal B, Chaudhary P, Shrestha S, Chhetri N (2019) Nutritive values of fodders at different seasons and altitudes in Gandaki River Basin of Nepal. J Agric Nat Resour 2(1):109–126
Singh B, Todaria NP (2012) Nutrients composition changes in leaves of *Quercus semecarpifolia* at different seasons and altitudes. Ann For Res 55(2):189–196
Singh B, Bhatt BP, Prasad P (2010) Altitudinal variation in nutritive value of adult-juvenile foliage of Celtisaustralis L.: a promising fodder tree species of central Himalaya, India. J Am Sci 6(2):108–112
Singh B, Kumar M, Cabral-Pinto MM, Bhatt BP (2022) Seasonal and altitudinal variation in chemical composition of *Celtis australis* L. tree foliage. Land 11(12):2271
Singh N, Riyal MK, Singh B, Khanduri VP, Rawat D, Singh C, Pinto, MMC Kumar M. (2024) Carbon sequestration potential of agroforestry versus adjoining forests at different altitudes in the Garhwal Himalayas. Atmos 15(3):313
Subrahmanya DJ, Kumar R, Tyagi N, Ram H, Singh M, Meena RK, Pandey AK (2017) Yield of fodder maize (*Zea mays*) and its chemical composition under varying plant densities and nutrient management. Indian J Animal Nutr 34(4):425–429
Toky OP, Kumar P, Khosla PK (1989) Structure and function of traditional agroforestry systems in the western Himalaya. II Nutrient cycling. Agrofor Syst 9:71–89
Tsombou FM, Saeed SulaimanJemei Al Dhanhani A, Mirza SB, Youssouf B, Ridouane FL (2024) Effect of harvest timing and plant parts on the nutritional and chemical profile of five potential fodder plants found in eastern coast of United Arab Emirates. Sci Rep 14(1):11371. https://doi.org/10.1038/s41598-024-62258-x
Untea AE, Saracila M, Vlaicu PA (2023) Feeding strategies and nutritional quality of animal products. Agriculture 13(9):1788

Verma T, Bhardwaj DR, Sharma U, Sharma P, Kumar D, Kumar A, Kumar A (2023a) Agroforestry systems in the mid-hills of the north-western Himalaya: A sustainable pathway to improved soil health and climate resilience. J Environ Manag 348:119264

Verma M, Pandove G, Kaur A (2023b) Role of PGPR as microbial inoculants in improving fodder crops productivity and quality: a review. Forage Res 48(4)

Yadav RP, Bisht JK (2013) Agroforestry: a way to conserve MPTs in North Western Himalaya. Res J Agric For Sci ISSN, 2320, 6063

Yadav N, Singh B, Khanduri VP (2023) Seasonal variation in nutrient composition in the leaves of two bauhinia species. Folia ForestaliaPolonica Ser A For 65(3):173

Chapter 10
Allelopathic Effects of Agroforestry Trees on Traditional Agricultural Crops of the Himalaya

Uday Bhanu Pratap, Naresh Singh, Bhupendra Singh, and Munesh Kumar

Abstract The Himalayan region, characterized by steep terrains and fragile ecosystems, relies heavily on agroforestry systems for sustainable agriculture, soil conservation, and livelihood support. However, the introduction and management of tree species within farmlands may have unintended consequences on traditional agricultural crops through allelopathic interactions. Allelopathy is the term used to describe how one plant chemically inhibits another through the release of allelochemicals through the leaves, roots, bark, or decomposing litter. This chapter explores the allelopathic effects of commonly used agroforestry tree species such as *Alnus nepalensis*, *Quercus leucotrichophora*, *Pinus roxburghii,* and *Grewia optiva* on traditional hill crops like maize (*Zea mays*), finger millet (*Eleusine coracana*), and paddy (*Oryza sativa*). Laboratory bioassays and field-level observations from various altitudinal zones in the Western and Central Himalayas reveal diverse impacts ranging from inhibition of seed germination and root–shoot growth to delayed crop maturation. Tree species like *Q. leucotrichophora* showed strong suppressive effects due to high concentrations of phenolics and other secondary metabolites, whereas *A. nepalensis* exhibited relatively neutral or even beneficial interactions. The chapter highlights how crucial it is to combine scientific research and traditional ecological knowledge in order to create tree–crop combinations that

U. B. Pratap (✉)
Department of Forest Products & Utilization, College of Forestry, Ranichauri, Veer Chandra Singh Garhwali Uttarakhand University of Horticulture and Forestry, Bharsar, Uttarakhand, India

N. Singh
Department of Forest Biology and Tree Improvement, College of Forestry, Ranichauri, Veer Chandra Singh Garhwali Uttarakhand University of Horticulture and Forestry, Bharsar, Uttarakhand, India

B. Singh
Department of Silviculture and Agroforestry, College of Forestry, Veer Chandra Singh Garhwali Uttarakhand University of Horticulture and Forestry, Tehri Garhwal, Uttarakhand, India

M. Kumar
Department of Forestry and Natural Resources, Hemvati Nandan Bahuguna Garhwal University (A Central University), Srinagar Garhwal, Uttarakhand, India

B. Singh et al. (eds.), *Himalayan Agroforestry Systems*, Advances in Agroforestry 15,
https://doi.org/10.1007/978-3-032-25096-4_10

are both ecologically harmonious and productive. It is essential to comprehend these allelopathic relationships in order to direct the design of agroforestry, increase crop yields, and guarantee the resilience of farming systems in the Himalayan region. Additionally, suggestions are given for future lines of inquiry and management techniques to lessen adverse allelopathic effects.

Keywords Agroforestry · Allelopathy · Himalayas · Traditional crops · Tree–crop interaction · Sustainable agriculture

1 Introduction

Allelopathy is a well-known ecological phenomenon that occurs naturally and strengthens the bonds between crops, animals, and trees (Zeng et al. 2008). The Greek terms "Allelon," which means "each other," and "Pathos," which means "to suffer," are the origins of the word "allelopathy" (Kundu et al. 2016), using a variety of techniques like volatilization, root exudation, and leaching (Bertin et al. 2003), decomposition of residues (Singh et al. 2001), exposure to stressful circumstances like high temperatures, droughts, and ultraviolet light (Rice et al. 2012) in agricultural and natural environments. Despite the fact that the term "allelopathie" was first used in 1937 by the Austrian scientist Hans Molisch (Willis 2007), the phenomenon of chemical interaction among plants has been understood for thousands of years. Different plant parts, including leaves, flowers, fruits, and buds, contain these inhibitory chemicals (Ashrafi et al. 2007) and are discharged into the surroundings, impeding the development of nearby plants.

Allelopathy represents a fundamental biological phenomenon by which an organism produces one or more biochemicals that influence the germination, growth, survival, and reproduction of other organisms (Rahaman et al. 2022). Originally conceptualized by Austrian professor Hans Molisch in 1937, the term derives from Greek compounds meaning "mutual harm" or "suffering," though modern understanding encompasses both beneficial and detrimental effects. These biochemicals, known as allelochemicals, are secondary metabolites that are not directly required for the metabolism of the allelopathic organism but serve crucial ecological functions in mediating interspecies interactions (Cheng and Cheng 2015). Allelochemicals are non-nutritive plant secondary metabolites that can have inhibitory or stimulatory effects on other plants and microorganisms, released through various pathways, including plant residues, roots, leaves, and volatilization (Mushtaq et al. 2020). The phenomenon extends beyond simple plant-to-plant interactions, encompassing chemical-mediated competition between any type of organisms, though it is most commonly studied in the context of plant communities where it serves as a biotic factor providing competitive advantages through direct interference with neighboring organisms' development (Kong et al. 2024).

The significance of allelopathy in agroforestry systems cannot be overstated, as these interactions fundamentally shape the productivity and sustainability of integrated tree–crop associations. Agroforestry systems, which deliberately combine trees with agricultural crops or livestock, create complex ecological environments where allelopathic interactions greatly influence crop production outcomes (Batish et al. 2008). When trees and crops are grown together, they interact through direct or indirect allelopathic mechanisms that can either inhibit or stimulate growth and yield, resulting in consequent effects including inhibited germination rates, reduced root and shoot extension, lack of root hairs, and swelling or necrosis of root systems (Devi 2017). Woody perennials in these systems release phytochemicals into the soil that adversely affect the germination and yield of understory crops, with multipurpose trees like *Leucaena leucocephala, Populus deltoides*, *Eucalyptus,* and *Acacia* species being particularly notable for their negative allelopathic effects on crop performance (Nair et al. 2021).

The relevance of studying allelopathy becomes particularly pronounced in the Himalayan context, where unique topographical, climatic, and ecological conditions create distinctive challenges and opportunities for agroforestry development. The Himalayan region's diverse altitudinal gradients, varying precipitation patterns, and complex microclimate variations influence the production and efficacy of allelochemicals, which are known to be affected by environmental variables, including nutrient availability, temperature, and soil pH (Kumar et al. 2020). Understanding allelopathic interactions in Himalayan agroforestry systems is crucial for developing sustainable agricultural practices that can address the region's specific challenges of soil conservation, slope stability, and climate adaptation while maintaining agricultural productivity. The mountainous terrain and fragile ecosystems of the Himalayas necessitate careful consideration of plant species selection and management practices to avoid negative allelopathic effects that could compromise crop yields or soil health. Furthermore, the rich biodiversity of the Himalayan region offers tremendous potential for discovering novel allelochemicals with beneficial properties for pest control and crop enhancement. A comprehensive understanding of allelopathy in this context can help develop more sustainable agroforestry systems that leverage the stimulatory influences of allelopathic plants while mitigating autotoxicity and other negative effects, ultimately contributing to the ecological integrity and economic viability of Himalayan agricultural landscapes (Rizvi et al. 1999).

2 Classification of Allelochemicals

Allelochemicals consist of various chemical families and are classified into the following categories based on chemical similarity (Rice 1974): water-soluble organic acids, straight-chain alcohols, aliphatic aldehydes and ketones; simple unsaturated lactones; long-chain fatty acids and polyacetylenes; benzoquinone, anthraquinone, and complex quinones; simple phenols, benzoic acid, and its derivatives; cinnamic acid and its derivatives; coumarin; flavonoids; tannins; terpenoids and steroids; amino acids and peptides; alkaloids and cyanohydrins; sulfide and glucosinolates; and purines and nucleosides (Cheng and Cheng 2015). Allelochemicals are also thought to include plant growth regulators such as ethylene, gibberellic acid, and salicylic acid (Fig. 10.1).

3 Concept of Allelopathy

The study of allelopathy traces its conceptual origins to ancient agricultural observations, though the term itself was formalized in 1937 by Austrian botanist Hans Molisch, who defined it as biochemical interactions where one plant inhibits another through released compounds (Xie et al. 2021). Historical records reveal that Theophrastus (300 BCE) documented pigweed's suppression of alfalfa, while the first -century Chinese text *Shennong Ben Cao Jing* cataloged 267 plants with pesticidal properties, hinting at early recognition of allelochemical effects (Willis 2007).

Allelopathic mechanisms operate through biochemical interference, where secondary metabolites disrupt physiological processes in target organisms. These compounds inhibit seed germination by destabilizing hormonal balances, impair root elongation through cellular membrane disruption, and suppress photosynthetic efficiency by blocking electron transport chains (Zhang et al. 2020). For instance, quinones like juglone (from *Juglans* species) and sorgoleone (from *Sorghum bicolor*) inhibit chloroplast oxygen evolution at micromolar concentrations, crippling ATP synthesis in both photosynthetic and mitochondrial pathways (Hejl 2001). Phenolic acids, such as cinnamic and benzoic derivatives, alter membrane permeability, induce oxidative stress via reactive oxygen species (ROS) accumulation, and interfere with nutrient uptake by disrupting ion channels (Hajam et al. 2023). Such multifaceted actions often result from synergistic effects of multiple allelochemicals, as seen in *Sorghum* species, where combined compounds target disparate cellular systems to amplify phytotoxicity (Hussain et al. 2021).

Allelochemical diversity spans several major classes, each with distinct ecological roles. *Phenolics*, derived from shikimic and phenylpropanoid pathways, constitute the largest group and include flavonoids, tannins, and coumarins, which inhibit enzyme activity and microbial symbionts (Shi et al. 2024). *Terpenoids*, volatile compounds like monoterpenes from *Eucalyptus*, disrupt membrane integrity and respiratory functions, while sesquiterpene lactones from *Ambrosia* species suppress auxin transport (El Amerany 2024). *Alkaloids*, nitrogen-containing metabolites such as caffeine (Coffea spp.) and nicotine (Nicotiana spp.), act as neurotoxicants by blocking acetylcholine receptors or disrupting DNA replication. Additionally, *glucosinolates* from Brassicaceae hydrolyze into isothiocyanates, inhibiting seed germination and microbial growth, whereas *cyanogenic glycosides release* hydrogen cyanide upon tissue damage, inducing cellular hypoxia (Awuchi 2019).

The release of allelochemicals occurs via four primary pathways, each shaped by ecological and physiological contexts. *Root exudation* delivers compounds like sorgoleone directly into the rhizosphere, creating chemical barriers against competitors (Chung 2018). *Leaching* from foliage, as observed in *Eucalyptus*, transfers water-soluble phenolics to soil through rainfall, while *volatilization* emits terpenoids and aldehydes into the air, affecting distant plants a strategy employed by *Artemisia* species to reduce neighboring seed viability (Chu et al. 2014). *Decomposition* of plant residues, such as juglone-rich walnut leaves, releases allelochemicals during microbial breakdown, inducing persistent soil toxicity. These pathways often

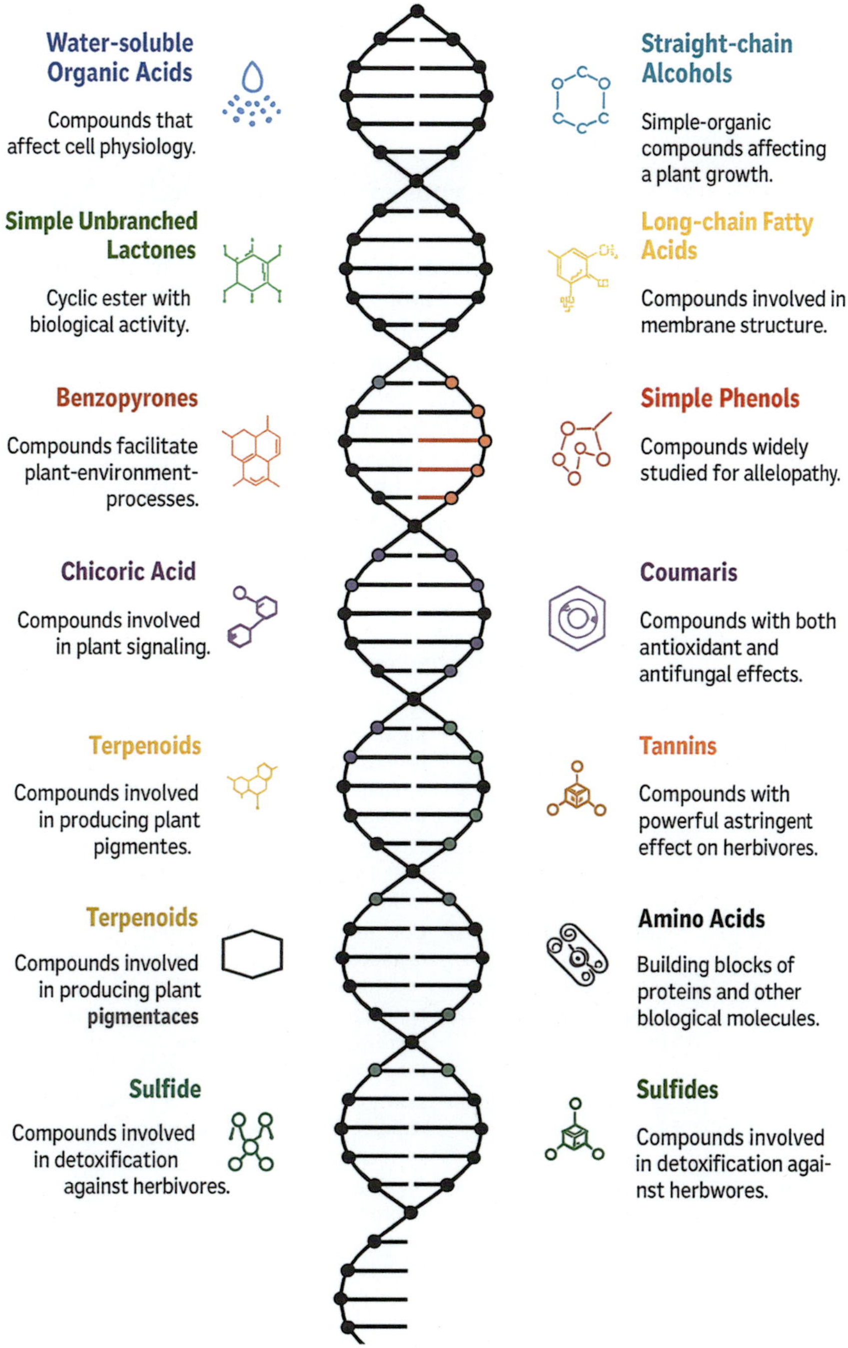

Fig. 10.1 Allelochemical classification. Putnam and Duke (1978)

operate synergistically; for example, *Centaurea maculosa* combines root exudates and litter decomposition to dominate invaded ecosystems through cumulative allelopathic pressure (Xu et al. 2023).

4 Traditional Agricultural Crops of the Himalayas

The Himalayan region, spanning across diverse altitudinal gradients and microclimates, hosts a rich tapestry of traditional crops that have sustained local communities for millennia. These crops, ranging from drought-resistant cereals like Finger millet (*Eleusine coracana*) and Barley (*Hordeum vulgare)* to protein-rich pulses such as Horse gram (*Macrotyloma uniflorum*) and Rajma (*Phaseolus vulgaris*), exemplify ecological adaptation and cultural significance. Recent studies highlight the critical role of traditional crops in combating climate change impacts, with farmers in Uttarakhand and Nepal reviving millets and drought-tolerant varieties to enhance food security (Pandey and Kumar 2024). The different traditional agricultural crops of the Himalaya are described in Table 10.1 (Fig. 10.2).

Table 10.1 Traditional crops of the Himalayas

Crop	Common name	Altitudinal range (m)	Key traits	Nutritional highlights	Sources
Zea mays	Maize	500–2500	Short growing period, drought-tolerant	High in carbohydrates, moderate protein	Salifu (2021)
Oryza sativa	Rice	600–2200	Flood-tolerant varieties in valleys	Rich in carbohydrates, B-vitamins	Negi and Joshi (2004)
Triticum aestivum	Wheat	1500–3000	Cold-tolerant landraces	High in protein, gluten content	Maikhuri et al. (1997)
Eleusine coracana	Finger Millet	1000–2500	Hardy, pest-resistant, long shelf life	Rich in calcium, iron, fiber	Rathi et al. (2024)
Panicum sumatrense	Little Millet	800–2000	Low input requirement, early maturing	High in fiber, iron, and micronutrients	Negi and Joshi (2004)
Vigna mungo	Black gram	600–1800	Short duration, improves soil fertility	High in protein and essential amino acids	Dhyani et al. (2011)
Phaseolus vulgaris	Common bean	1200–2400	Rich landrace diversity, cool-tolerant	Excellent protein source, rich in fiber and minerals	Rana et al. (2009)
Amaranthus hypochondriacus	Amaranth	1200–2800	Tolerant to drought and poor soils	Rich in lysine, protein, calcium	Nazeer et al. (2022)

Fig. 10.2 Landscape-scale traditional agroforestry models in the Himalayan mid-hills demonstrating integration of food crops, multipurpose trees, and homestead farming systems for biodiversity conservation and livelihood security

The common agroforestry tree species with their allelopathic effects are shown in Table 10.2 (Fig. 10.3).

Weeds in the Garhwal Himalaya exert significant *allelopathic effects* on crops, mainly through the release of allelochemicals that inhibit seed germination, seedling growth, and overall crop vigor. Species such as *Grewia oppositifolia* and *Ficus auriculata* have been shown to negatively impact traditional agricultural crops in this region, leading to reduced crop productivity. Toxic leachates from weeds like *Parthenium hysterophorus* further decrease germination rates and seedling development by disrupting nutrient uptake and cellular processes in crops. Proper *weed management* is critical in the region to prevent agricultural losses and maintain crop yields (Table 10.3).

5 Medicinal Plants in Different Agroforestry Systems

In India, diverse agroforestry practices have emerged in response to varying ecological and socioeconomic backgrounds across different regions. The adoption of agroforestry by marginal farmers is seen as a pathway to strengthen resistance to the effects of climate change. This practice not only diversifies livelihood sources but

Table 10.2 Common agroforestry tree species with allelopathic potential

Tree species	Local name	Allelopathic traits	Observed effects on crops	References
Quercus leucotrichophora	Banj oak	Tannins and phenolic compounds in leaf litter	Moderate suppression of seed germination in pulses	Pratap et al. (2020)
Alnus nepalensis	Uttis	Nitrogen-fixing, low allelopathic potential	Neutral to stimulatory effects on maize and millets	Parmar et al.(2022)
Eucalyptus globulus	Safeda	Volatile oils, high phenolic content	Strong inhibition of germination in wheat and legumes	Babu et al. (1997)
Grewia optiva	Bhimal	Moderate phenolics in leaves and bark	Minimal to mild suppression on finger millet	Sharma et al. (2025)
Populus deltoides	Poplar	Allelochemicals in leaf litter and bark leachates	Reduces germination and growth in vegetable crops	Singh et al. (2001)
Acacia catechu	Khair	Catechins, tannins, and polyphenols	Inhibitory to gram and lentil germination	Hossen and Hisashi (2020)
Melia azedarach	Daikan	Leaf and fruit leachates with bioactive compounds	Reduces seedling vigor in sensitive crops like spinach	Kumar et al.(2017)
Ficus auriculata	Timla	Phytotoxic compounds in leaves and bark	Inhibits growth of traditional food crops (*Echinochloa frumentaceae, Eleusine coracana, Amaranthus caudatus,* and *Vigna umbellata*)	Satapathy et al. (2022)
Ficus semicordata	Khaina	Secondary metabolites with allelopathic activity	Inhibits growth of traditional food crops (*Echinochloa frumentaceae, Eleusine coracana, Amaranthus caudatus,* and *Vigna umbellata*)	Satapathy et al. (2022)
Celtis australis	Khirk/Khari	Leaf litter and aqueous extracts releasing phenolics	Reduced seed germination growth of *Triticum aestivum, Hordeum vulgare, Brassica rapa, Lens culinaris,* and *Vigna umbellate*	Pant et al. (2018)

(continued)

Table 10.2 (continued)

Tree species	Local name	Allelopathic traits	Observed effects on crops	References
Rhododendron arboreum	Burans	Bioactive compounds in flowers and bark	Potential allelopathic effects, mainly found in *Eleusine coracana* and *Glycine max*	Rana et al. (2024)
Myrica esculenta	Kaphal	Phenolics, flavonoids, tannins, glycosides, saponins, terpenoids, alkaloids etc.	Inhibited germination of (*Eleusine coracana Setaria italica, amaranthus Amaranthus caudatus, Vigna umbellate, and Panicum miliaceum*)	Granthi (2023)
Ficus subincisa	Not specifically given (a multipurpose boundary tree in Garhwal Himalaya)	Leaf and bark leachates contain water-soluble allelochemicals; effect concentration-dependent	Strong inhibitory effect on germination of *Triticum aestivum* and *Hordeum vulgare*	Singh et al. (2009)
Bauhinia purpurea	Kachnar	Leaf and bark leachates exert both inhibitory and stimulatory effects; bark leachates particularly stimulatory at low concentration	Least toxic overall; inhibitory at 2% on *T. aestivum*	Singh et al. (2009)
Toona hexandra	Toon	Leaf and bark leachates contain phytotoxic substances; toxic at low concentration, stimulatory at higher concentration	Most toxic leaf leachate: reduced germination of *Brassica campestris*	Singh et al. (2009)
Phyllanthus emblica L.	Aonla (Indian Gooseberry)	Ttannins and phenolics; toxicity concentration-dependent	Reduction in germination and growth of *Amaranthus caudatus*; *coracana*, *Phaseolus mungo*)	Singh et al. (2010)

(continued)

Table 10.2 (continued)

Tree species	Local name	Allelopathic traits	Observed effects on crops	References
Bauhinia variegata	Kachnar	Leaves and bark contain phenolics, flavonoids, tannins, saponins, alkaloids, terpenoids	Aqueous extracts inhibited germination and seedling growth of *Triticum aestivum* (wheat); bark extracts stronger than leaves	Yadav et al. (2024)
Bauhinia retusa	Koiral	Similar secondary metabolites (phenolics, tannins, flavonoids, saponins, terpenoids)	Showed both inhibitory and stimulatory effects on wheat germination and seedling growth depending on extract concentration	Yadav et al. (2024)

Fig. 10.3 Field-level traditional agroforestry practices in the Himalayan region illustrating crop establishment, intercropping patterns, and soil–water conservation functions

Table 10.3 Allelopathic effects of weeds on crops

Species	Local name	Allelopathic traits	Observed effects on crops	References
Poa annua	Annual bluegrass	Phenolics, flavonoids, terpenoids, nitrogenous compounds	At 5% extract: ~5.66% reduction in crop germination; mild inhibition overall	Padu et al. (2023)
Erigeron annuus	Annual fleabane	Flavonoids, terpenoids, phenolics, benzoxazinoids	Strongest inhibitor: Up to −39.97% nutrient uptake; reduces germination by ~17.55% at 5% extract	Padu et al. (2023)
Stellaria media	Chickweed	Apigenin glycosides, tannins, coumarins, phenolics	Inhibited germination by ~18.86% at 5% extract; reduced chlorophyll content up to −63.74%	Padu et al. (2023)

also has a vital function in providing shade during heat waves and buffering soil moisture during dry spells (Singh et al. 2025). Medicinal plants play a vital role in different agroforestry systems of the Himalayan region by enhancing biodiversity, ecological sustainability, and rural livelihoods. These systems integrate medicinal and aromatic plants (MAPs) such as *Aconitum heterophyllum* (Atis), *Picrorhiza kurroa* (Kutki), *Swertia chirayita* (Chirayita), *Nardostachys jatamansi* (Jatamansi), and *Ocimum sanctum* (Tulsi) with multipurpose tree species. In the Western and Central Himalayas, MAPs are cultivated under the canopy of trees like oak (*Quercus leucotrichophora*) and poplar (*Populus deltoides*), creating microclimates conducive to shade-loving herbs. Traditional agroforestry practices include intercropping, multilayer cropping, and alley cropping, which optimize land use, conserve soil, and stabilize ecosystems. The cultivation of medicinal plants in agroforestry not only supports biodiversity conservation but also generates diversified income sources for local communities, preserving cultural heritage. Recent efforts focus on blending traditional knowledge with scientific techniques to ensure sustainable cultivation and effective marketing of medicinal plants, promoting environmental health and socioeconomic benefits across the Himalayas (Table 10.4).

6 Management Strategies

In the Western Himalayas, managing allelopathy requires a combination of strategies to ensure tree–crop compatibility and sustained productivity. Careful selection of compatible species such as *Grewia optiva* and *Bauhinia variegata*, while avoiding strongly allelopathic trees like *Eucalyptus*, helps reduce chemical stress (Bhatt and Singh 2009). Pruning, lopping, and timely removal or composting of leaf litter prevent toxic metabolite accumulation in soils (Bhatt et al. 2009). Spatial arrangements like buffer zones and contour planting further minimize root exudate spread and chemical leaching (Honfy et al. 2023). Additionally, crop rotation with tolerant species and adjusting sowing schedules allow natural degradation of allelochemicals, ensuring sustainable yields (Khamare et al. 2022).

Table 10.4 Agroforestry system types in the Western Himalaya showing key tree species and their associated medicinal plants along with their major uses

Agroforestry system type	Key tree species	Associated medicinal plants	Remarks/uses
Agrisilviculture	*Grewia optiva*	*Aloe vera*	Fodder tree; Aloe leaves used for medicinal gel
Agrisilviculture	*Toona ciliata*	*Rauvolfia serpentina*	Timber species; Sarpagandha roots used for hypertension treatment
Hortisilviculture	*Juglans regia*	*Ocimum sanctum*	Nut crop; Tulsi leaves valued for religious and medicinal purposes
Silvipastoral	*Bauhinia variegata*	*Asparagus racemosus*	Fodder and edible flowers; Shatavari roots used as a female tonic
Silvipastoral	*Celtis australis*	*Phyllanthus emblica*	Fodder species; Amla fruits rich in vitamin C, used in Ayurveda
Agrisilvipastoral	*Populus deltoides*	*Terminalia bellirica*	Timber species; Baheda fruits used for cough and digestive issues
Agrisilvipastoral	*Sapindus mukorossi*	*Withania somnifera*	Soapnut fruits used for cleaning; Ashwagandha roots as health tonic
Agrisilvipastoral	*Pinus roxburghii*	*Andrographis paniculata*	Resin source; Kalmegh leaves used for liver ailments

6.1 Selection of Compatible Tree–Crop Combinations

In the context of agroforestry in the Western Himalayas, careful selection of tree and crop species plays a crucial role in minimizing allelopathic stress. Native or locally adapted tree species that have neutral or positive interactions with traditional crops should be prioritized. For instance, species like *Grewia optiva*, *Bauhinia variegata,* and *Morusalba* have shown compatibility with crops such as wheat, barley, and mustard. Avoiding tree species known to release strong allelochemicals, like *Eucalyptus* or *Populus deltoides*, in close proximity to sensitive crops can help maintain productivity. Scientific screening of tree–crop combinations through bioassays and field trials can guide farmers in selecting combinations that ensure mutual growth benefits rather than chemical competition (Bhatt and Singh 2009).

6.2 Use of Pruning, Lopping, and Leaf Litter Management

Pruning and lopping of tree branches at regular intervals help reduce shading and also minimize the deposition of allelopathic leaf litter on agricultural fields. In the Western Himalayan context, where deciduous and broadleaf trees are common, leaf fall is significant and, if unmanaged, can lead to the accumulation of toxic secondary metabolites in the soil. Timely removal or composting of leaf litter before it decomposes can mitigate its allelopathic effects. Additionally, controlled use of tree biomass as mulch after composting (rather than fresh application) ensures that the decomposed organic matter supports soil fertility without introducing harmful phytotoxins (Bhatt et al. 2009).

6.3 Spatial Arrangement and Buffer Zones

Optimizing the spatial layout of trees and crops is another essential strategy. Proper spacing between tree rows and crop beds can reduce the spread of allelopathic chemicals through root exudates and fallen litter. Establishing buffer zonesstrips of neutral or tolerant vegetation such as grasses (e.g., *Cymbopogon* spp. or *Vetiver*) between allelopathic tree species and main crop areas can act as a biological barrier. In sloping terrains of the Himalayas, contour planting and strip cropping are also effective in controlling not just soil erosion, but also chemical leaching from trees to crops. Such designs help balance resource sharing and reduce the impact of chemical interference (Honfy et al. 2023).

6.4 Crop Rotation and Timing to Reduce Allelopathic Stress

Practicing crop rotation with tolerant species and adjusting sowing times can significantly reduce the impact of allelopathic interactions. Crops that are less sensitive to allelochemicals, such as maize or certain legumes, can be included in rotation cycles to naturally break the allelopathic cycle in the soil. Also, staggering planting times to allow allelopathic compounds to degrade before sowing sensitive crops helps mitigate negative effects. In the cooler climates of the Western Himalayas, understanding seasonal decomposition rates and aligning cropping schedules accordingly can optimize productivity while minimizing chemical stress on crops (Khamare et al. 2022).

6.5 Perceptions of Farmers and Indigenous Knowledge

In the Himalayan region, indigenous knowledge and traditional ecological wisdom of farmers play an important role in managing complex agroforestry systems, including the mitigation of allelopathic effects. Local farmers, through generations of observation and experience, have identified which tree–crop combinations are harmonious and which cause yield reduction due to chemical interference. For example, many avoid sowing sensitive crops near *Ailanthus altissima* or *Eucalyptus* due to their known adverse effects on soil and plant growth. Instead, they prefer trees like *Ficus roxburghii* or *Grewia optiva*, which have more neutral or beneficial interactions. Farmers also use traditional practices like drying or composting leaf litter before use, rotating crops and timing planting after heavy leaf fall to reduce allelopathic stress. Such local strategies, passed down orally and refined through trial and error, reflect a deep understanding of ecological balance and offer valuable insights for sustainable land management in the Western Himalayas (Islam et al. 2017).

Agroforestry in the Himalayan region often integrates multipurpose trees with high-value cash crops such as Turmeric (*Curcuma longa*), Ginger (*Zingiber officinale*), and Cardamom (*Elettaria cardamomum*). These combinations are adapted to the varied altitudes and microclimates of the Himalayas, maximizing land productivity and income diversification.

For example, farmers commonly plant turmeric and ginger in the shade of poplar (*Populus deltoides*) or oak (*Quercus leucotrichophra*), which provide partial canopy cover that protects the delicate rhizomes from direct sunlight while improving soil moisture retention. Similarly, cardamom thrives under forest canopies or agroforestry systems with shade trees like alder or species of *Alnus*. This mixed cropping enhances soil fertility through leaf litter and organic matter added by trees, conserves soil on steep slopes, and supports biodiversity.

6.6 Research Gaps and Future Directions

Contemporary allelopathy research stands at a critical juncture where significant advances in laboratory understanding require translation into practical agricultural applications. Current research reveals substantial gaps between controlled laboratory conditions and field implementations, highlighting the urgent need for methodological improvements and technological innovations. The field faces challenges in establishing clear evidence of allelopathic effects under natural conditions, inadequate chemical extraction and identification methodologies, and limited understanding of allelochemical mechanisms in complex environmental systems (Cheng and Cheng 2015). These limitations underscore the necessity for comprehensive approaches that bridge laboratory findings with real-world agricultural applications through enhanced research methodologies and interdisciplinary collaboration.

6.7 Need for Long-Term Field Studies

Long-term field studies represent the most critical gap in allelopathy research, as laboratory bioassays frequently yield different results compared to field assessments due to environmental variability and biological decomposition of allelochemicals. Current research demonstrates that many studies showing promising laboratory results fail to translate effectively to field conditions, where complex soil–plant interactions, microbial activity, and environmental factors significantly influence allelochemical bioavailability and effectiveness (Sahrir et al. 2023).

6.8 Identification and Quantification of Specific Allelochemicals

The identification and quantification of specific allelochemicals remain a fundamental challenge, limiting the practical application of allelopathy in agriculture. Recent advances using gas chromatography-mass spectrometry (GC-MS) combined with bioassays have successfully identified numerous compounds, including phenolic acids, organic acids, and aldehydes, with studies revealing up to 61 different chemical substances in soil samples (Bachheti et al. 2020).

6.9 Integration of Allelopathy in Agroforestry Planning

The integration of allelopathy principles into agroforestry planning represents an underexplored but promising avenue for sustainable agricultural development. Current research indicates that allelopathic plants can be strategically incorporated into farming systems through crop rotation, intercropping, and mulching practices to achieve natural weed suppression and enhance crop productivity (Rizvi et al. 1999).

6.10 Role of Biotechnology and Molecular Tools

Biotechnology and molecular tools offer unprecedented opportunities to advance allelopathy research and application, yet their potential remains largely untapped in current agricultural practices. Molecular breeding approaches have shown promise in developing allelopathic crop varieties, with studies identifying quantitative trait loci on specific chromosomes related to weed competitive ability in wheat (Akhtar et al. 2024).

7 Conclusion

The allelopathic interactions between agroforestry trees and traditional agricultural crops in the Himalayan region are complex and species-specific. While some tree species like *Alnus nepalensis* demonstrate minimal or even beneficial effects on crop growth, others such as *Pinus roxburghii* and *Grevillea robusta* exhibit strong inhibitory impacts due to the presence of allelochemicals like phenolics and terpenoids. These interactions can significantly influence seed germination, seedling vigor, and overall crop productivity. Therefore, understanding the allelopathic behavior of tree species is essential for designing ecologically harmonious and economically viable agroforestry systems. Integrating indigenous knowledge with scientific assessments can help in selecting compatible tree–crop combinations that support both ecological sustainability and agricultural resilience. For sustainable land use in the fragile Himalayan ecosystem, it is crucial to conduct long-term field studies, isolate active allelochemicals, and promote awareness among farmers. Such informed approaches will ensure balanced agroforestry practices and food security in the region.

References

Akhtar N, Shadab M, Bhatti N, SajidAnsarì M, Siddiqui MB (2024) Biotechnological frontiers in harnessing allelopathy for sustainable crop production. Funct Integr Genomics 24(5):155

Ashrafi ZY, Sadeghi S, Mashhadi HR, Hassan MA (2007) Allelopathic effects of sunflower (*Helianthus annuus*) on germination and growth of wild barley (*Hordeum spontaneum*). J Agric Technol 3(2):63–78

Awuchi GC (2019) The biochemistry, toxicology and uses of the pharmacologically active phytochemicals: alkaloids, terpenes, polyphenols and glycosides. J Food Pharm Sci 7(3):131–150

Babu RC, & Kandasamy OS (1997) Allelopathic effect of *Eucalyptus globulus* Labill. on *Cyperus rotundus* L. and *Cynodon dactylon* L. Pers. Journal of Agronomy and Crop Science, 179(2), 123–126.

Bachheti A, Sharma A, Bachheti RK, Husen A, Pandey DP (2020) Plant allelochemicals and their various applications. In: Co-evolution of secondary metabolites, pp 441–465

Batish DR, Singh HM, Kohli RK (2008) Allelopathic tree–crop interactions under agroforestry systems. In: Ecological basis of agroforestry. CRC Press, Florida

Bertin C, Yang X, Weston LA (2003) The role of root exudates and allelochemicals in the rhizosphere. Plant Soil 256(1):67–83. https://doi.org/10.1023/A:1026290508166

Bhatt BP, Singh JK (2009) Allelopathic effects of tree species on crops in Eastern Himalaya, India. Allelopathy J 23(2):333–344

Bhatt BP, Hussain S, Walling I, Singh JK (2009) Allelopathic effects of agroforestry trees on field crops in Eastern Himalaya, India. Allelopathy J 24(2)

Cheng F, Cheng Z (2015) Research progress on the use of plant allelopathy in agriculture and the physiological and ecological mechanisms of allelopathy. Front Plant Sci 6:1020

Chu C, Mortimer PE, Wang H, Wang Y, Liu X, Shixiao Y (2014) Allelopathic effects of Eucalyptus on native and introduced tree species. For Ecol Manag 323:79–84

Chung HY (2018) The influence of sorghum physiology on rhizosphere interactions and their effect on root disease. Doctoral dissertation, University of the Free State

Devi M (2017) Allelopathy in agroforestry: a review. J Pharmacogn Phytochem 6(3):686

Dhyani BL, Madhu M, Singh C, Kumar A, Bankey B, Mandal D, Muruganandam (2011) Participatory dissemination and assessment of land and water management technologies for livelihood security in rainfed areas of north western Himalayas. Annual report, 2011–2012, CSWCRTI, Dehradun, pp 99–100

El Amerany F (2024) The role of terpenoids in plant development and stress tolerance. In: Molecular and physiological insights into plant stress tolerance and applications in agriculture-part 2. Bentham Science Publishers, pp 71–98

Granthi D (2023) Allelopathic effect of leaf and bark extract of *Myrica esculenta* on agricultural test crops under laboratory and nursery conditions. Master's dissertation, College of Forestry, Ranichauri

Hajam YA, Lone R, Kumar R (2023) Role of plant phenolics against reactive oxygen species (ROS) induced oxidative stress and biochemical alterations. In: Plant phenolics in abiotic stress management. Springer Nature Singapore, Singapore, pp 125–147

Hejl AM (2001) Mechanisms of sorgoleone and juglone growth inhibition in soybean (Glycine max L. Merr) and corn (Zea mays L.). University of South Dakota

Honfy V, Pödör Z, Keserű Z, Rásó J, Ábri T, Borovics A (2023) The effect of tree spacing on yields of alley cropping systems—a case study from Hungary. Plants 12(3):595

Hossen K, & Hisashi KN (2020) Determination of allelopathic properties of *Acacia catechu* (Lf) Willd. Notulae Botanicae Horti Agrobotanici Cluj-Napoca, 48(4), 2050–2059.

Hussain MI, Danish S, Sánchez-Moreiras AM, Vicente Ó, Jabran K, Chaudhry UK et al (2021) Unraveling sorghum allelopathy in agriculture: concepts and implications. Plants 10(9):1795

Islam MA, Qaisar KN, Bhat GM (2017) Indigenous knowledge in traditional agroforestry systems of Kashmir valley: current challenges and future opportunities. Int J For Crop Improv 8:68

Khamare Y, Chen J, Marble SC (2022) Allelopathy and its application as a weed management tool: a review. Front Plant Sci 13:1034649

Kong CH, Li Z, Li FL, Xia XX, Wang P (2024) Chemically mediated plant–plant interactions: allelopathy and allelobiosis. Plants 13(5):626

Kumar D, Thakur NS, & Gunaga RP (2017) Allelopathic influence of leaf aqueous extract and leaf litter of Indian lilac (*Melia azedarach* L.) on germination, growth, biomass and grain yield of green gram (*Vigna radiata* L.) and black Chickpea (*Cicer arietinum* L.). International Journal of Current Microbiology and Applied Sciences, 6(10), 2669–2683.

Kumar N, Kumar A, Jeena N, Singh R, Singh H (2020) Factors influencing soil ecosystem and agricultural productivity at higher altitudes. In: Microbiological advancements for higher altitude agro-ecosystems & sustainability, pp 55–70

Kundu R, Ghosh S, Mukherjee A (2016) Allelopathic potential of medicinal plants: a review. Int J Herb Med 4(3):88–93

Maikhuri RK, Semwal RL, Rao KS, Nautiyal S, Saxena KG (1997) Eroding traditional crop diversity imperils the sustainability of agricultural systems in Central Himalaya. Curr Sci 73:777–782

Mushtaq W, Siddiqui MB, Hakeem KR (2020) Allelopathy. Springer International Publishing

Nair PR, Kumar BM, Nair VD, Nair PR, Kumar BM, Nair VD (2021) Multipurpose trees (MPTs) and other agroforestry species. In: An introduction to agroforestry: four decades of scientific developments, pp 281–351

Nazeer S, Zubair Akram M, Ali M (2022) Amaranth as nutrition-rich and climatic resilient crop: a review. Agric Conspec Sci 87(4):281–293

Negi GCS, Joshi V (2004) Rainfall and spring discharge patterns and relationships in two small catchments in the western Himalayan Mountains, India. The Environmentalist 24:19–28

Padu K, Khanduri VP, Singh B, Rawat D, Riyal MK, Kumar KS (2023) Phytotoxicity of common weeds on germination, seedling growth, NPK uptake and chlorophyll content of four hill crops of Garhwal Himalaya. J Agric Food Res 12:100539

Pandey M, Kumar N (2024) Diversity of traditional grain legumes of Himalayan region of Uttarakhand: a review. Agric Rev 45(1):96–102

Pant N, Khanduri VP, Bali RS, Rawat P, Gopal R (2018) Allelopathic effect of Celtis australis on germination and growth parameters of some important food crops of the Garhwal Himalayas. J Appl Nat Sci 10(3):1003

Parmar B, Vishwakarma A, Padbhushan R, Kumar A, Kumar R, Kumari R, Kumar Yadav B, Giri SP, Kaviraj M. & Kumar U (2022) Hedge and alder-based agroforestry systems: Potential interventions to carbon sequestration and better crop productivity in Indian sub-himalayas. Frontiers in Environmental Science, 10, 858948.

Pratap UB, Singh B, Khanduri VP, & Pant KS (2020) Allelopathic propensity of Quercus species on the performance of traditional food crops of the mid-Himalayan region. Journal of Pharmacognosy and Phytochemistry, 9(5), 2521–2525.

Putnam AR, Duke WB (1978) Allelopathy in agroecosystems

Rahaman F, ShukorJuraimi A, Rafii MY, Uddin K, Hassan L, Chowdhury AK et al (2022) Allelopathic potential in rice-a biochemical tool for plant defence against weeds. Front Plant Sci 13:1072723
Rana DK, Rawat SS, Uniyal SP, Mishra AC (2009). Effect of sowing dates, phosphorus levels and seed treatment with rhizobium culture on growth and yield of french bean. Vegetable Sci 36(1):116-118
Rana K, Dhanai CS, Negi A (2024) Influence of *Rhododendron arboreum* Sm. leaf and bark aqueous extract on germination and growth of *Eleusine coracana* L., *Echinochloa frumentacea* Link, *Glycine max* L. and *Vigna umbellata* Thunb. Allelopath J 63(2):197
Rathi BK, Jain AK, Sharma N, Jain RK (2024) Millets as nutri-cereals: global and Indian scenario and its nutritional and health benefits. Urban India 44(2):143–158
Rice EL (1974) Allelopathy. Academic Press, NY
Rice PJ, Pang T, Coats JR (2012) Phenolic root exudates as allelochemicals and their impacts on soil microbial communities. J Chem Ecol 38(6):746–757. https://doi.org/10.1007/s10886-012-0138-7
Rizvi SJH, Tahir M, Rizvi V, Kohli RK, Ansari A (1999) Allelopathic interactions in agroforestry systems. Crit Rev Plant Sci 18(6):773–796
Sahrir MAS, Yusoff N, Azizan KA (2023) Allelopathy activity under laboratory, greenhouse and field conditions: a review. AIMS Agric Food 8(1):78
Salifu M (2021) The Effects of Water Supply on the Viability, Growth, Yield, Physiology, and Quality of Maize (*Zea Mays* L.) (Doctoral dissertation, Debreceni Egyetem (Hungary).
Satapathy SR, Khanduri VP, Singh B, Riyal MK, Kumar S, Kumar P, Rawat D (2022) Allelopathic potential of Ficus auriculata and Ficus semicordata on growth of four traditional food crops of Garhwal Himalaya. J Agric Food Res 9:100352
Sharma M, Kaler NS, Sharma S, Sharma R, Sharma S, & Ghabru A (2025) Allelopathic influence of *Morus alba* and *Grewia optiva* on growth and performance of pulse crops. Forest Ecology and Management, 596, 123113.
Shi R, Jin G, Shen S, Xu G, Zheng F, Clements DR et al (2024) Allelopathic molecular mechanisms of the two main allelochemicals in sweet potato. Curr Issues Mol Biol 46(11):11890–11905
Singh H, Kohli R, Batish D (2001) Allelopathic interference of *Populus deltoides* with some winter season crops. Agronomie 21(2):139–146
Singh HP, Batish DR, Kohli RK (2001) Allelopathy in agroecosystems: an overview. J Crop Prod 4(2):1–41. https://doi.org/10.1300/J144v04n02_01
Singh B, Jhaldiyal V, Kumar M (2009) Effects of leaf and bark leachates on some multipurpose trees on germination and growth of field crops. Estonian J Ecol 58(1):38–46
Singh B, Bhatt S, Uniyal P (2010). Allelopathic behavior of *Phyllanthus emblica* on germination and growth of field crop. Range Manag Agrofor 31(1):27–30
Singh N, Bali RS, Thakur P., Lal P, Gopal R, Kumar A, Kumar A (2025) Livelihood improvement through the integration of medicinal plants in agroforestry. In: Kumar A, Shukla G, Gopal R, Dinesha S (eds) Agroforestry as a nature-based solution: a holistic approach to food security and climate change. Cambridge Scholars Publishing, pp 327–358
Willis RJ (2007) The history of allelopathy. Springer Science & Business Media
Xie Y, Tian L, Han X, Yang Y (2021) Research advances in allelopathy of volatile organic compounds (VOCs) of plants. Horticulturae 7(9):278
Xu Y, Chen X, Ding L, Kong CH (2023) Allelopathy and allelochemicals in grasslands and forests. Forests 14(3):562
Yadav N, Khanduri VP, Singh B, Rawat D, Riyal MK (2024) Phytotoxicity of two Bauhinia species on four *Triticum aestivum* varieties in laboratory bioassay. Int J Plant Biol 15(3):599–606
Zeng RS, Mallik AU, Luo SM (2008) Allelopathy in sustainable agriculture and forestry. Springer Science & Business Media
Zhang S, Sun SW, Shi HL, Zhao K, Wang J, Liu Y, Wang W (2020) Physiological and biochemical mechanisms mediated by allelo chemicalisoliquiritigenin on the growth of lettuce seedlings. Plants 9(2):245

Chapter 11
Soil Carbon Storage Potential of Different Land-Use Systems in a Srinagar Valley of Garhwal Himalaya, India

Arjun Tomar, Munesh Kumar, and Tarun Kumar

Abstract Land-use changes occur due to degradation and shifts of ecosystems, accomplishing the anthropogenic demands of development and population explosion, especially in the developing world. The changes in land use and land cover (LULC) from forest land to urban cause a decrease in carbon sink. With this perspective, this study was conducted to estimate the carbon storage potential of four different land uses, that is, forest land (FL), agriculture land (AL), habitation land (HL), and river land (RL) in the lower Himalaya, India. Soil organic carbon (SOC) and bulk density (BD) were estimated based on the standard protocol by collecting samples from the area under study. BD of soil was estimated to be the maximum in RL, followed by HL, and the minimum in AL. Similarly, SOC stock was also estimated to be the maximum in RL and the minimum in HL. Further, SOC storage potential in RL is found to be the maximum. Thus, we recommend that RL may be optimally utilized by suitable tree planting matching with site conditions to support nature-based solutions to climate change by improving its carbon sink potential.

Keywords Afforestation · Carbon sink · LULC · Soil organic carbon

1 Introduction

Different land uses play an important role in controlling the amount and quality of litter input, decomposition rate, and stabilization of soil organic carbon (SOC), which further influences carbon dynamics, both source and sink (Padhan et al. 2016;

A. Tomar · M. Kumar (✉)
Department of Forestry and Natural Resources, Hemvati Nandan Bahuguna Garhwal University (A Central University), Srinagar Garhwal, Uttarakhand, India

T. Kumar
Department of Environmental Science, Indira Gandhi National Tribal University, Amarkantak, Madhya Pradesh, India
e-mail: tarun.thakur@igntu.ac.in

B. Singh et al. (eds.), *Himalayan Agroforestry Systems*, Advances in Agroforestry 15,
https://doi.org/10.1007/978-3-032-25096-4_11

Sahoo et al. 2019). Any change in land use and land cover (LULC) influences SOC, which alters the carbon fluxes to the atmosphere (Schlesinger 1997; Guggenberger et al. 1995). In the soil, changes in SOC and nutrients due to LULC also change the plant diversity and soil carbon stock (Ostle et al. 2009). The change in land use accounts for variation in vegetation structure, besides nutrient status, bringing the spatial variability in the SOC. Therefore, strategic management practices for carbon sequestration should be managed as per the land- use type by accounting for their potential for carbon capturing (Wang et al. 2010).

It is reported that the changes in land use from annual to perennial in progressive to SOC level and carbon sequestering in the soil (Gebhart et al. 1994). It is also evident that anthropogenic activity in tropical forests is the main carbon source, which not only brings change to its vegetation and ecological processes but also to its soil properties (Fan et al. 2016; Iqbal and Tiwari 2016). Benbi et al. (2014) highlighted that cultivated soils lose about 21–36% organic carbon compared to uncultivated soils. The change in the soil structure and other activities leads to SOC loss (Golchin and Asgari 2008). Thus, the changes in land use can be the most important CO_2 sources that cause a rise in atmospheric temperature and aid in climate change. The change of natural forests to other purposes such as plantations, croplands, and others, particularly in the tropics and subtropics, is of great concern.

Soil is an important carbon pool, and therefore estimation of SOC would provide baseline information on the carbon storage potential of different land uses and contribute to the nationally determined carbon contribution (INDC) and effective planning on land-use practices. Forests and agriculture are two of the major land-use sectors in India, and there has been large-scale conversion of forest land to agriculture in India due to population pressure and to provide alternate sources of livelihood. However, during the land-use conversion, the consequences of this to climate change are mostly ignored. However, when the degraded lands are brought under effective agriculture, the soil starts to build up, and there occurs a progressive SOC change in soil (Singh et al. 2018). SOC being very vital to crop growth, its spatial variation in cropping systems would affect crop productivity. Therefore, it would be most desirable to know the SOC status for soil fertility management and decide if supplementation of residue input is desirable for optimum crop productivity (Kumar and Sharma 2015, 2016).

The change in the SOC stock in different land uses has a direct impact on the global carbon stock and land productivity, and the changes in SOC may be crucial for carbon dynamics (sink and source) and greenhouse gases (GHGs) for mitigation purposes. The study of different land-use patterns has been carried out by different workers such as Abera and Belachew (2011) in Ethiopia; Sabina Yeasmin et al. (2020) in Bangladesh; Chenglong Tu et al. (2018) in China; Kenye et al. (2019) in Mizoram, India; and Sharma et al. (2019) in Rajasthan, India. However, so far SOC studies in different land-use patterns, in particular of the study region, Srinagar Garhwal of Uttarkhand, have not been reported; thus, the gap of information in SOC under different land uses is being accounted for in the study by generating and estimating the SOC stock in different land uses (e.g., habitation land, agriculture land, river land, and forest land) in the Srinagar Valley of Uttarakhand, India, with the hypothesis that the different land-use patterns have differential capabilities of storing the SOC stock. The findings of this study contribute to the National Strategy for adaptation to climate change, the United Nations Framework Convention on Climate Change (UNFCCC), and other spatial levels of estimates of carbon stock. Thus, to understand the potential of soil organic carbon stock in different land-use patterns, with the objective to understand SOC's storing potential in different land-use patterns in the Srinagar Valley of Garhwal Himalaya.

2 Materials and Methods

2.1 Study Area

The study was carried out in the Srinagar Valley of Garhwal Himalaya in District Tehri Garhwal (Uttarakhand, India). The valley is located between latitude 30° 12′ 45″ to 30° 14′ 42″ N and longitude 78° 44′ 51″ to 78° 49′ 30″ E at altitudes between 564 and 624 m a.m.s.l and falls under a subtropical climate (Fig. 11.1, location

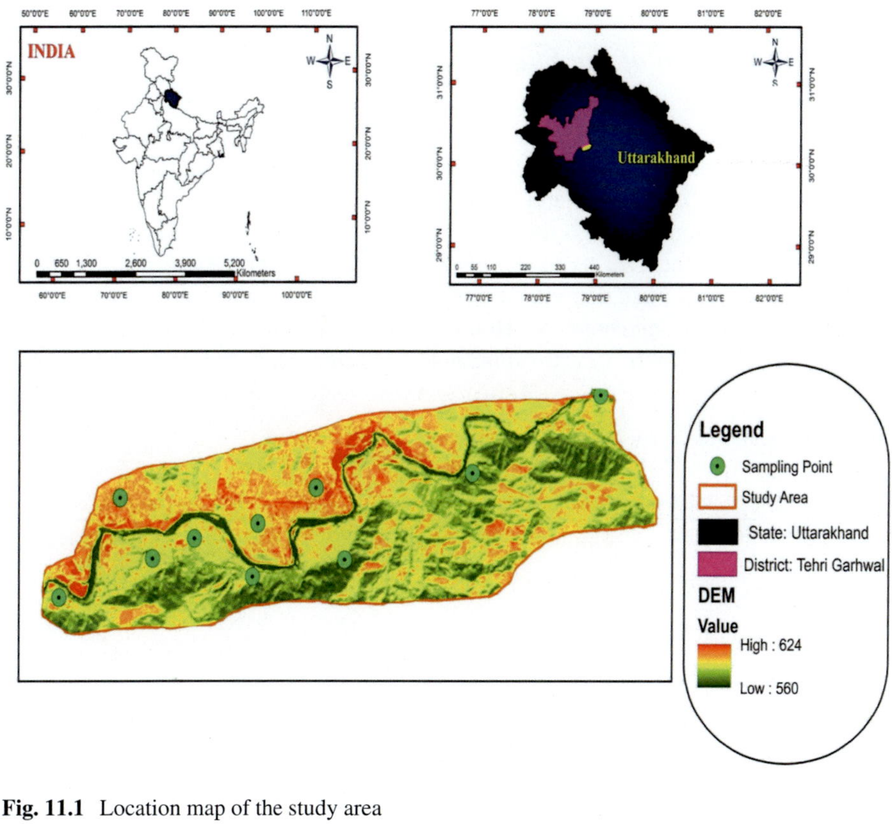

Fig. 11.1 Location map of the study area

map). The climate is hot in summer and cold in winter with three distinct seasons (i.e., summer, rainy, and winter). The annual mean temperature during 2017–18 (study period) ranges from 10.53 to 28.90 °C and rainfall of 151 mm (Department of Rural Technology 2018). The River Alakananda flows across the Srinagar Valley, which is a major city of the Himalayan state of Uttarakhand. The total population of Srinagar city is about 20,000 (census, 2011). The livelihood of city dwellers revolves around service and business; however, villagers are primarily dependent on agriculture, livestock, and forests. The soil of the area falls between sandy loams and sandy clayey loam. The forests are moderately dense and act as a good source of forest products for local communities, besides providing fodder for livestock and a source of grazing land. The common trees found in the forest land area are Chir pine (*Pinus roxburghii* Sarg.), Dhaora (*Anogeissus latifolia* Wall ex Bedd.), Indian ash tree [*Lannea coromendlica* (Houtt.) Merr., J. Arnold Arbor.], and Amla (*Emblica officinalis* Gaertner). The trees in agricultural land and habitation land are Bhimal (*Grewia optiva* J.R. Drumm.ex Burret), Mulberry (*Morus alba* L.), Kharik (*Celtis australis* L.), and orchid tree (*Bauhinia variegata* L. Benth.); however, along the river, the major species grown in the region are Sissoo (*Dalbergia sissoo* L.) and Khair (*Acacia catechu* L.f. Willd.).

2.2 *Stratification of the Study Area*

The valley area, after a reconnaissance survey, was classified into four land-use types based on their use. These land uses were forest land (FL), agriculture land (AL), habitation land (HL), and river land (RL). Sampling sites were selected randomly for the study and distributed across the area. Table 11.1 provides a detailed description of the study area. Under FL type, *Pinus roxburghii* Sarg., and *Anogeissus latifolia* Wall ex Bedd. forests were dominant trees with associated species of *Acacia catechu* L.f. Willd., *Lannea coromendalica* (Houtt.) Merr., J. Arnold Arbor, in the tree layer and *Lantana indica* Roxb., *Carrisa spinarum* L., *Rhus parviflora* Roxb., and *Murrya*

Table 11.1 Details of the sites, samples, and location of the study area

Land use	Total site[a]	Total sample[b]	Altitude (m asl)	Latitude	Longitude
Habitation land	3	12	555–656	30° 12′ 871″ N 30°14′ 46″ N	78° 44′ 951″ E 78° 49′ 29″ E
Agriculture land	3	9	550–624	30° 13′ 0.300″ N 30° 14′ 42″ N	78° 45′ 559″ E 78° 49′ 25″ E
River land	2	6	530–545	30° 13′ 39.99″ N 30° 13′ 48.36″ N	78° 48′ 37.44″ E 78° 49′ 17.60″ E
Forest land	3	6	687–627	30° 13′ 212″ N 30° 14′ 40″ N	78° 45′ 028″ E 78° 49′ 12″ E

[a]Each site has three depths
[b]Number of samples at one depth

koenigii (L.) Spreng. in the shrub layer. In AL use, the dominant crops grown were wheat (*Triticum aestivum* L.), pea (*Pisum sativum* L.), and rai (*Brassica compestris* L.) for Ravi crop and rice (*Oryza sativa* L.), millets for the Kharif crop. The HL use consists of built-up areas where people reside and are densely populated. The soil samples in the habitation were collected randomly across the movement path to ensure the regular anthropogenic activities led to almost no growth of vegetation. The RL along the riverbank with floodplains were selected randomly close to the riverbed so that the impacts of stream change and flood could be accounted for. The sample points used in the study area are mentioned in Fig. 11.1.

2.3 *Methodology Adopted*

In the present study, Srinagar Valley was surveyed to collect the soil samples at random from different land-use covers, that is, HL, AL, RL, and FL. The collected soil samples during 2017–2018 from each land use and three depths (i.e., 0–30, 30–60, and 60–90 cm) were brought to the laboratory for further analysis. A total of 36 soil samples (12 samples × 3 depths) were collected from HL, 27 (9 samples × 3 depths) from AL, and 18 (6 samples × 3 depths) samples both from RL and FL. Thus, from all land-use systems, a total of 99 soil samples were collected. The variation in soil sampling from different land uses was based on variability within the land use. The collected samples were used in the laboratory for analysis. The soil bulk density (BD) was estimated using core sample methods (Kumar and Kumar 2020). The SOC was estimated by Walkley and Black methods (1934) using the following formula:

$$\mathrm{SOC}(\%) = \{10(\mathrm{B}-\mathrm{T})/\mathrm{B}\} \times \{0.003 \times 100 / \mathrm{wt.of\ soil\ sample\ (g)}\}$$

The actual soil organic carbon was estimated using the following formula:

$$\mathrm{SOC}(\%) = \mathrm{Estimated\ soil\ organic\ carbon} \times 1.32(\mathrm{correction\ factor})$$

The correction factor was used to compensate for the incomplete oxidation of organic matter (OM), as only 77% carbon was recovered under the process (Kumar et al. 2017). SOC stock (t ha^{-1}) was estimated using the following equation (Kumar et al. 2017):

$$\mathrm{SOC\ Stock}\left(\mathrm{t\,ha^{-1}}\right) = \mathrm{Soil\ BD}\left(\mathrm{g.cm^{-3}}\right) \times \mathrm{Soil\ depth}(\mathrm{cm}) \times \%\mathrm{OC}$$

2.3.1 Land Use and Vegetation Analysis

In the study area, land cover and vegetation were analyzed using RS and GIS techniques with a few field measurements.

2.3.2 Satellite Data Collection and Preprocessing of Satellite Data

The data covers the entire study area and its surrounding environments. Sentinel 2A satellite data from 2020 was freely obtained from the United States Geological Survey (http://glovis.usgs.gov). The digital image analysis for the LULC classes of the procured data was performed on ERDAS Imagine software on a personal computer, and the ancillary data were collected through the field sampling technique as ground control points with the help of GPS and direct local field measurements for ground verification, which was performed for image classification. The coordinates of the sample points in the field were taken, and a total of 25 samples were selected for ground survey in various land-use types.

2.3.3 Supervised Classification and Accuracy Assessment

The digital classification method of Sentinel 2A satellite data for the analysis of LULC and vegetation was done using supervised classification in ERDAS Imagines software 2014 (Kumar et al. 2017; Thakur et al. 2021). Four LULC classes, viz., habitation land, agriculture land, forest land, and river land, were delineated (Fig. 11.2, Table 11.2). The accuracy of classification was estimated between ground truth and classified pixels (Congalton 1991). The kappa coefficient was calculated to assess overall accuracy of classification.

2.3.4 Normalized Difference Vegetation Index (NDVI)

NDVI is used to assess the contribution of vegetation to multispectral observations (Pearson and Miller 1972) and estimated by the following:

$$\mathrm{NDVI} = \frac{\mathrm{NIR} - \mathrm{R}}{\mathrm{NIR} + \mathrm{R}} \text{ or } \frac{\mathrm{Band\,3} - \mathrm{Band\,2}}{\mathrm{Band\,3} + \mathrm{Band\,2}}$$

Here, NIR indicates near-infrared light and R means visible red light; NDVI range of +1.0 to -1.0, of barren rock, sand, or snow shows very low NDVI (i.e., 0.1 or less). The NDVI map is shown in Fig. 11.3.

2.4 Statistical Tools Used

The descriptive statistics were estimated for the collected data. ANOVA was used to test the difference in SOC between the different land uses. Analyses were done using SPSS 20.0 and Microsoft Word, 2007 version.

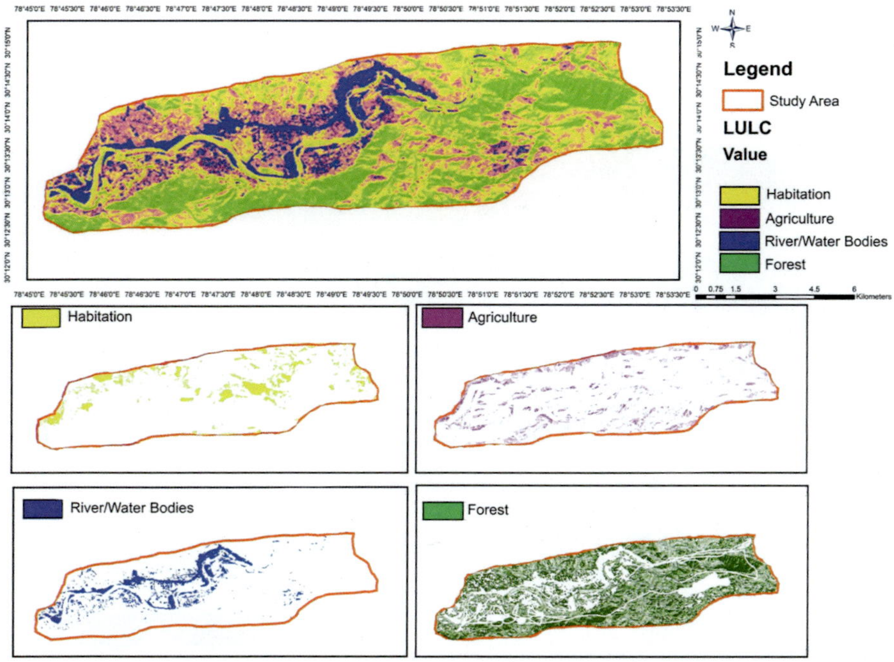

Fig. 11.2 Classified image of different land-use patterns in the study area

Table 11.2 LULC with classification accuracy of the study area

Class name	Area (km^2)	Distribution percent	Classification accuracy
Habitation land	2.01	4.820144	87.25
Agriculture land	5.36	12.85372	91.19
River land	11.25	26.97842	98.21
Forest land	23.08	55.34772	93.56
Total	41.7	100	

3 Results and Discussion

This study was conducted in the subtropical region of the Himalayas under four different land uses. Among all four land-use types, BD of soil increased with increasing soil depth. The range of BD in each land-use pattern was from 1.21 to 1.35 g cm^{-3} (HL), 1.12 to 1.18 g cm^{-3} (AL), 2.71 to 2.83 g cm^{-3} (RL), and 1.16 to 1.38 g cm^{-3} (FL) (Fig. 11.4). Among the land uses, the maximum average BD was reported to be 2.77 ± 0.21 g cm^{-3} in RL followed by HL (1.3 ± 0.22 g cm^{-3}), FL (1.28 ± 0.15 g cm^{-3}), and AL (1.16 ± 0.21 g cm^{-3}) (Table 11.3). The maximum value of BD in RL could be because of regular deposition of the layer of soil with boulders, granules, and gravel at the river bed, and with a low occurrence of vegetation, it creates compactness. The BD of HL may be accounted for due to the compaction

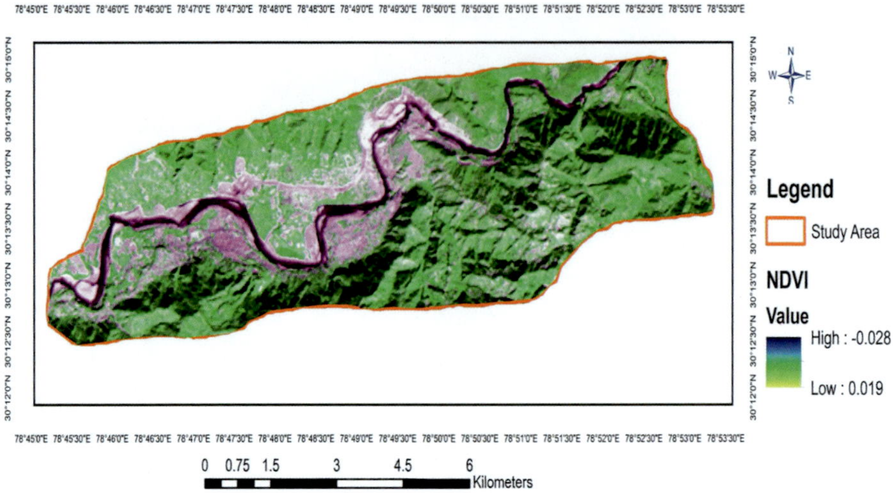

Fig. 11.3 NDVI map of the study area

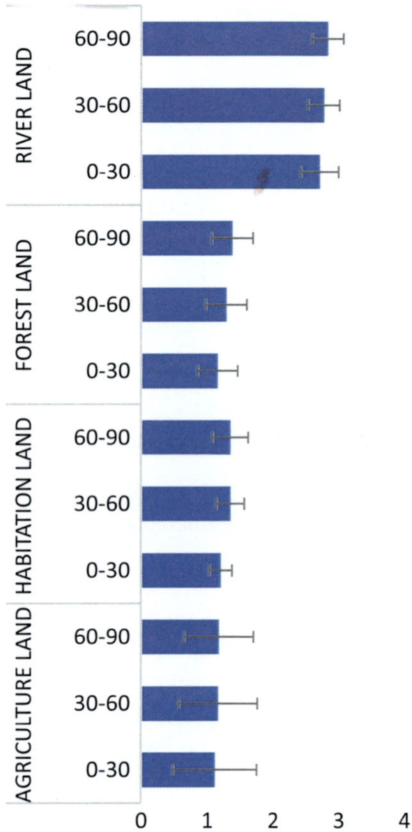

Fig. 11.4 BD (g.cm^{-3}) (mean ± SD) of soil in different land-use types at different depths (cm)

Table 11.3 Descriptive statistics (mean ± SD) for BD, SOC, and SOC stock in different land--use types

Land-use types	BD (g/cm^3)	SOC (%)	SOC stock (t ha^{-1})
Habitation land	1.30 ± 0.22	1.17 ± 0.47	46.2 ± 18.58
Agriculture land	1.16 ± 0.21	1.52 ± 0.25	53.68 ± 13.55
River land	2.77 ± 0.21	2.01 ± 0.18	168.08 ± 12.56
Forest land	1.28 ± 0.15	2.52 ± 0.80	96.38 ± 24.14

BD bulk density, *SOC* soil organic carbon concentration

of soil and regular anthropogenic activities leading to compacted clay and minerals. The low BD in FL is attributed to regular decomposition of leaf litter and other organic matter in the soil due to various microbial activities. The lowest BD in AL is because of regular plowing of farmland and crop cultivation, which restricts the higher compactness in the soil and high OM with clay contents.

This difference in BD among the land uses was found to be significant ($p < 0.05$) between HL and RL; AL and RL; and RL and FL. The BD in forests was similar to the present study in a community-managed forest of Garhwal Himalaya (Mahato et al. 2016). BD under AL was also reported to be similar for cultivated land in other parts of the world, such as in Nigeria (Ufot et al. 2016), Thailand (Swangjang 2015), and Iran (Abad et al. 2014). The BD for different land-use types, that is, open barren shrub land (4.50 g cm^{-3}), closed species-rich shrub land (1.46 g cm^{-3}), secondary forest (1.43 g cm^{-3}), and primary forest (1.22 g cm^{-3}), is reported for subtropical India by Purakayastha and Patra (2007). BD is highly correlated with the SOC stock in the Himalayan region (Kumar and Sharma 2016; Kumar et al. 2018) mainly due to different vegetation patterns and soil, climatic conditions, and root hairs.

Land-use changes play an important role in carbon sequestering in soil. The carbon pools depend on land-use patterns due to variability in the source material (quality and/or quantity), physicochemical characteristics of soil, and microbial activities, which in turn affect the degradation rate of OM (Kumar et al. 2017). The ranges of SOC concentration estimated in different land-use types were 1.04 to 1.44% (HL), 1.39 to 1.65% (AL), 1.90 to 2.09% (RL), and 2.07 to 2.96% (FL) (Fig. 11.5). The effect of SOC on different land-use patterns and depths was significantly different ($p < 0.05$). The significant difference in SOC concentration was between 0–30 cm and 60–90 cm and a nonsignificant difference was reported between 0–30 cm and 30–60 cm and 30–60 cm and 60–90 cm depths. The SOC concentration was reported highest in FL (2.52 ± 0.80%), followed by RL (2.01 ± 0.18%), AL (1.52 ± 0.25%), and HL (1.17 ± 0.47%) (Table 11.3).

In a community-managed forest in Garhwal Himalaya, India, the average SOC concentration was 1.28%. The low SOC under AL and HL may be attributed to anthropogenic disturbances and therefore indicate that human activities were an important contributor to the variability in the concentration of SOC levels. SOC concentrations lie between 5.2% and 6.5% for mangrove land and 1.2–5.3% in wetlands studied in West Bengal, India. The high concentration of SOC is attributed to large carbon sequestration of carbon in soil in wetland. The mean SOC

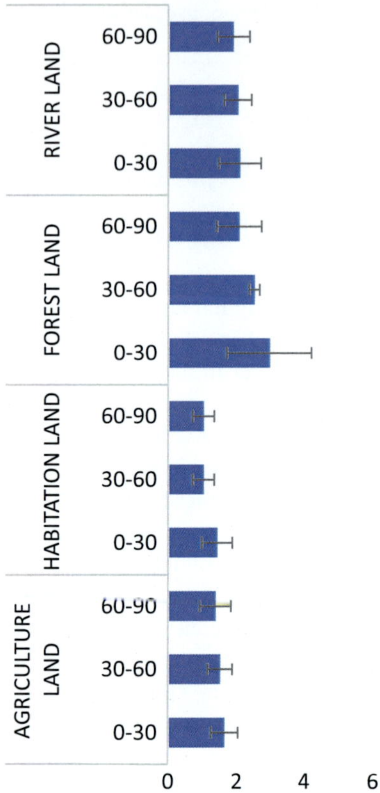

Fig. 11.5 SOC % (mean ± SD) in different land-use types at different depths (cm)

concentration in the forest soil under subtropical regions is estimated to be 2.47% (agroforestry), 2.38% (vegetable field), 2.09% (sewage-irrigated rice–wheat), 2.01% (tube-well-irrigated rice–wheat), and 1.86% (uncultivated soil) (Purakayastha and Patra 2007). In northeast India, SOC concentration reported in different land uses varied from 0.53% (mango plantation) to 3.72% (dense forest) at 0–15 cm soil depth and 0.38% (mango plantation) to 2.13% (dense forest) at 15–30 cm soil depth (Singh et al. 2018). The SOC concentrations in various plantations in northeast India were oil palm (0.96–1.15%), orange (0.95–1.30%), teak (0.90–1.41%), coffee (0.97–1.43%), arecanut (1.35–1.65%), jhum fallows (0.83–1.26%), and wet rice cultivation (0.69–1.28) (Singh et al. 2018). The SOC reported for various land uses in other countries like Thailand were 2.01% (mangrove), 1.37% (paddy land), 0.94% (field crop), and 0.89% (horticulture) (Swangjang 2015). These reported values are also close to the SOC estimates of the present study. The SOC reported in the Nigerian soil under different land-use patterns are 1.03% for fallow land, 1.01% for palm plantation, 0.75% for Yam cultivation land, and 0.39% for cassava cultivation land (Ufot et al. 2016). However, in the vegetation zone, SOC differs from 1.78% (forest), 1.48% (plantation), and 1.27% (cropland) (Olorunfemi et al. 2018). In the river basin of China, SOC is estimated for grassland, woodland, cropland, and unused land to the tune of 5.15%, 2.15%, 0.85%, and 0.45%, respectively (Yang et al. 2018). In Mediterranean climate with different land-use patterns, the SOC differ from 2.7% (field crop) to 4.5% (forest), to 3.4% (orchard) (Olorunfemi et al. 2018). In South Ethiopia in different land use, SOC was the maximum in agroforestry (2.40%) and the minimum (1.49%) in cropland (Worku et al. 2004). The SOC for other land use reported was 2.06% (shrubland) and 1.71% (grazing land).

Understanding the vertical pattern of SOC would enhance our understanding of the carbon dynamics in a given soil profile and its potential for climate change. In general, the SOC concentration in each land use decreased with an increase, as also reported by others (Dinakaran and Krishnayya 2008; Alamgir and Amin 2008; Singh et al. 2018). The higher percentage of SOC in the top oil could be because higher inputs of organic matter and higher microbial activities lead to rapid decomposition of forest litter in a favorable environment. The reduction of SOC with depth indicated lower biological activity in deeper layers (Singh et al. 2018; Jobbágy and Jackson 2000; Sheikh et al. 2020). Higher activities of microbes in upper soil layers lead to accelerating the decomposition of litter (Upadhyay and Singh 1989). Further, the upper soil layer has rich root tips, favoring porous and nutritious soil (Samina et al. 1999). Temperature and precipitation influence the decomposition and SOC stored (Sheikh et al. 2009).

The mean values of carbon stock of different depths are shown in Table 11.3. Among the land uses, the SOC stock ranged from 42.23 to 54.14 t ha^{-1} (HL), 50.63 to 55.83 t ha^{-1} (AL), 161.72 to 172.15 t ha^{-1} (RL), and 88.19 to 102.70 t ha^{-1} (FL) (Fig. 11.6). The differences in SOC at different land uses were found to be significant. The cumulative SOC up to 90 cm depth in different land-use types (Fig. 11.7) reported the highest in RL (504.24 t ha^{-1}), followed by FL (289.13 t ha^{-1}), AL (161.03 t ha^{-1}), and the lowest in HL (138.60 t ha^{-1}). The SOC of coniferous subtropical and broadleaf temperate forests of Garhwal Himalaya (Sheikh et al. 2009) ranges from 141.6 to 124.8 t C ha^{-1} (subtropical forest) to 185.6 to 160.8 t C ha^{-1} (temperate forest), which is lower than the estimated SOC of the study. Moreover, it is reported that the SOC for plantation (eucalyptus, popular, shishm, and teak) is more than the barren land in Uttarakhand and Haryana states (Gupta and Pandey 2008). In Garhwal Himalaya, India, SOC stock was 218.57 t ha^{-1} in community-managed forests (Mahato et al. 2016). In the Garhwal Himalaya region, SOC is the maximum (78.49 t ha-1) in forests, followed by 75.76 t ha^{-1} (in grasslands), 54.33 t ha^{-1} (in orchards), and 46.13 t ha^{-1} (in plantations) (Gupta and Sharma 2011). The SOC is affected by the microbial activity in the soil, and its rate is directly proportional to soil respiration (Murthy et al. 2002; Kumar and Kumar 2020; Kumar et al. 2017).

A balance between organic matter (OM) decomposition and biomass debris driven by various factors controls the SOC stocks and quantity of OM (Guggenberger et al. 1995; Ostle et al. 2009; Sheikh et al. 2009). In general, residence times of SOC vary according to the soil pools, but most of the SOC have a long degradation rate (e.g., semi-labile and refractory organic carbon) and are thus considered cycling carbon compared to labile organic carbon (Nandwa 2001; Kumar and Sharma 2015; Kumar et al. 2018). SOC stock is influenced by soil carbon inputs through the

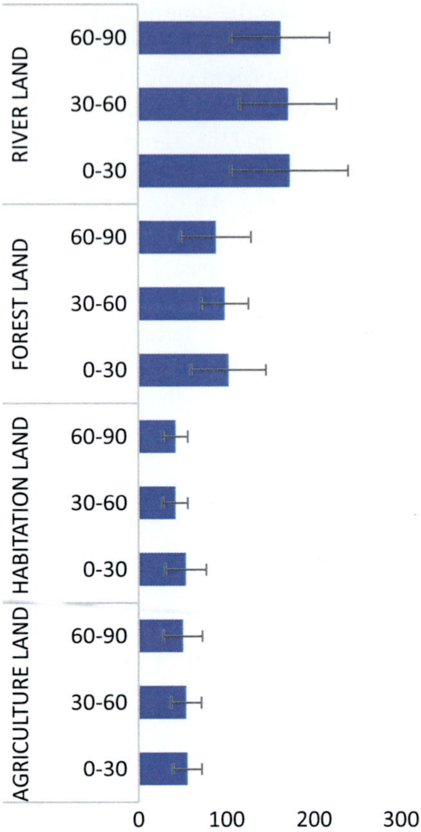

Fig. 11.6 SOC stock (t ha^{-1}) (mean ± SD) in different land-use types at different depths (cm)

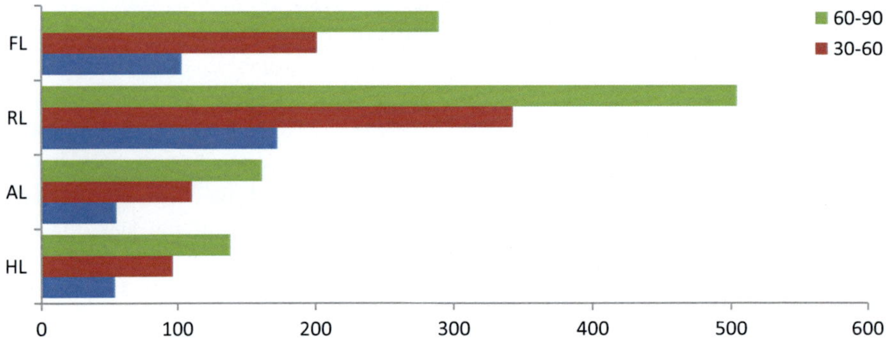

Fig. 11.7 Depth (cm)-wise carbon stock (t ha^{-1}) in different land-use types

addition of plant residues (quality and quantity) and outputs (alteration in decomposition processes) (Sahoo et al. 2019). The variations in SOC in different land uses occur when one vegetation is replaced by others due to LULC changes. Conversion of agriculture to forest and agroforestry lands has been reported to significantly

Fig. 11.8 Soil organic carbon (SOC) map of the study area

increase SOC stock, while the conversion of agroforestry to forest, agroforestry to agriculture and plantation, and grasslands has drastically reduced the SOC stock (Singh et al. 2018). Therefore, rampant conversion of land use should be discouraged, and before any land-use conversion, the implication of such conversion to SOC stock and their contribution to climate change should be envisaged. The soil organic carbon (SOC) map of the study area is shown in Fig. 11.8.

3.1 Land Use and Vegetation Classification

Four land-cover/vegetation classes, viz., habitation land, agriculture land, forest land, and river land, classify the satellite image, where FL occupied the most area and HL the least area. Land-use and land-cover changes are depicted, and according to the LULC results, the FL covered 55.34% (23.08 km^2) of the total study area, followed by RL in 26.97% (11.25 km^2) and AL in 12.85% (5.36 km^2) of the total area. HL were only found in 4.82% (2.01 km^2) area. The classification accuracy ranged between 87.25% and 98.21%. The highest accuracy was in RL and FL, followed by AL and HL.

Earlier studies demonstrated the use of remote sensing satellite data for characterizing land-use types (Thakur et al. 2014; Park and Lee 2016; Soha and El-Raey 2019; Thakur et al. 2020). Many workers (El-Tantawi et al. 2019; Thakur et al. 2019) have done land use and vegetation mapping through satellite remote sensing in different environments.

4 Conclusion

The present evaluation was conducted at the Lower Himalaya region of Uttarakhand, India, under four (FL, AL, HL, and RL) different land uses. Land-use changes due to anthropogenic activity in the studied catchment cause a change in SOC stocks and therefore contribute to the mitigation of climate change. Results reveal that the BD of soil was found to be the maximum in RL, followed by HL and the minimum in AL. Similarly, SOC stock was also estimated to be the maximum in RL and the minimum in HL. This may provide an option and additional inputs for the inclusion of built-up area and river land in mitigating carbon for the National Communication to UNFCCC. This information is also useful for accounting for the shifting of land uses and their relevance in the mitigation mechanism to climate change. Moreover, the SOC estimate may also provide an input for understanding the productivity of the land, especially the farmland, and therefore assist in planning for food security by the policy planner and thus facilitate adaptation options to climate change to the communities of the area. Similarly, a study in a broader domain would facilitate a precise estimate of carbon flow and thereby assist climate actions.

Authors' Contributions AT and MK developed a draft of the manuscript; TK improved the manuscript by adding new content. All authors approved the manuscript.

References

Abad JRS, Khosravi H, Alamdarlou EH (2014) Assessment the effects of land use changes on soil physicochemical properties in Jafarabad of Golestan province, Iran. Bull Environ Pharmacol Life Sci 3(III):296–300

Abera Y, Belachew T (2011) Effects of landuse on soil organic carbon and nitrogen in soils of bale, southeastern Ethiopia. Trop Subtrop Agroecosyst 14:229–235

Alamgir M, Amin MA (2008) Storage of organic carbon in forest undergrowth, litter and soil within geoposition of Chittagong (south) forest division, Bangladesh. Int J Usufruct Manage 9:175–191

Benbi DK, Brar K, Toor AS, Singh P (2014) Total and labile pools of soil organic carbon in cultivated and undisturbed soils in northern India. Geoderma 237–238:149–158

Congalton RG (1991) A review of assessing the accuracy of classifications of remotely sensed data. Remote Sens Environ 37:35–46

Department of Rural Technology, HNB Garhwal University, Srinagar Garhwal (2018)

Dinakaran J, Krishnayya NSR (2008) Variation in type of vegetal cover and heterogeneity of soil organic carbon in affecting sink capacity of tropical soils. Curr Sci 94:1–9

El-Tantawi AM, Bao A, Chang C, Liu Y (2019) Monitoring and predicting land use/cover changes in the Aksu-Tarim River Basin, Xinjiang-China (1990–2030). Environ Monit Assess 191:480. https://doi.org/10.1007/s10661-019-7478-0

Fan S, Guan F, Xu X, Forrester DI, Ma W, Tang X (2016) Ecosystem carbon stock loss after land use change in subtropical forests in China. Forests 7:142. https://doi.org/10.3390/f7070142

Gebhart DL, Johnston HB, Mayeux HS, Polley HW (1994) The CRP increases soil organic carbon. J Soil Water Conserv

Golchin A, Asgari H (2008) Land use effects on soil quality indicators in northeastern Iran. Aust J Soil Res 46:27–36

Guggenberger G, Zech W, Thomas RJ (1995) Lignin and carbohydrate alteration in particle size separates of an Oxisol under tropical pastures following native savanna. Soil Biol Biochem 27:1629–1638

Gupta MK, Pandey R (2008) Soil organic carbon pool under different plantation in some districts of Uttarakhand and Haryana. Indian J For 31(3):369–374

Gupta MK, Sharma SD (2011) Sequestrated carbon: organic carbon pool in the soils under different forest covers and land uses in Garhwal Himalayan Region of India. Int J Agric For 1(1):14–20

Iqbal S, Tiwari SC (2016) Soil organic carbon pool under different land uses in Achanakmar biosphere Reserve of Chhattisgarh, India. Curr Sci 110:771–773

Jobbágy EG, Jackson RB (2000) The vertical distribution of soil carbon and its relation to climate and vegetation. Ecol Appl 10:423–436

Kenye A, Sahoo UK, Singh SL, Gogoi A (2019) Soil organic carbon stock of different land uses of Mizoram, Northeast India. AIMS Geosci 5(1):25–40

Kumar A, Kumar M (2020) Estimation of biomass and soil carbon stock in the hydroelectric catchment of India and its implementation to climate change. J Sustain For 39. https://doi.org/10.1080/10549811.2020.1794907

Kumar A, Sharma MP (2015) Review of methodology for estimation of labile organic carbon in reservoirs and lakes for green house gas emission. J Mater Env Sci 5:653–660

Kumar A, Sharma MP (2016) Estimation of soil organic carbon in the forest catchment of two hydroelectric reservoirs in Uttarakhand, India. Hum Ecol Risk Assess 22:991–1001

Kumar A, Mahendra P., Sharma A, Taxak K (2017) Effect of vegetation communities and altitudes on the soil organic carbon stock in Kotli Bhel-1A Catchment, India. Clean Soil Air Water 45(8):1600650. https://doi.org/10.1002/clen.201600650

Kumar A, Sharma MP, Yang T (2018) Estimation of carbon stock for greenhouse gas emissions from hydropower reservoirs. Stoch Env Res Risk A 32:3183–3319

Mahato S, Dasgupta S, Todaria NP (2016) Tree and soil carbon stock by community managed forests in Garhwal Himalaya. Plant Arch 6(2):805–811

Murthy D, Kirschbaum MUF, McMurtrie RE, McGilvray H (2002) Does conversion of forest to agricultural land change soil carbon and nitrogen? A review of the literature. Global Change Biol 8:105–123

Nandwa S (2001) Soil organic carbon (SOC) management for sustainable productivity of cropping and agro-forestry system in Eastern and Southern Africa. Nutr Cycl Agroecosyst 61(1):143–158

Olorunfemi IE, Fasinmirin JT, Akinola FF (2018) Soil physic-chemical properties and fertility status of long term land use and cover changes: a case study in forestry vegetative zone of Nigeria. Eurasian J Soil Sci 2:133–150

Ostle NJ, Levy P, Evans CD, Smith P (2009) UK land use and soil carbon sequestration. Land Use Policy 26:274–283

Padhan D, Pradhan AK, Chakraborty M, Sen A (2016) Assessment of the effects of land use pattern on distribution of sulphur fractions in soil. J Appl Nat Sci 8(3):1685–1691

Park J, Lee J (2016) Detection of land use/land cover changes using remotely-sensed data. J For Res. https://doi.org/10.1007/s11676-016-0270-x

Pearson RL, Miller LD (1972) Remote mapping of standing crop biomass for estimation of the productivity of the short grass prairie. ed. Remote Sens Environ VIII:1355

Purakayastha T, Patra AK (2007) Long-term effects of different land use and soil management on various organic carbon fraction in an lnceptisol of subtropical India. Aust J Soil Res 45:33–40

Sahoo UK, Singh SL, Gogoi A, Kenye A, Sahoo SS (2019) Active and passive soil organic carbon pools as affected by different land use types in Mizoram, Northeast India. PLoS One 14(7):e0219969. https://doi.org/10.1371/journal.pone.0219969

Samina U, Singh SP, Rawat YS (1999) Fine root productivity and turnover in two evergreen central Himalayan forests. Ann Bot 84:87–94

Schlesinger WH (1997) Biogeochemistry an analysis of global change. Academic, San Diego

Sharma G, Sharam LK, Sharma KC (2019) Assessment of land use change and its effect on soil carbon stock using multitemporal satellite data in semiarid region of Rajasthan, India. Ecol Process 8:42

Sheikh MA, Kumar M, Bussmann RW (2009) Altitudinal variation in soil organic carbon stock in coniferous subtropical and broadleaf temperate forests in Garhwal Himalaya. Carbon Balance Manag 4:1–6

Sheikh MA, Kumar M, Todaria NP, Pandey R (2020) Biomass and soil carbon along altitudinal gradients in temperate Cedrus deodara forests in Central Himalaya, India: implications for climate change mitigation. Ecol Indic 111:101625

Singh SL, Sahoo UK, Gogoi A, Kenye A (2018) Effect of land use changes on carbon stock dynamics in major land use sectors of Mizoram, Northeast India. J Environ Prot 9:1262–1285. https://doi.org/10.4236/jep.2018.912079

Soha AM, El-Raey EM (2019) Land cover classification and change detection analysis of Qaroun and Wadi El-Rayyan lakes using multi-temporal remotely sensed imagery. Environ Monit Assess 191:229. https://doi.org/10.1007/s10661-019-7339-x

Swangjang K (2015) Soil carbon and nitrogen ratio in different land use. Int Conf Adv Environ Res 87:36–40

Thakur T, Swamy SL, Nain AS (2014) Composition, structure & diversity analysis of dry tropical forest of Chhattisgarh using satellite data. J For Res 25 (4:819–825

Thakur TK, Padwar GK, Patel DK (2019) Monitoring land use, species composition and diversity of dry tropical environ in AABR, India using satellite data. Biodivers Int J 3(4):162–172

Thakur TK, Patel DK, Bijalwan A, Dobriyal MJ, Kumar A, Thakur A, Bohra A, Jahangeer B (2020) Land use land cover change detection through geospatial analysis in an Indian Biosphere Reserve. Trees For People 2:100018. https://doi.org/10.1016/j.tfp.2020.100018

Thakur TK, Patel DK, Dutta J, Kumar A, Kaushik S, Bijalwan A et al (2021) Assessment of decadal land use dynamics of upper catchment area of Narmada River, the lifeline of Central India. J King Saud Univ Sci 33:101322. https://doi.org/10.1016/j.jksus.2020.101322

Tu C, Liu C, Quine TA, Jones MW, Liu T, Li L, Liu W (2018) Dynamics of soil organic carbon following land-use change: insights from stable C-isotope analysis in black soil of Northeast China. Acta Geochim 37:746–757

Ufot OU, Iren BO, Njoku Chikere UC (2016) Effects of land use on soil physical and chemical properties in Akokwa Area of Imo State, Nigeria. Int J Life Sci Scienti Res 2(3):273–278

Upadhyay VP, Singh JS (1989) Nitrogen release pattern in Nainital hills, India. Indian For 115:320–326

Walkley AE, Black JA (1934) An examination of the Degtiga vet. Method for determining soil organic matter and proposed modification of the chronic acid titration method. Soil Sci 37:29–38

Wang ZM, Zhang B, Song KS, Liu DW, Ren CY (2010) Spatial variability of soil organic carbon under maize monoculture in the Song Nen plain, Northeast Cjina. Pedospbere 20(1):80–89

Worku G, Bantider A, Temesgen H (2004) Effects of land use/land cover change on some soil physical and chemical properties in Ameleke micro-watershed, Gedeo and Borena zones, South Ethiopia. J Environ Earth Sci 4(11):1–12

Yang S, Sheng D, Adamowski J, Yifan G, Zhang J (2018) Effect of land use change on soil carbon storage over the last 40 years in the Shi Yang River Basin, China. Land 7(11):2–9

Yeasmin S, Jahan E, Molla MA, Mominul Islam, AKM, Anwar MP, Rashid MHO, Chungopast S (2020) Effect of land use on organic carbon storage potential of soils with contrasting native organic matter content

Chapter 12
Bamboo-Based Livelihood Development in Agroforestry Under Changing Climate

Indra Singh, Gaddam Sudharshan Reddy, Modala Rakesh, M. V. Durai, and Aman Dabral

Abstract Bamboo is economically and ecologically significant because it is a fast-growing, sustainable resource. Historically, timber has been overly relied upon, but bamboo offers a greener alternative from an Indian and global perspective owing to its remarkable adaptability and high carbon sequestration capacity. Agroforestry systems based on bamboo combine ecological restoration and income generation, thus contributing to soil and water conservation, as well as biodiversity and rural livelihoods, promoting land-use sustainability. Hill bamboos store biomass quickly, thereby serving as nature-based solutions to climate change by providing landslide control, soil stabilization, and rehabilitation of degraded lands, thus modifying the mountain ecosystem and economy. There remain many bottlenecks/barriers to bamboo-based livelihoods (lack of planting material, insecure land tenure, limited technical knowledge, highly fragmented market, etc.) Focused education with interlocking land reform, cooperative development, and infrastructure development may facilitate the scaling of bamboo-based livelihoods. Innovations in tissue culture and the development of value-added products for the industry will enhance global competitiveness while aiding in the biodiversity conservation of bamboo-based industries. Looking further into the Himalayan region, incorporating bamboo into agrotourism or nature-based solutions can strengthen climate-friendly growth. Women and young people can also be meaningfully employed through community-led enterprises, self-help groups (SHGs), and public–private collaborations. This

I. Singh (✉)
Department of Silviculture and Agroforestry, Veer Chandra Singh Garhwali Uttarakhand University of Horticulture and Forestry, Tehri Garhwal, Uttarakhand, India

G. S. Reddy · M. V. Durai
ICFRE – Institute of Wood Science and Technology, Bengaluru, Karnataka, India

M. Rakesh
ICFRE – Forest Research Institute, Dehradun, Uttarakhand, India

A. Dabral
Department of Forest Biology and Tree Improvement, Veer Chandra Singh Garhwali Uttarakhand University of Horticulture and Forestry, Ranichauri, Tehri Garhwal, Uttarakhand, India

B. Singh et al. (eds.), *Himalayan Agroforestry Systems*, Advances in Agroforestry 15,
https://doi.org/10.1007/978-3-032-25096-4_12

further increases their resilience to poverty and provides sustainable livelihood opportunities. In these vulnerable areas, bamboo is an essential catalyst for rural transformation and climate-smart development.

Keywords Bamboo · Sustainable resource · Agroforestry systems · Bamboo-based livelihoods · Self-help groups

1 Introduction

1.1 Bamboo: A Sustainable Resource with Global Importance and Distribution

Bamboo, a fast-growing grass, belongs to the Poaceae family and the Bambusoideae subfamily (Chaudhary et al. 2024). Its exceptional growth rate, high yield, wide availability, and structural strength make it an ideal raw material for various applications. One of bamboo's most striking features is its rapid growth—reaching full height within 2–4 months, typically between 15 and 30 m, and occasionally up to 35 m. Daily growth rates can vary from 20 cm to 100 cm (Liese 1987).

Bamboo is often described as "green gold," "cradle to coffin timber," and "poor man's timber" due to its diverse advantages and wide applicability (Tewari et al. 2019). Its multifunctional uses make it one of the most versatile natural resources. Its uses span numerous sectors, including animal fodder, raw material for paper, pulp production, and sustainable building material. Additionally, bamboo is utilized for producing charcoal and edible products, and forms the backbone of many rural and cottage industries. The rising demand for wood and wood-based products has placed considerable strain on the limited forest resources available. In this scenario, bamboo presents itself as a practical and sustainable alternative to conventional timber. Its fast growth rate, renewable nature, and comparable structural qualities make it an effective substitute, thereby reducing the burden on natural forests while simultaneously catering to the expanding requirements of multiple industries across the globe. Some of the important ecological and economic benefits associated with bamboo are shown in Fig. 12.1. Beyond its economic uses, bamboo offers numerous environmental benefits. It plays a significant role in climate change mitigation by producing more oxygen and sequestering greater amounts of carbon dioxide than many other plant species (Yadav and Mathur 2021). The growing demand and application of bamboo also offer the potential for widespread employment generation, which in turn could contribute to poverty alleviation and socioeconomic progress.

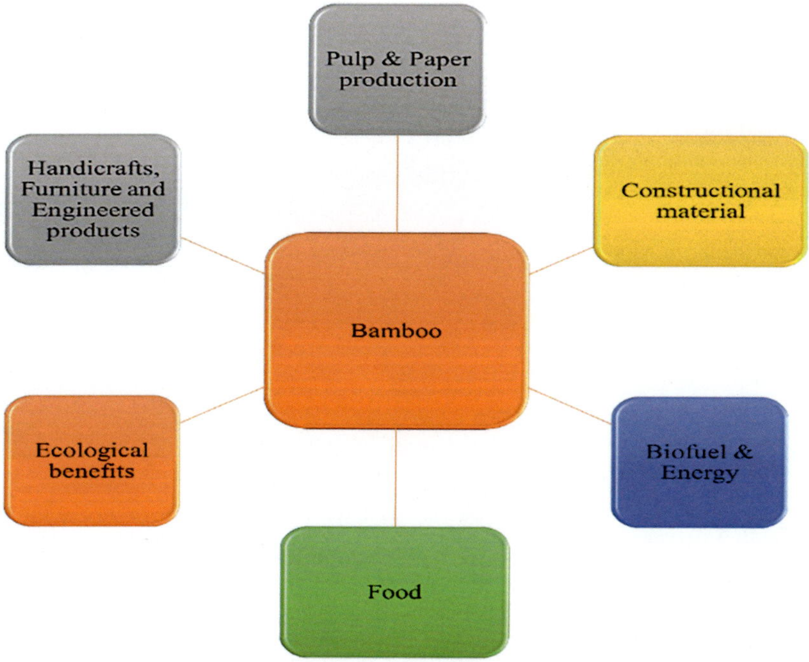

Fig. 12.1 Major ecological and socioeconomic benefits of bamboo

1.2 Bamboo Area in the World

Bamboo demonstrates exceptional adaptability to various climatic and soil conditions, thriving in nutrient-poor to mineral-rich soils, and tolerating a wide range of moisture levels, making it suitable for rehabilitating degraded lands (Chaudhary et al. 2024).

The Global Forest Resource Assessment (2020) report published by the Food and Agriculture Organization states that there are 35 million hectares of bamboo forests growing globally (FAO 2020). The detailed statistics on regional bamboo distribution are presented in Table 12.1.

1.3 Bamboo Distribution Across the Indian States

Asian regions are rich in bamboo resources. India is rich in bamboo biodiversity and is home to nearly 125 native and 11 introduced species spread across 23 genera (ISFR 2021). The country's bamboo-bearing area is estimated to be 154,670 km^2 (ISFR 2023), reflecting a growth of 5227 km^2 since the previous ISFR assessment. Some of the most common bamboo genera found in India are *Bambusa*,

Table 12.1 Areas of bamboo, by region and subregion

Region/subregion	Bamboo area (1000 ha)
Eastern and Southern Africa	3984
Northern Africa	30
Western and Central Africa	634
Total Africa	**4648**
East Asia	**7005**
South and Southeast Asia	17,872
Western and Central Asia	0
Total Asia	**24,877**
Total Europe	**0**
Caribbean	125
Central America	0
North America	0
Total North and Central America	**125**
Total Oceania	0
Total South America	**5389**
World	**35,040**

Source: FAO (2020)

Table 12.2 Distribution of bamboo area in recorded forest areas (Km^2)

S. no.	State/union territory	Bamboo-bearing area (km^2)	S. no.	State/union territory	Bamboo-bearing area (km^2)
1.	Andhra Pradesh	6370	16.	Meghalaya	5347
2.	Arunachal Pradesh	18,424	17.	Mizoram	4772
3.	Assam	11,246	18.	Nagaland	3980
4.	Bihar	1109	19.	Odisha	12,328
5.	Chhattisgarh	11,139	20.	Punjab	311
6.	Goa	235	21.	Rajasthan	1706
7.	Gujarat	3895	22.	Sikkim	1010
8.	Haryana	**42**	23.	Tamil Nadu	3217
9.	Himachal Pradesh	1154	24.	Telangana	3801
10.	Jharkhand	3746	25.	Tripura	4466
11.	Karnataka	7334	26.	Uttar Pradesh	1598
12.	Kerala	2443	27.	Uttarakhand	1307
13.	Madhya Pradesh	**20,421**	28.	West Bengal	754
14.	Maharashtra	13,572	29.	A & N Islands	1426
15.	Manipur	7517	**Total**		**1,54,670**

Source: ISFR (2023)

Dendrocalamus, *Arundinaria*, *Chimonobambusa*, *Dinochloa*, and *Ochlandra* (ISFR 2021). The distribution of bamboo-bearing areas within the recorded forest regions of the country, as reported in the ISFR (2021), is summarized in Table 12.2. Among all states, Madhya Pradesh accounts for the largest bamboo-bearing extent, covering 20,421 km^2.

1.4 *Bamboo Potential in the Himalayan State (Uttarakhand)*

In Uttarakhand, the bamboo-bearing area extends over **1307 km^2.** The state is home to a diverse range of bamboo flora, with **eight naturally occurring species** distributed under **five genera.** The prominent species include *Bambusa bambos* (L.) Voss, *Dendrocalamus patellaris* Gamble (syn. *Ampelocalamus patellaris*), *Dendrocalamus somdevae* H.B. Naithani, and *Dendrocalamus strictus*. Among the hill bamboos, *Arundinaria falcate* Nees (syn. *Drepanostachyum falcatum*) and *Chimonobambusa jaunsarensis* (Gamble) Bahadur & H.B. Naithani (syn. *Yushania anceps*), *Thamnocalamus falconeri* Hook.f. ex-Munro (syn. *Himalayacalamus falconeri*), and *Thamnocalamus spathiflorus* Munro (Sundriyal and Joshi 2015). Within the state, Sundriyal and Sundriyal (2011) identified Uttarkashi as having the largest bamboo-covered area in Uttarakhand, followed by Rudraprayag, Haridwar, Nainital, Tehri, Udham Singh Nagar, Dehradun, Almora, Champawat, Pauri, Pithoragarh, Chamoli, and Bageshwar districts. The abundance of bamboo resources in the Himalayan region provides a plethora of opportunities for utilizing these vast resources, which can support livelihood development and contribute toward tackling the global issue of climate change.

1.5 *Bamboo-Based Agroforestry*

Bamboo-based agroforestry is emerging as a sustainable, eco-friendly, and economically viable land-use system, particularly in tropical and subtropical regions (Dutta et al. 2025). It involves integrating bamboo cultivation with other agricultural practices, such as growing food crops and medicinal plants or rearing livestock. This system promotes ecological balance and provides livelihood opportunities for rural communities (Tiwari et al. 2017). The two characteristics that make bamboo an ideal candidate for agroforestry systems are its versatility coupled with its high growth rate and low human input demands (Dev et al. 2020).

2 Environmental Importance of Bamboo Agroforestry

The greatest advantage of bamboo-based agroforestry is that it unconsciously contributes to the protection of the environment (Solomon et al. 2020). The greatest advantage of bamboo-based agroforestry is that it unconsciously contributes to the protection of the environment (Solomon et al. 2020). Bamboo is of paramount importance for soil conservation because its fibrous and zigzag root system helps in soil binding, thereby minimizing soil erosion and stabilizing the land, especially sloped and degraded lands (Shen et al. 2025; Kaushal et al. 2021). This feature is particularly valuable in regions that are vulnerable to landslides and sudden flash floods. Bamboo also contributes to carbon sequestration by absorbing high levels of carbon dioxide from the atmosphere and releasing oxygen at a higher rate than many tree species (Yiping et al. 2010). Consequently, bamboo cultivation aids in climate change mitigation and environmental restoration. In terms of water conservation, bamboo helps improve water infiltration and prevents runoff, which, in turn, recharges groundwater tables (Singh et al. 2013). Moreover, it increases biodiversity, as it is a source of shelter and food for birds, insects, and small animals. Bamboo forests when incorporated in other native vegetative seeding bring about mini-ecosystems, making the forests have that eco-balance desired.

2.1 Economic and Livelihood Benefits

Bamboo is an economically versatile venture for rural and small-scale farmers (Lin et al. 2019). It is a multiuse crop, with its shoots being eaten as delicacies in most Asian foods, and its culms utilized in construction, furniture-making, arts, fencing, and even in paper manufacturing (Rane et al. 2016). Bamboo grows much faster than many hardwood trees, matures within 3–5 years, and can be harvested once every year without the necessity to replenish, which makes it a renewable and sustainable resource (Scheba et al. 2017). Bamboo is a wonderful crop with a low maintenance cost, which is suitable to marginal as well as poor farmers. With minimal investment, it offers high returns, especially if integrated into value chains such as bamboo handicrafts, bamboo charcoal, agarbatti (incense stick) production, and furniture manufacturing (Rao et al. 2009). These industries also create rural employment opportunities, particularly for women and artisans. The bamboo industry is associated with high labor intensity and provides livelihoods to large numbers of people in and out of farms. Farming and processing of bamboo play an important role in the rural economy; it provides revenues and employment in harvestings, handicrafts, industries, and construction (Mekonnen et al. 2014). To indicate the potential livelihood suitability of bamboo and stimulate more investment in the industry, this economic influence should be measured.

2.2 Social Inclusion in Bamboo Livelihoods

Bamboo-based livelihoods in the Himalayas involve most social strata, such as tribes, women, and young people (Banerjee et al. 2021). Bamboo resource and benefits distribution should become a priority toward inclusive development. Marginalized groups can be given a voice through training and support, and such efforts can strengthen their roles and stimulate community welfare.

2.3 Uses in Agroforestry Systems

Bamboo can also be integrated into different agroforestry systems, such as alley cropping, in which bamboo is planted in lines between food or medical crops; silvopasture, as bamboo grows with grazing livestock and home gardens, where it grows with edible and fruit trees (Viswanath and Lubina 2018; Chandewar 2025). It is also a great windbreak or shelterbelt capable of shielding other crops against strong winds and minimizing evapotranspiration (Kumar 2016). In places where the utilization of bamboo is conventional, farmers mix it with crops, such as ginger, turmeric, yams, and legumes (Sharma 2023). Bamboo shoots are edible and can be harvested during the early seasons, providing an easy source of income, whereas the culms take a few years to mature (Zhang et al. 2024). Bamboo leaves may be used as animal feed, agricultural mulch, or even as compost to increase soil fertility (Sharma et al. 2022). Furthermore, bamboo plantations are potentially useful for restoring waste, such as newly abandoned mine sites or heavily grazed lands.

2.4 Climate Resilience and Adaptability

Bamboo can also grow in various soils and climatic conditions and is especially drought resistant, making it an appropriate measure in areas with climate variability (Naskar et al. 2025). It can be cultivated in areas with poor nutrients and degraded/rocky arable lands where most crops cannot be produced. Therefore, covering farmlands with bamboo improves the climate resilience of farming systems, as agroforestry allows them to remain productive even under unfavorable environmental conditions (Partey et al. 2017). This feature is vital amid climate-related risks, such as surges in droughts, sporadic rainfalls, and land erosion.

2.5 Government Policies and Initiatives

Governments in most countries, including India, are encouraging the growth of bamboo by organizing programs such as the National Bamboo Mission, which facilitates the establishment of bamboo nurseries, training, and access to the market (Ghode 2023; Arjumand and Rani 2023). Policy changes have facilitated the process of farmers growing and harvesting bamboo, which was previously considered a forest product, thereby requiring the permission of the forest department. These changes have also mainstreamed the uptake of bamboo into farming systems and stimulated farmers to embrace bamboo agroforestry as a livelihood option.

3 Challenges and Considerations

Despite its many benefits, bamboo-based agroforestry faces several challenges (Addo-Danso and Amankwaa-Yeboah 2021). These include small-capacity processing, farmer ignorance, and poor market connections in Sudan. Moreover, monoculture bamboo plantation threaten to emerge, impacting biodiversity much more if they are adequately combined with other species (Wang et al. 2019). The solutions to these problems are training, awareness programs, and investment in bamboo-based industries in rural areas.

3.1 Genetic Diversity and Conservation of Bamboo

Himalayan bamboo species are also extremely genetically diverse, which is essential for maintaining resilient ecosystems and sustainable use (Meena et al. 2023). Conservation of such genetic resources implies ensuring that they are not lost through habitat destruction or excessive harvesting. In this way, we can prevent the extinction of bamboo in the future.

3.2 Future Prospects and Innovations

Bamboo is recognized as one of the fastest-growing plants globally, and its future role in agriculture will largely depend on effective value addition, strong research support, and continuous innovation. Production can be increased by cultivating more productive and disease-resistant bamboo varieties. Meanwhile, businessmen and new companies are discovering new and novel ways to use bamboo such as they are making it into textiles, floors, biocomposites, and even fully biodegradable materials, which is adding to its commercial value. Bamboo-based bioenergy (such as bamboo charcoal and pellets) is also gaining interest as a clean energy source in rural areas.

4 Bamboo's Role in Climate Change Mitigation and Adaptation

4.1 Carbon Sequestration Potential of Hill Bamboos

Carbon sequestration refers to the process by which plants capture and store atmospheric carbon dioxide (CO_2) to mitigate global warming (Lal 2009). Hill bamboos are especially effective due to their fast growth rate, substantial biomass yield, and ability to thrive on sloping and marginal lands. Species such as *Dendrocalamus hamiltonii*, *Dendrocalamus strictus*, and *Bambusa nutans* thrive in hilly regions, growing quickly even in poor soils with minimal care (Nath et al. 2009). These clump-forming bamboos mature in just 3–5 years, much faster than traditional hardwoods, and can reach heights of up to 20 m. They store carbon both above ground in culms, branches, and leaves, and below ground in rhizomes and roots, making them effective for both short- and long-term carbon storage (Tripathi et al. 2024). Bamboo plantations also enhance soil organic carbon through dense root networks and leaf litter, improving soil health. Studies have shown that hill bamboo plantations can sequester between 5 and 12 tons of CO_2 per hectare annually, depending on species, age, and management (Ni et al. 2021; Sohel et al. 2015). In addition, harvested bamboo products continue to hold carbon in furniture and building materials for many decades, and they replace even more carbon-intensive materials. In addition to carbon storage, bamboo stabilizes soils and watersheds, prevents erosion on landslide-prone hills, and grows on degraded lands, thereby contributing to ecological restoration. The difficulties include interspecific variability in the ability to sequester, and sustainable management is required. Climate policies are increasingly acknowledging the role of hill bamboos, integrating them into frameworks like the Nationally Determined Contributions (NDCs) under the Paris Agreement, as well as in carbon offset initiatives such as REDD+. The enhancement of bamboo forests, enterprise education, and the implementation of bamboo for carbon accounting will address this better in the climate mitigation frame (Pan et al. 2023). In summary, hill bamboos present a nature-based scalable solution to favor carbon sequestration, ecological well-being, and socio-financial gains within prone mountainous areas.

4.2 Soil Conservation and Slope Stabilization

Soil conservation and slope stabilization are essential in hilly and mountainous regions to prevent erosion, landslides, and land degradation (Acharya 2018). One of the most effective nature-based approaches to mitigating soil erosion and slope instability is the establishment of vegetative cover, with hill bamboos playing a particularly important role. Their dense, fibrous root systems interlock soil particles, strengthening slope stability. By forming an extensive underground network, these roots reinforce soil structure, minimize surface runoff, and significantly reduce the likelihood of mass movement during periods of heavy rainfall. Bamboo is also a natural dam that decelerates the flow of water and facilitates important water infiltration, which helps maintain the hydrological balance. Its wide canopy catches rainfall, which reduces the extent to which soil is directly affected and helps reduce erosion. Bamboo is also adaptable to poor lands and infertile sloping lands, and it is therefore the right species to engage in reforestation and environmental restoration. Optimized by structural measures such as terraces, retaining walls, and check dams, bamboo-based interventions provide a cost-effective and environmentally friendly approach to prevent soil erosion and stabilize erosion-prone slopes to preserve ecosystems and local populations.

4.3 Restoring Degraded Lands and Increasing Farm Resilience

One of the key advantages of hill bamboos is their ability to reclaim degraded lands and make farms more resistant, especially in high-risk mountainous regions (Sawarkar et al. 2025). Bamboo is highly suited for the rehabilitation of degraded lands, primarily due to its remarkable ability to thrive in nutrient-poor and marginal soils. Unlike many conventional crops or tree species that struggle in such conditions, bamboo can establish and grow successfully, making it an ideal choice for restoring degraded sites. The dense root system not only stabilizes the soil and avoids severe erosion and run-off, but also greatly enhances the soil structure and fertility over time, as the continuous deposition of organic matter by litter fall and root turnover increases fertility. This work of restoring fertile land can directly assist in restoring productive agricultural terrain and local means of subsistence. Furthermore, bamboo's rapid growth and year-round canopy offer farmers diverse economic products, such as fodder, poles, and handicraft materials, thereby diversifying income sources and reducing over-reliance on climate-sensitive crops. This inherent diversification enhances farm resilience, equipping communities with alternative resources and income streams to better withstand adverse weather events and crop failures. By offering both ecological and livelihood benefits, hill bamboos contribute significantly to sustainable land management while enhancing the resilience of farming communities to the challenges posed by climate change.

5 Challenges to Bamboo-Based Livelihood Promotion

Livelihood promotion based on bamboo has several challenges that limit the growth and sustainability of livelihood development (Alamerew et al. 2024). Inadequate quality planting material, indisputable land tenure, and lack of technical knowledge are limits to good cultivation and management. Moreover, the markets are divided, the necessary funding is insufficient, and the existing regulations have become outdated, which interferes with value addition and the creation of a stable supply chain. Addressing these challenges with the help of enhanced training, market organization, and favorable policies is important for maximizing the economic and bamboo utilization potential of inland rural communities.

5.1 Access to Quality Planting Material and Land Tenure Issues

Effective bamboo cultivation hinges on overcoming two pivotal challenges: ensuring access to high-quality planting material and addressing land tenure insecurity (Adewolu 2024). The limited availability of certified, disease-free, and genetically superior bamboo saplings directly impacts plantation success; inferior stock leads to poor establishment, slow growth, and reduced yields, eroding farmer confidence and economic viability. Simultaneously, ambiguous or temporary land rights significantly discourage long-term investment in bamboo, a crop requiring several years to mature. Without secure tenure, farmers are understandably hesitant to commit the necessary resources and adopt sustainable management practices. Addressing these interconnected issues by improving the supply chain for quality planting stock and clarifying land ownership is therefore crucial for scaling bamboo-based livelihoods and realizing their full environmental and economic potential.

5.2 *Inadequate Awareness, Limited Training Opportunities, and Insufficient Technical Expertise*

One of the basic slowdowns on the way to bamboo-based livelihoods is the negligence of awareness and thorough training and the essential technical experience. Most of the rural societies and farmers are hardly aware of the tremendous economic and environmental potential of bamboo, and thus they do not value it as a wild plant. The lack of this knowledge relates to key areas such as the best choice of species, current cultivation procedures, and sustainable harvesting activities, which do not allow efficient use of these resources. Also, low awareness to the market forces and sophisticated processing techniques does not allow artisans and entrepreneurs to involve themselves in the production of higher value products, thereby limiting the generation of income. Closing the wide gap by introducing special educational programs, skill-enhancing training sessions, and convenient extension services is the key to unleashing bamboo as the potentially resource and potential contributor to sustainable livelihood upgrading and increased climate resilience.

5.3 *Market Fragmentation and Absence of Cooperatives*

Market fragmentation and the lack of strong cooperatives significantly hinder the growth of bamboo-based livelihoods (Ghimire 2008). Across many regions, bamboo growers and artisans work in isolation, resulting in inconsistent product quality, limited market access, and an inability to scale operations. This disorganized structure weakens bargaining power, prevents investment in shared infrastructure, and makes it difficult to meet the demands of bulk or high-value markets. Without coordinated efforts, producers face challenges in standardization, market linkage, and accessing credit or government support. Establishing well-functioning cooperatives can address these issues by promoting joint marketing, resource sharing, and collective training, ultimately fostering a more resilient, efficient, and competitive bamboo industry.

5.4 *Innovations, Research, and Development Opportunities*

Tissue Culture and Clonal Propagation
Plant biotechnology is enormous in terms of tissue culture and clonal propagation. Their use enables scientists to produce enormous quantities of genetically identical, disease-free bamboo plants (Lai and Lai 2019). Using small tissue sections (explants) in sterile, controlled environments, these methods allow for rapid, reliable propagation compared to traditional techniques. This innovation is especially valuable for conserving rare bamboo species and enhancing traits such as growth rate, fiber quality, and pest resistance. It also reduces pressure on natural forests by supplying consistent planting material through lab-raised saplings. Beyond conservation, this approach benefits farmers and entrepreneurs with predictable yields and supports the development of climate-adapted, region-specific bamboo varieties. Research opportunities include optimizing media compositions, growth regulators, and regeneration protocols for diverse and often recalcitrant bamboo species. Integration of molecular markers can aid in early trait screening and genetic fidelity checks, ensuring high-quality outputs. Additionally, the automation of micropropagation processes and the exploration of cryopreservation techniques can improve cost-efficiency and enable long-term germplasm storage. Scalable technology provides a means of expanding larger projects, and when the biotech industry collaborates with it, the outcome can elevate commercial nurseries, provide employment, and assist in keeping the biodiversity. That is, the combo is a major component of the progress of sustainable bamboo farming and forestry studies.

5.5 *Development of High-Value Bamboo Products*

Bamboo has been renowned as simple things like chopsticks or scaffolding, but now it is becoming a high-value product that combines sustainability, innovation, and economic development and relies on emerging material science and design (Sun and Liu 2022). Raw bamboo can be transformed into engineered products such as laminated bamboo lumber (LBL), cross-laminated bamboo (CLB), and durable composites—positioning bamboo as a green alternative to timber, plastic, and even steel in construction (Hosseini et al. 2025). Additional applications include biodegradable packaging, textiles, furniture, flooring, and artistic crafts, which appeal to both domestic and global eco-conscious markets. Advanced technologies, including 3D printing, nanofiber extraction, and bamboo-based activated carbon production, further diversify the product potential. There is also promising research into extracting pharmaceutical compounds and integrating bamboo materials with smart technology, like embedded sensors for structural or environmental monitoring. These products are capable of competing in the global marketplace with clever advertisement, stylish design concepts, and the support of the government and research institutions. The region also enjoys most of the money earned through the sale of the goods due to the fact that more effort is placed in terms of converting the raw materials to useful goods and hence economic gains remain within the region. Overall, this industry demonstrates that the environment can be safeguarded, and, at the same time, to generate high-quality employment and new plants in the countryside.

6 Integration with Agrotourism and Nature-Based Solutions (NbS)

The synergetic combination of the bamboo with agrotourism and nature-based solutions (NbS) is an all-embracing concept of sustainable rural development since it combines the efforts of ecological restoration with income generation and community participation (Chen et al. 2024). Bamboo plantations can be transformed into ecotourism destinations featuring guided educational trails, bamboo craft workshops, farm stays, and farm-to-table culinary experiences centered on bamboo shoots providing diverse revenue streams and cultural value (Rana et al. 2024). The reason why bamboo is awesome is that it is fast growing, has good root systems, and is adaptive. All these characteristics render it ideal in natural-based solutions (NbS) like erosion control, riverbank stabilization, carbon sequestration, and habitat creation. The scientists are also able to research the role of bamboo in phytoremediation, land reclamation, and cleansing of the water. When bamboo is used in greening urban areas, watershed, and climate-resilience interventions, it can only increase its environmental worth. With bamboo, resilient agroforestry systems can be developed that increase biodiversity and soil health as well as serve as living laboratories of eco-innovation. Such initiatives ensure green jobs and local economies and fulfill global sustainability. Concisely, this integration leads to a symbiotic model in which economic, environmental, and social goals can be integrated to advance sustainable green growth that is inclusive and resilient.

7 Case Studies from States and Districts of the Himalayas and Uttarakhand

The Himalayan belt, particularly Uttarakhand, is characterized by diverse communities living in remote and difficult mountain landscapes. Over the years, numerous community-led initiatives have emerged as shining examples of sustainable development. In Uttarakhand, several villages have successfully revived traditional water harvesting systems such as the *khal*, ensuring a perennial water supply for agriculture and daily needs. This recovery has greatly cut off distress migration to stabilize living in that area. The other notable one is conservation of the medicinal plants whereby people in various districts such as Chamoli district and Uttarkashi have formed organizations in order to preserve and harbor herbal gardens and nurseries. Such activities transmit favorable harvesting procedures and guarantee a reasonable value to any collector, thus guaranteeing both an ecological equilibrium and a powerful economy. Himachal Pradesh, an area in India, Kullu, and Kinnaur districts have made organic farming bloom and encouraged cash crops like apples and vegetables. This is to maintain the population healthier and the farming system as a whole more efficient and combining local knowledge that has existed over a long period of time with current farming research. These cases demonstrate that the combination of teamwork and native knowledge can meet local needs and establish sustainable resource management.

7.1 Role of SHGs, Cooperatives, and Forest-Based Microenterprises

Self-help groups (SHGs), cooperatives, and forest-based microenterprises have transformed the Himalayan communities, particularly in the remote and marginalized regions, enormously. SHGs, which are mainly initiated by women, provide access to credit, increase savings and entrepreneurship in activities such as handloom weaving, knitting and food processing where products such as jams and pickles are produced using locally available produce. Cooperatives have enabled farmers and artisans to pool resources, collectively market their products, and secure better prices by bypassing exploitative middlemen. Notable examples include dairy cooperatives ensuring fair milk prices and craft cooperatives that market traditional Himachal woolen products and shawls. Microenterprises typically established in a forest-based are established by self-help groups (SHGs) or other community organizations. The main objective is the harvesting and marketing of non-timber forest products (NTFPs), including medicinal plants, wild edible species, and bamboo-based crafts. These enterprises not only generate income but also contribute to forest conservation by creating local employment that depends on the sustainable management and continued health of forest ecosystems.

8 Impact on Women's Empowerment and Youth Employment

SHGs, cooperatives, and microenterprises have enhanced the empowerment of women and employment of the youth in the Himalayan region to a considerable extent as community-led activities. Women who have always been at the periphery in terms of making economic decisions have become financially independent and have acquired a voice of their own because of their participation in SHGs. This involvement has enhanced their confidence, leadership abilities, and decision-making roles both within their households and in the wider community. Many women, previously confined to domestic roles, have become successful entrepreneurs contributing substantially to family incomes and local development. For youth, these initiatives have opened new employment avenues, reducing the pressure to migrate to urban centers for work. Traditional crafts, ecotourism, organic farming, and sustainable forestry projects have developed skills among the youths that are market oriented. Local microenterprises, including value chains, have also created potential in terms of youth entrepreneurship, marketing, and management to form a strong sense of ownership and a connection with their ancestral lands. These all are making a strong society where women and young people establish an important role in sustainable development.

8.1 Strategies for Sustainable Expansion and Upscaling

Sustainable growth and upscaling of the bamboo industry, especially in areas such as Himalayas, ought to be multiprong, where human capital, strong supply chains, and collaborative governance are dealt with. These are the major strategies to go into detail.

8.2 Capacity Building and Vocational Training

Effective and sustainable growth in the bamboo sector relies heavily on strengthening human capital through targeted capacity building and vocational training (Zhaohua and Wei 2018). Equipping local communities with comprehensive skills across the bamboo value chain—from sustainable cultivation and harvesting to innovative processing and product diversification—is crucial (Samdin et al. 2024). Farmer training programs focus on promoting best practices in nursery management, plantation establishment, pest and disease management, and selective harvesting methods that facilitate natural regeneration. Meanwhile, artisans and entrepreneurs benefit from vocational courses focused on contemporary design, advanced manufacturing technologies, quality assurance, and effective marketing approaches for premium bamboo products. In addition to technical skills, business management, understanding of financial literacy, and building of organizations, especially in the establishment and management of self-help groups and cooperatives, are also part of training. Such initiatives do not only increase productivity and quality of products but also promote innovativeness and entrepreneurship, which eventually result in higher earnings and sustainable livelihoods. Frequent transfer of knowledge by the workshops or extension services is a way of making the communities sustainable and adaptive and competitive to promote sustainable development within the bamboo sector.

8.3 Strengthening Bamboo Supply Chains and Institutions

A robust and efficient bamboo supply chain is paramount for the sustainable growth of the industry, creating seamless connections from the source of raw materials to the final market (Iroegbu and Ray 2021). Key strategies for achieving this involve establishing local collection and aggregation centers to reduce transportation costs and postharvest losses. Investing in primary processing units (like cutting, splitting, and initial treatment) at the local level is crucial for adding value at the origin and generating local employment opportunities (Bramsiepe et al. 2012). Beyond the physical chain, strengthening institutional support is vital. This includes bolstering research and development institutions focused on new bamboo species and innovative products, as well as developing supportive policy frameworks that promote sustainable bamboo cultivation and trade practices. Implementing market information systems, facilitating market access through platforms like trade fairs and online channels, and establishing clear quality standards and certification mechanisms are also essential for enhancing competitiveness and building consumer trust in bamboo products. Finally, an effective bamboo chain supply is all about producers, processors, and consumers being linked together effectively. This entails maximizing the methods of harvesting, storing correctly to reduce wastage and better transportation mechanism to transport goods on time. More importantly, the cooperation and producer groups pool capital together, have a say in the prices, and gain access to credit, technology, and training, which helps the small producers compete favorably. The creation of conducive policies and infrastructure like high-dedicated markets and processing centers through the effort of governments and nongovernmental organization or NGOs also plays a key role. Such an integrated solution can directly result in minimized waste, high returns, and the sustainability of bamboo businesses in the long run.

8.4 Public/Private Community Partnerships

Establishment of partnership between the public/private/community (PPCPs) is necessary to create a healthy and sustainable bamboo industry, utilizing individual capabilities of each of the stakeholders. The foundation is established by the government through a friendly policy environment, development of infrastructure, financing (including research funding), extension services, and other financial incentives, which provide an enabling environment to growth. The private sector drives commercial viability by providing investment, technological innovation, market access, and entrepreneurial expertise, particularly in scaling processing units and developing high-value bamboo products. Crucially, community involvement ensures that initiatives remain socially inclusive, environmentally sustainable, and culturally appropriate by incorporating traditional knowledge and local stewardship of bamboo resources. These pairings may assume various prototypes of either joint ventures, contractual farming, or community-based enterprising that is complete with government subsidization and mentorship by the private sector. PPCPs promote risk sharing, collective rewards, and conservation, along with economic development through collaborations in research, capacity building, and marketing. Also, they incubate new models of financing and foster startups, catalyzing development in the industry. In sum, such a combined system develops a profitable yet long-term sustainable and inclusive structure of bamboo, which can work as an example of integrated sector development.

9 Conclusion and Way Forward

Bamboo-based agroforestry is not a traditional practice but represents a contemporary approach to address modern challenges such as climate change, rural unemployment, and soil erosion. The prospects of developing go green livelihoods and ecological upscaling in the Himalayan region are enormous. This can only be achieved by supportive policy, climate-adaptive research, investment in processing and infrastructure, and better market access. The entrepreneurial activities will help improve the value chains of bamboo with the establishment of community–government–private relationship. Bamboo in climate and rural development policies can contribute to the idea of a greener, sustainable future of the Himalayan communities.

References

Acharya MS (2018) The use of bamboo for erosion control and slope stabilization: soil bioengineering works. In: Bamboo: current and future prospects, p 105

Addo-Danso A, Amankwaa-Yeboah P (2021) The potentials of bamboo-based agroforestry systems in improving the productivity of tropical African agricultural systems. Agric Food Sci J Ghana 14(1):1468–1482. https://doi.org/10.4314/afsjg.v14i1.11

Adewolu AO (2024) Green and sustainable construction techniques in South-West Nigeria

Alamerew AN, Kozak R, Shrestha AK, Zhu Z, Wang G (2024) A way for livelihood improvement: systematic review on bamboo forest research. Trees For People:100571. https://doi.org/10.1016/j.tfp.2024.100571

Arjumand T, Rani BR (2023) The Case study on Bamboo farming, Centre for Climate Change and Adaptation (CCA), National Institute of Agricultural Extension Management (MANAGE), Hyderabad, India

Banerjee S, Basak M, Dutta S, Chanda C, Dey A, Das M (2021) Ethnobamboology: traditional uses of bamboos and opportunities to exploit genomic resources for better exploitation. In:

Biotechnological advances in bamboo: the "Green Gold" on the Earth, pp 313–352. https://doi.org/10.1007/978-981-16-1310-4_14
Bramsiepe C, Sievers S, Seifert T, Stefanidis GD, Vlachos DG, SchnitzerH, SchembeckerG (2012) Low-cost small scale processing technologies for production applications in various environments—mass produced factories. Chem Eng Process Process Intensif 51:32–52
Chandewar N (2025) Agroforestry and sustainable land management. Addition Publishing House
Chaudhary U, Malik S, Rana V, Joshi G (2024) Bamboo in the pulp, paper and allied industries. Adv Bamboo Sci:100069
Chen WJ, Jan JF, Chung CH, Liaw SC (2024) Evaluating the economic viability of agro-ecotourism as a nature-based solution for a climate adaptation strategy: a case study of Yuanshan Township, Taiwan. Sustainability 16(18):8267. https://doi.org/10.3390/su16188267
Dev I, Ram A, Ahlawat SP, Palsaniya DR, Singh R, DhyaniSK PJ (2020) Bamboo-based agroforestry system (Dendrocalamus strictus+ sesame–chickpea) for enhancing productivity in semi-arid tropics of central India. Agrofor Syst 94:1725–1739. https://doi.org/10.1007/s10457-020-00492-8
Dutta S, Gorain S, Roy J, Das R, Banerjee S, Gorai SK, Das S (2025) Bamboo for global sustainability: a systematic review of its environmental and ecological implications, climate action, and biodiversity contributions. Environ Rev. https://doi.org/10.1139/er-2025-0032
FAO (2020) Global Forest resources assessment 2020: main report, Rome
Ghimire A (2008) An assessment of the dependency of farmers on bamboo resource for rural livelihood in Lalitpur District, Nepal
Ghode N (2023) Manual on bamboo cultivation: bamboo cultivation in Chhattisgarh, India: a sustainable and economically viable solution. Shashwat Publication
Hosseini M, Gaff M, Wei Y, Tu C (2025) Engineered bamboo building materials: Types, production, and applications. Forests 16(4):662. https://doi.org/10.3390/f16040662
Iroegbu AOC, Ray SS (2021) Bamboos: from bioresource to sustainable materials and chemicals. Sustainability 13(21):12200. https://doi.org/10.3390/su132112200
ISFR (India State of Forest Report) 2021 by Forest Survey of India. Ministry of Environment and Forests and Climate Change, Government of India. Dehradun, India
ISFR (India State of Forest Report) 2023 by Forest Survey of India. Ministry of Environment and Forests & Climate Change, Government of India. Dehradun, India
Kaushal R, Kumar A, Patra S, Islam S, Tomar JMS, Singh DV, Durai J (2021) In-situ soil moisture conservation in bamboos for the rehabilitation of degraded lands in the Himalayan foothills. Ecol Eng 173:106437. https://doi.org/10.1016/j.ecoleng.2021.106437
Kumar V (2016) Multifunctional agroforestry systems in tropics region. Nat Environ Pollut Technol 15(2):365
Lai R, Lai S (2019) Role of Tissue culture in rapid clonal propagation and production of pathogen–free plants. In: Crop improvement utilizing biotechnology. CRC Press, pp 73–116
Lal R (2009) Sequestering atmospheric carbon dioxide. Crit Rev Plant Sci 28(3):90–96. https://doi.org/10.1080/07352680902782711
Liese W (1987) Research on bamboo. Wood Sci Technol 21:189–209
Lin J, Gupta S, Loos TK, Birner R (2019) Opportunities and challenges in the Ethiopian bamboo sector: a market analysis of the bamboo-based value web. Sustainability 11(6):1644. https://doi.org/10.3390/su11061644
Meena RK, Negi N, Shankhwar R, Bhandari MS, Sharma R (2023) Population genetic analysis illustrated a high gene diversity and genetic heterogeneity in Himalayacalamus falconeri: a socio-economically important Indian temperate woody bamboo taxon. J Plant Biochem Biotechnol 32(3):438–450. https://doi.org/10.1007/s13562-022-00824-y
Mekonnen Z, Worku A, Yohannes T, Alebachew M, Kassa H (2014) Bamboo Resources in Ethiopia: their value chain and contribution to livelihoods. Ethnobot Res Appl 12:511–524. https://ethnobotanyjournal.org/index.php/era/article/view/923
Naskar J, Kumar Jha A, SinghTN, Aeron S (2025) Climate change and soil resilience: a critical appraisal on innovative techniques for sustainable ground improvement and ecosystem pro-

tection. J Hazard Toxic Radioact Waste 29(4):03125002. https://doi.org/10.1061/JHTRBP.HZENG-1465
Nath AJ, Das G, Das AK (2009) Above ground standing biomass and carbon storage in village bamboos in North East India. Biomass Bioenergy 33(9):1188–1196
Ni H, Su W, Fan S, Chu H (2021) Effects of intensive management practices on rhizosphere soil properties, root growth, and nutrient uptake in Moso bamboo plantations in subtropical China. For Ecol Manag 493:119083. https://doi.org/10.1016/j.foreco.2021.119083
Pan C, Zhou G, Shrestha AK, Chen J, Kozak R, Li N, Wang G (2023) Bamboo as a nature-based solution (NbS) for climate change mitigation: biomass, products, and carbon credits. Climate 11(9):175. https://doi.org/10.3390/cli11090175
Partey ST, Sarfo DA, Frith O, KwakuM, Thevathasan NV (2017) Potentials of bamboo-based agroforestry for sustainable development in Sub-Saharan Africa: a review. Agric Res 6(1):22–32. https://doi.org/10.1007/s40003-017-0244-z
Rana VS, Raina A, Bathla G (eds) (2024) Global sustainable practices in gastronomic tourism. IGI Global
Rane AD, Chandramouli S, Viswanath S (2016) Can Dendrocalamus stocksii (munro.) be the ideal multipurpose bamboo species for domestication in Peninsular India. J Bamboo Rattan 15(1–4):23–32
Rao IR, KumarA, RezaS, Motokuri B (2009) A pathway out of poverty. Bamboo Incense Sticks Production as a Livelihood Option for Rural Women in Tripura, India. International Network for Bamboo and Rattan (INBAR) and Beijing and Centre for Indian Bamboo Resource and Technology (CIBART), New Delhi, pp. 7–39
Samdin Z, Abdullah Zawawi NI, Saadun N, Kamaruddin N (2024) A review of the potential socio-economic impacts of bamboo plantation on local community. Pertanika J Sci Technol 32
Sawarkar AD, Shrimankar DD, Manekar SC, Kumar M, Garlapati PK, SinghL (2025) Bamboo as a sustainable crop for land restoration in India: challenges and opportunities. Environ Dev Sustain 27(1):157–189. https://doi.org/10.1007/s10668-023-03911-9
Scheba A, Blanchard R, Mayeki S (2017) Bamboo for green development. The opportunities and challenges of commercialising bamboo in South Africa, Human Sciences Research Council (HSRC)
Sharma M (2023) Studies on ginger based intercropping systems for higher yield and income. Doctoral dissertation, Dr Yashwant Singh Parmar University of Horticulture and Forestry
Sharma U, Bhardwaj DR, SharmaS, Sankhyan N, Thakur CL, Rana N, Sharma S (2022) Assessment of the efficacy of various mulch materials on improving the growth and yield of ginger (Zingiber officinale) under bamboo-based agroforestry system in NW-Himalaya. Agrofor Syst 96(5):925–940. https://doi.org/10.1007/s10457-022-00753-8
Shen J, Zeng X, Fan S, LiuG (2025) Impacts of intensive management practices on the long-term sustainability of soil and water conservation functions in bamboo forests: a mechanistic review from silvicultural perspectives. Forests 16(5):787. https://doi.org/10.3390/f16050787
Singh AK, Kal S, Dubey SK, Mahopatra KP, Rao BK, Gaur ML (2013) Impact of resources conservation treatments on runoff, soil loss and growth performance of bamboo for reclamation of Yamuna ravine systems of India. In: Natural resource conservation: emerging issues and future challenges, pp 361–369
Sohel MSI, Alamgir M, Akhter S, Rahman M (2015) Carbon storage in a bamboo (Bambusa vulgaris) plantation in the degraded tropical forests: implications for policy development. Land Use Policy 49:142–151. https://doi.org/10.1016/j.landusepol.2015.07.011
Solomon T, Moon H, Abebe S, Minale AS, Teketay D (2020) Promoting bamboo-based agroforestry for enhancing ecosystem services from degraded lands. In: Agroforestry for degraded landscapes: recent advances and emerging challenges, vol 2, pp 423–444. https://doi.org/10.1007/978-981-15-6807-7_16

Sun Y, Liu X (2022) How design technology improves the sustainability of intangible cultural heritage products: a practical study on bamboo basketry craft. Sustainability 14(19):12058. https://doi.org/10.3390/su141912058

Sundriyal M, Joshi K (2015) Bamboo status and trade vulnerability: a central Himalayan case study. In: 10th World Bamboo Congress, Korea, pp 1–9

Sundriyal M, Sundriyal RC (2011) Bamboo trade and future prospects in the central Himalaya: a case study of the traditional artisans of Uttarakhand, India. Ethnobotany Res Appl 9:445–454

Tewari S, Negi H, Kaushal R (2019) Status of bamboo in India. Int J Econ Plants 6:030–039

Tiwari P, Kumar R, Thakur L, Salve A, Parmar Y (2017) Agroforestry for sustainable rural livelihood: a review. Int J Pure Appl Biosci 5(1):299–309. https://doi.org/10.18782/2320-7051.2439

Tripathi A, Yadav S, Nishtha, Nkengnamai M, Thakur A (2024) Bamboo: a fast-growing species to mitigate carbon footprint. In: Forests and climate change: biological perspectives on impact, adaptation, and mitigation strategies. Springer, Singapore, pp 469–487. https://doi.org/10.1007/978-981-97-3905-9_23

Viswanath S, Lubina PA (2018) Traditional agroforestry systems. In: Agroforestry: anecdotal to modern science. Springer, Singapore, pp 91–119. https://doi.org/10.1007/978-981-10-7650-3_3

Wang X, Hua F, Wang L, Wilcove DS, Yu DW (2019) The biodiversity benefit of native forests and mixed-species plantations over monoculture plantations. Divers Distrib 25(11):1721–1735. https://doi.org/10.1111/ddi.12972

Yadav M, Mathur A (2021) Bamboo as a sustainable material in the construction industry: an overview. Mater Today Proc 43:2872–2876

Yiping L, Yanxia L, Buckingham K, Henley G, Guomo Z (2010) Bamboo and climate change mitigation: a comparative analysis of carbon sequestration. Int Network Bamboo Rattan 30:1–47

Zhang Y, Wu L, Li Y, Yang J, Yang H, Zhao Y, Chen G (2024) Bamboo shoot and its food applications in last decade: an undervalued edible resource from forest to feed future people. Trends Food Sci Technol:104399. https://doi.org/10.1016/j.tifs.2024.104399

Zhaohua Z, Wei J (2018) Sustainable bamboo development. CABI

Chapter 13
Wild Edible Plants for Economic Development in Agroforestry Systems

Vatika Sharma, Garima Bhatt, Bhupendra Singh, Deepa Rawat, Vinod Prasad Khanduri, and Munesh Kumar

Abstract Wild edible plants (WEPs), or uncultivated species used for food, have been essential to human sustenance since before agriculture was invented. Although contemporary diets depend on a small selection of cultivated crops, WEPs continue to be an important food source, especially in developing countries such as India, where they augment diets and enhance food security, particularly in rural areas. Numerous WEP species may be found in the Indian Himalayan state of Uttarakhand, which is known for its rich biodiversity, including beneficial wild fruits that are crucial for the livelihoods of local communities, providing key nutrients and income sources. In spite of a global trend favoring cultivated foods, WEPs possess the capability to improve nutrition, offer economic benefits, and support the sustainable intensification of food systems, emphasizing the necessity for their ongoing acknowledgment and use, especially in areas where they are essential to local cultures and economies. Wild edible plants (WEPs) serve as essential natural resources vital for the survival of many, especially in areas like Uttarakhand, India. These plants, which include highly nutritious species such as *Prunus persica* (peach), *Myrica esculenta* (box myrtle), and *Rubus ellipticus* (golden Himalayan raspberry), as well as those with medicinal benefits like *Ficus auriculata* (elephant ear fig), *Berberis asiatica* (Indian barberry), *Punica granatum* (pomegranate), and *Solanum nigrum* (black nightshade), contributed significantly to the Kumaun and Garhwal regions' rural communities' food security, nutritional well-being, and revenue source. Conservation initiatives that blend traditional and scientific methods with community participation are essential to protect these invaluable resources, ensuring

V. Sharma · G. Bhatt · B. Singh (✉) · V. P. Khanduri
Department of Silviculture and Agroforestry, College of Forestry, Veer Chandra Singh Garhwali Uttarakhand University of Horticulture and Forestry, Tehri Garhwal, Uttarakhand, India

D. Rawat
Department of Forests Resource Management, College of Forestry, Veer Chandra Singh Garhwali Uttarakhand University of Horticulture and Forestry, Tehri Garhwal, Uttarakhand, India

M. Kumar
Department of Forestry and Natural Resources, Hemvati Nandan Bahuguna Garhwal University (A Central University), Srinagar Garhwal, Uttarakhand, India

B. Singh et al. (eds.), *Himalayan Agroforestry Systems*, Advances in Agroforestry 15,
https://doi.org/10.1007/978-3-032-25096-4_13

ecological balance, future food security, and the economic empowerment of Himalayan communities through the creation of functional foods and nutraceuticals.

Keywords Conservation · Natural resources · Indian Himalaya subsistence agriculture · Wild plant

1 Introduction

The term "wild edible plants" (WEPs) describes plant species that are available from a variety of natural habitats and are consumed as food but are not grown or domesticated. There is at least one edible part in wild edible plants, such as a leaf, fruit, or root. Fruit is a flower's ripened ovary, which can be either nut-like or fleshy. Generally, there are two kinds of fruits: wild and domesticated. Fruits that are grown are planted by farmers for profit, but wild fruits only exist in their natural environment. With an increasing population, fewer plants are found in the plantation and forest regions used for human well-being; consequently, wild plants lose their individuality and become less common. Since their challenging geology and climate make them unsuitable for human survival, wild fruiting plants predominate there and are unaffected by human interaction (Tiwari et al. 2010a). Around the world, many traditions have collected WEPs from a variety of environments, including forests, arable fields, and even anthropogenically damaged areas like wastelands and roadside ditches. Wild edible plants (WEPs) were a major source of food for humans before civilization developed and modern fruits were domesticated. Incorporating these wild fruits into people's diets has been crucial. Wild food plants have been a vital part of human life since ancient times. Wild edible plants have historically supported human populations in all of the populated continents (Khyade et al. 2009). The majority of people living in rural areas of India rely on wild plants to supplement their diet (Tiwari et al. 2010a). The variety of wild plant species improves food security in the home by adding diversity to the family's diet. Although a relatively small number of crops now provide the majority of human plant food, it is evident that the usage of wild plants is not insignificant in many regions of the world (Prescott-Allen and Prescott-Allen 1990; Pieroni et al. 2007).

Uttarakhand is renowned for both its wide variety of ethnomedicinal plants and its long history of wild food plant systems. According to Arora and Anjula's 1996 report, Uttarakhand is home to 97 species of edible plants, including 8 crops of cereals and pseudocereals, 16 crops of millets and minor millets, 11 crops of oilseeds, 28 crops of vegetables, 10 crops of spices and condiments, and 19 crops of fruits. Uttarakhand's diverse eco-geographical and eco-climatic circumstances produce a highly enriched vegetation, including wild edible fruits (Saklani and Chandra 2011). Compared to cultivated fruits, some wild fruits have been found to have superior nutritional content (Maikhuri et al. 1994). Wild edible fruit consumption satisfies the needs of poor residents in Uttarakhand's hilly regions for proteins, carbs, lipids, vitamins, and minerals (Sundriyal and Sundriyal 2001). Due to its high economic worth, people in the hills without a source of income wait for the ripening season. The fact that the residents of the hills may make more money every season by selling the wild fruits is a good indicator of the species' popularity (Bhatt et al. 2000). Because the Himalayas are so abundant in natural resources, the Garhwal Himalayas region is home to numerous stunning holy sites, valleys, and slopes. The forest resources are essential to the local communities' survival. The region's tremendous diversity is used by the people for a wide range of things, including as lumber, agricultural implements, food, fuel, fodder, and medicine. Among these, edible wild plants offer the local inhabitants a crucial food source in times of famine. The locals gather a variety of wild edible plants for sustenance due to their subsistence agriculture and limited land holdings. A range of biological resources are used by people in the Himalayan countryside to support themselves. Their primary sources of income are marginal agriculture and animal husbandry (Kala 2005, 2007). Wild plants can occasionally have better nutritional content than a number of well-known conventional vegetables and fruits (Nordeide et al. 1996; Orech et al. 2007). The landscape, vegetation, people, and customs of the Garhwal Himalaya are all unique. The livelihoods of the local populations in the area are significantly influenced by the forest resources. They still rely on the forest's natural resources to survive (Tiwari et al. 2010b). In many developing countries, wild plants provide humans with a sufficient amount of nourishment by serving as sources of food and other necessities. In many other developing nations, millions of people lack access to enough food to meet their daily needs and are lacking in one or more nutrients (FAO 2004). Humans have used around 7000 WEPs in the past, but there are still a lot of useful plants and food sources that need to be discovered (Mohan 2000). The majority of societies still rely heavily on agricultural crops. The use of WEPs has not completely stopped.

According to a study by the Food and Agriculture Organization (FAO), one billion or more individuals consume wild foods in their diet. It is now commonly known that all of the current cultivated plants were once wild edible plants (WEPs). Humans have been cultivating plants for thousands of years, choosing those with desirable characteristics to create the domesticated crops we are familiar with today. Only a few plant and fungal species have become economically significant, but some are marketed in city and local markets. Useful plants and fungus may be the most significant ecosystem provision service available to rural communities, with the potential to meet fundamental population needs for food, medicine, and photochemicals, among other things. Numerous species are suitable for the sustainable intensification of crop production and are essential components of agroforestry systems. Since ancient times, edible wild plants and fungus have been utilized by human societies. Although the use of these species has mainly been discontinued in developed nations, there is currently a growing movement to revalue them. In contrast, gathering and using wild plants and mushrooms is still common in less developed areas. A number of these species are also commercialized and have significant socioeconomic value. Although the majority of plant-based foods consumed by humans today are derived from a relatively small number of crops, it is evident that the use of wild plants is not insignificant in many regions of the world. Wild plants can occasionally have better nutritional content than a number of well-known, everyday fruits and vegetables. In India, a total of 1532 species of wild edibles were reported (Bhatia et al. 2018), in which 675 species were from the Indian Himalayas region and 800 species were only used by tribal communities (Bandyopadhyay and Mukherjee 2009) (Tables 13.1, 13.2, and 13.3).

2 Economic and Nutritional Values of Wild Edibles

According to surveys conducted on a few key species in certain agro-ecologies and niche regions in Uttarakhand, wild plants can yield INR 5000–500,000 per group. In terms of energy and nutritional content, many wild food sources are frequently thought to be superior to those from numerous cultivated species.

It should be noted that "malnutrition is a major health burden in developing countries and use of wild plant foods can be enlisted as policy support to secure wild food use and preserve habitats for wild edible species" (Bharucha and Pretty 2010). Therefore, comprehensive food composition is crucial, particularly for communities most at risk for malnutrition (Misra et al. 2008). However, compared to cultivated species, we know very little about the macro- and micronutritional characteristics of wild foods. Essential micronutrients, including Fe, P, Ca, Zn, Na, K, and others, are found in a wide variety of naturally occurring food plants. Numerous natural food sources found in the Uttarakhand highlands are high in micronutrients (Negi and Gaur 1994). Commonly consumed edible parts of agroforestry systems (CPRs) and a variety of wild backyard plants are significant sources of antioxidants, vitamins A, B_2, and C, especially during lean seasons. Many are essential foods in famine that ensure nutrition stability all year long in spite of possible scarcity of food. Wild fruits are currently gaining more attention globally as a potential dietary supplement or less costly alternative to commercial fruits. In addition to their well-established nutritional worth, there is evidence of the health benefits of consuming wild fruits, and there is typically sufficient information available regarding the antioxidant capacity of fruits from different species. *Ficus carica, Randia echinocarpa, Ficus deliciosa, Kadsura coccinea, Myrciaria dubia, Ficus racemosa, Actinidia eriantha Phyllanthus emblica, Punica granatum, Ziziphus mauritiana,* and *Randia echinocarpa* are a few instances. In addition to fruits, other plant sections are also known to possess antioxidant qualities. The economic worth of wild foods cannot be fully estimated (Bharucha and Pretty 2010), quantitative evaluations encounter methodological challenges, and the trade in wild foods is frequently informal or takes place at local markets, meaning it is not taken into account by traditional accounting methods (Jaarsveld et al. 2005) (Table 13.4).

Table 13.1 List of northwestern Himalayan wild edible trees

S. no.	Botanical name	Common name	Family	Local distribution	Altitudinal range	Plant part used	Major chemical constituents	Availability Status	Ethnobotanical and medicinal uses
1	*Aesculus indica*	Horse chestnut, Pangar	Hippocastanaceae	Himachal Pradesh, Uttar Pradesh, Meghalaya, Sikkim, Nepal,	1500–2800	Seeds are edible	Saponins, aesculine	++	Hamorrhoids, thorombosis
2.	*Aesandra butyracea*	Indian butter tree, Baehni	Sapotaceae	Uttarakhand, Nepal, Bhutan, and parts of China	300–1500	Seed bark fruit pulp, leaves	α-Spinasterol, β-sitosterol, α-spinasterone	+	Emollient, antirheumatic, skin affection
3	*Aegle marmelos*	Bel	Rutaceae	Punjab, Bihar, Madhya Pradesh, Jharkhand Chhattisgarh, Uttar Pradesh, Odisha, and West Bengal	Up to 1200	Ripe fruits edible	Alkaloids, flavonoids, carotenoids, and phenolic compounds	+	Juice used to treat respiratory, cardiac, and digestive disorders
4	*Bauhinia variegata*	Kachnar, orchid tree, and mountain ebony	Caesalpinaceae	Uttarakhand, Assam, and Myanmar	1500–1830 m	Flowers, leaves	Saponins, terpenoids, and tannins, reducing sugars, steroids, flavonoids, and cardiac glycosides	+++	Flowers are roasted and eaten. Leaves as fodder. Wood is used for building and agricultural tools. Using dried leaf ash to cure cough

(continued)

Table 13.1 (continued)

S. no.	Botanical name	Common name	Family	Local distribution	Altitudinal range	Plant part used	Major chemical constituents	Availability Status	Ethnobotanical and medicinal uses
5	*Celtis australis*	Kharik, nettle tree	Ulmaceae	Western Himalayas, particularly in Uttarakhand, Himachal Pradesh, and Jammu and Kashmir	500–2500	Fruit, leaves, wood, bark, mucilage	Tanin phenolic compounds like chlorogenic acid and caffeic acid derivatives, flavonoids like vitexin and isovitexin, and other compounds like tannins, antioxidants, and antimicrobial compounds	+++	Reduce peptic ulcers and urinary bleeding. Bark is a source of color, leaves are feed, wood is used as fuel and to make small agricultural tools, and its paste is used to treat joint and bone discomfort
6	*Cinnamomum tamala*	Dalchini	Lauraceae	Himachal Pradesh, Uttarakhand, Sikkim, Assam, Mizoram, and Meghalaya	500–2200	Bark and leaves	Myrcene, limonene, methyl ether of eugenol, and alfapinene	++	Bark and dried leaves are eaten to relieve throat irritation and dyspepsia, and they are used to flavor tea and other food preparations

7	*Cornus capitata*	Bhamora, Himalayan Dogwood	Cornaceae	Western Himalayas, including states like Sikkim, West Bengal, Arunachal Pradesh, Nagaland, and Himachal Pradesh	1500–2500	Fruits, wood	Tannins, anthocyanins, flavonoids, lignans, and aromatic acids, along with terpenoids such as iridoids, and rich in minerals like sodium calcium, magnesium, and iron	++	Edible fruits that wildlife prefers, timber utilized in farming equipment
8	*Embilica officinalis*	Awnla	Euphorbiaceae	Maharashtra, Gujarat, Uttar Pradesh, Rajasthan, Andhra Pradesh, Himachal Pradesh, Tamil Nadu, and Karnataka	150–2300 meters	Fruits, leaves, and bark	Tannins, flavonoids, vitamin C, and phenolic acids	+	Edible fruits that are high in vitamin C. Fruits used as pickles and as a component of trifala, a remedy for a number of illnesses. Tannin-producing bark and leaves for feed

(continued)

Table 13.1 (continued)

S. no.	Botanical name	Common name	Family	Local distribution	Altitudinal range	Plant part used	Major chemical constituents	Availability Status	Ethnobotanical and medicinal uses
9	*Ficus auriculata*	Timla, elephant ear fig	Moraceae	Jammu and Kashmir, Assam, Arunachal Pradesh, West Bengal, Bihar, Maharashtra, Mizoram, Manipur, Meghalaya, Orissa, Sikkim, and South India	600–1800	Fruits, leaves, and wood	Flavonoids, phenols, tannins, glycosides, and terpenoids, with some studies also identifying alkaloids, saponins, and resins	++	Fruits (figs) can be eaten, ripe fruits can be eaten, unripe fruits can be turned into vegetables, leaves can be converted into cups and plates, and they make ideal cattle feed. White latex prevents diabetes, and wood is utilized as fuel
10	*Ficus palmata*	Bedu, wild Himalayan fig	Moraceae	Andhra Pradesh, Odisha, Tamil Nadu, Madhya Pradesh, Uttarakhand, Himachal Pradesh, Delhi, Uttar Pradesh, Punjab, and Jammu and Kashmir	600–1800	Fig, fruit, leaves, and twigs	Flavonoids, alkaloids, tannins, terpenoids, and cardiac glycosides, as well as minerals, vitamins, and sugars	++	Figs are a useful agroforestry plant that can be eaten raw with salt or baked. The fruit helps with digestive issues, while the leaves and twigs make wonderful fodder

11	*Ficus racemosa*	Umra	Moraceae	from the outer Himalayan ranges to the Deccan and South India	100 and 1700 meters	Fruits leaves	Flavonoids, triterpenoids, tannins, alkaloids, and steroids	++	Edible fruits, even cooked and fried immature fruits. Using leaves as feed
12	*Ficus semicordata*	Khaina	Moraceae	Andhra Pradesh, Assam, Bihar, and Uttarakhand and other regions in South and Southeast Asia	Up to 2000 m	Fruit, wood, and leaves	Flavonoids, phenols, tannins, saponins, anthraquinone, and glycosides	+	Fruit are edible, wood used as a fuel and leaves as fodder
13	*Flacourtia indica*	Kangu	Flacourtiaceae	Himachal Pradesh, Gujarat, Goa, Kerala, Karanatka, Andaman & Nicobar Islands, Andhra Pradesh, Bihar, West Bengal, Tamil Nadu, Chhattisgarh, Madhya Pradesh, Maharashtra, and Rajasthan	Up to 1200 m	Leaves, fruit	Phenolic glycosides, tannins, saponins, alkaloids, and flavonoids, as well as vitamins like ascorbic acid (vitamin C) and beta-carotene	++	Fruits are edible, leaves used as fodder, and fruit are used in hepatitis fever, dysentery, and diarrhea

(continued)

Table 13.1 (continued)

S. no.	Botanical name	Common name	Family	Local distribution	Altitudinal range	Plant part used	Major chemical constituents	Availability Status	Ethnobotanical and medicinal uses
14	*Grewia optiva*	Bhimal	Tiliaceae	Hills of northwestern states like Jammu and Kashmir, Himachal Pradesh, and Uttarakhand	500–2500 m	Fruit, leaves, fiber	Cellulose, hemicellulose, and lignin, with flavonoids and other bioactive compounds also present	++	The fiber from sticks is used to make soap and shampoo, the fruit is edible, and the leaves are used as fodder. In order to help with urinary problems, ropes, nets, brushes, brooms, and cattle tie ropes made from fiber and fruits
15	*Juglans regia*	Akhrot	Juglandaceae	Dry, cold, and alpine regions of Himachal Pradesh, including the hills of Uttarakhand and Jammu and Kashmir	1600–2700	Fruits, wood, and bark	Rich in vitamins, minerals fats, primarily unsaturated fatty acids like oleic acids and linoleic acid, along with protein, carbohydrates, and various micronutrients	++	Fruits are edible and produce veneers, oil, and wood for furniture, carvings, and gun stocks. Bark is used as medication and coloring. Used for tanning, coloring, and intoxicating fish. Leaves are used as insecticides and fungicides

16	*Mangifera indica*	Aam	Anacardiaceous	Uttar Pradesh, Andhra Pradesh, Karnataka, and Maharashtra	Up to 1200 m	Fruits, wood, and flowers	Polyphenols, flavonoids, and terpenoids, with mangiferin being a prominent xanthone	++	Fruits can be eaten, wood can be used to build and furnish, and flowers are a good source of food for bees
17	*Morus alba*	Sehtoot	Sehtoot	Karnataka, Andhra Pradesh, West Bengal, Jammu and Kashmir, Uttar Pradesh, Tamil Nadu, Kerala, and to a lesser extent in Madhya Pradesh, Bihar, Orissa, Assam, and Manipur	200–1800	Fruits and leaves	Bioactive compounds, including polyphenols, flavonoids, and other phytochemicals like chlorogenic acid, rutin, quercetin, and anthocyanins	++	Edible fruits, leaves that have been chopped for sheep and goat feed, and leaves that are fed to silkworms
18	*Myrica esculenta*	Kaphal, Bay Berry	Myricaceae	Uttarakhand, Jammu and Kashmir, Himachal Pradesh, Mizoram, Manipur, Sikkim, Arunachal Pradesh, Assam, Meghalaya, and Nagaland	1200–2400	Fruits, bark, root wood	Tripterpenes, Myrictrin	+++	Used in painful throat and cold. Wood is used as fuel and tools, fruits are consumed raw or prepared into cool beverages, bark is used to make yellow dye and intoxicate fish, and wood is used to cure wounds and oral diseases

(continued)

Table 13.1 (continued)

S. no.	Botanical name	Common name	Family	Local distribution	Altitudinal range	Plant part used	Major chemical constituents	Availability Status	Ethnobotanical and medicinal uses
19	*Phoenix humilis*	Khajoor, mountain date palm	Arecaceae	Native to the Himalayas and Northeast India, with its distribution extending from Uttarakhand to the northeastern states	Up to 1500	Fruits and leaves	Flavonoids, saponins, sterols, and terpenoids	++	Fruit is edible, and leaves are used for brooms, hats, and mats. Additionally, dry leaves are utilized to prepare mud homes' sturdy roofs
20	*Pinus roxburghii*	Chir	Pinaceae	Jammu and Kashmir, Himachal Pradesh, Uttarakhand, and Sikkim	1000–2400	Seeds, resin wood saw dust	Terpenes like α-pinene, β-pinene, and δ-3-carene, as well as resin acids, tannins, longifolene, and caryophyllene	++	Seeds can be eaten. Turpentine, which is extracted from the resin, has antibacterial, diuretic, rubefacient, and vermifuge properties. Wood is utilized in building. Resin is found in turpentine, paints, and varnishes. Using sawdust and honey to treat bronchitis and asthma

21	*Prunus armeniaca*	Chulu, apricot	Rosaceae	Uttarakhand, Himachal Pradesh, Ladakh, Jammu and Kashmir	700–3000	Fruits, flowers	Carbohydrates, proteins, vitamins (A, C), minerals (calcium, magnesium, potassium), and antioxidants like flavonoids and carotenoids, with oleic acid being the predominant fatty acid in the oil	++	Edible fruits and seed oil are also utilized as medicines for fevers and as a form of massage for body aches. Locally produced drinks made from dried fruit pulp. Flowers are used as feed by bees in apiculture
22	*Prunus cerasoides*	Payain, Himalayan cherry	Rosaceae	Jammu and Kashmir, Himachal Pradesh, Uttarakhand Sikkim, Manipur, Assam, and Mizoram	1200–2400	Fruits, branch leaves bark	Flavones, isoflavones, glycosides, terpenoids, and sterols	+++	Edible ripe fruits. Branches used in walking sticks. Bark in psycho medicine. Use leaves as feed. Bark juice is used to treat bodily edemas and disorientation
23	*Punica granatum*	Dalimu/Dadim	Punicaceae	Maharashtra, Karnataka, Andhra Pradesh, and Gujarat in wild in Himachal Pradesh and Uttarakhand	1000 and 2775	Fruit	Flavonoids, phenolic acids, ellagic acid, and ellagitannins	++	Fruits are edible, High source of minerals and fruits are used in cough and cold

(continued)

Table 13.1 (continued)

S. no.	Botanical name	Common name	Family	Local distribution	Altitudinal range	Plant part used	Major chemical constituents	Availability Status	Ethnobotanical and medicinal uses
24	*Pyrus pashia*	Melu, wild Himalayan pear	Rosaceae	Himachal Pradesh, Uttarakhand, Jammu and Kashmir, as well as parts of the northeastern hills and Punjab	1000–2400	Fruits, leaves, twigs wood, and flowers	Phytochemicals including flavonoids, tannins, alkaloids, glycosides, steroids, saponins, and tannins, as well as useful polyphenolic therapeutic constituents like arbutin, flavan-3-ols, and chlorogenic acids	+++	Leaves are utilized as fodder, and fruits are edible and blood-purifying. Edible ripe fruits are also utilized to treat stomach issues. Twigs and leaves for food. Sticks fashioned from wood. Flowers that are beneficial to apiculture
25	*Rhododendron arboreum*	Burans	Ericaceae	Uttarakhand, Sikkim, Arunachal Pradesh, and Nagaland	1200–3300	Flowers	Alkaloids, glycosides, terpenoids, steroids, anthraquinones, flavonoids, saponins, and tannins, with quercetin, rutin, and coumaric acid being prominent in flowers	+++	People in Uttarakhand use flowers to get their juice. Flowers that serve as food for bees. Bark and flowers are used as remedies for respiratory and intestinal conditions

26	*Rhododendron barbatum*	Burans, bearded rhododendron	Ericaceae	Uttarakhand and Sikkim, as well as in Darjeeling and West Bengal	2000–3300	Flowers	Grayanotoxins, and romedotoxin, and arbutin	+	Flowers of trees are used to prepare juice
27	*Rhus parviflora*	Tungula Toxicodendron parviflorum	Anacardiaceae	Uttar Pradesh, Uttarakhand, Himachal Pradesh, Arunanchal Pradesh, Madhya Pradesh, Sikkim, Kerala, Jammu and Kashmir, Karnataka, and Tamil Nadu	700–1100 m	Fruits, leaves stem, and wood	Flavonoids like myricetin, quercetin, and kaempferol, along with their 3-O-rhamnosides, and other compounds like hentriacontane, hentriacontanol, beta-sitosterol, lignoceric acid, and iso-rhamnetin-3-alpha-L-arvinoside	+	Ayurvedic traditional medicine Edible fruits and tobacco-infused leaves that are smoked. Leaves are frequently utilized as bio-fence. Fuel made from wood, toothbrush stems, and toothaches
28	*Rubus paniculatus*	Kali Hinsar, heart leaved raspberry	Rosaceae	Uttarakhand, Himachal Pradesh, and Sikkim	1500–2900 m	Fruits and flowers	Anthocyanins, phenolic compounds, and triterpenoids, including ursolic acid	++	Energy-producing fruit and apiculture-useful flowers

(continued)

Table 13.1 (continued)

S. no.	Botanical name	Common name	Family	Local distribution	Altitudinal range	Plant part used	Major chemical constituents	Availability Status	Ethnobotanical and medicinal uses
29	*Solanum nigrum*	Makoi, black nightshade	Solanaceae	Andhra Pradesh, Telangana, Karnataka, Maharashtra, Odisha, Tamil Nadu, and Jammu and Kashmir	Up to 2200 m	Young shoot fruit stem and leaf	Organic acids, polysaccharides, steroidal saponins, steroidal alkaloids, flavonoids, coumarin, lignin, and volatile oils	++	Fruit and stem extract are used to treat liver fever and diarrhea. Fruits used to treat diarrhea and digestive disorders. Leaf extract is used to treat ear discomfort. Herbs. Ripe fruits are edible, and young shoots and leaves are cooked as vegetables
30	*Syzygium cumini*	Jamun, black plum	Myrtaceae	Karnataka, Andhra Pradesh, Maharashtra, Tamil Nadu, Uttar Pradesh, and Himachal Pradesh	Up to 1200 m	Fruit and bark	Alkaloids, flavonoids, tannins, saponins, phenols, terpenoids, and steroids, as well as essential oils rich in compounds like τ-cadinol, globulol, and caryophyllene	++	Edible fruit, multivitamins, diabetic treatment, tanning, and dyeing bark

31	*Taxus wallichiana Zucc.*	Thuner	Taxaceae, Himalayan yew	Jammu and Kashmir, Uttarakhand, Himachal Pradesh, Arunachal Pradesh, Sikkim, Manipur, Meghalaya, Nagaland, and West Bengal	2000–3600	Bark and seeds	Taxane alkaloids (like taxol), diterpenoids, lignans, biflavonoids, steroids, and sugar derivatives, with taxol being a notable anticancer agent	+	Tree bark and seeds are edible
32	*Ziziphus glaberrima*	Ber	Rhamnaceae	Karnataka, Maharashtra, and northeast India	Up to 1000	Fruit root leaves	Jujube fruit is rich in vitamins A, B, and C, as well as various compounds such as alkaloids, flavonoids, sterols, tannins, saponins, and fatty acids	++	The root is helpful and applied to old wounds and ulcers, while the fruit is more nutritive and edible. It is also a good source of calcium and potassium. Leaf extract helps with fever, asthma, and liver issues

Source: Bharti and Lata (2018), Radha et al. (2013), Chauhan et al. (2020), Agarwal and Chandra (2019), and Bhandari et al. (2021)
+ uncommon, ++ common, +++ abundant

Table 13.2 List of wild edible shrubs of northwestern Himalayas

S. no.	Botanical name	Common name	Family	Local distribution	Altitudinal range	Plant part used	Major chemical constituents	Availability status	Ethnobotanical and medicinal uses
1.	*Berberis aristata*	Darul haldi	Berberidaceae	Himachal Pradesh, Jammu and Kashmir, Tamil Nadu, Sikkim, and Uttarakhand	1800–2700	Edible fruits	Alkaloids, phenolics	+++	Roots and bark are consumed orally to treat jaundice
2.	*Berberis asiatica*	Kilmora	Berberidaceae	Uttarakhand, Assam, Bihar, Arunanachal Pradesh, and Himachal Pradesh	1200–2500	Fruit used raw or dried	Phenolics, alkaloids, tannins, and sugars	++	The fruit has laxative and cooling properties. Roots used in jaundice, fevers etc. Tender leaf buds are chewed for dental caries. Bark and wood produce yellow dye
3.	*Cannabis sativa*	Bhaang	Cannabinaceae	Himachal Pradesh, Uttarakhand, Jammu and Kashmir, Rajasthan, and Odisha	1300–2100	Roasted seeds are used as condiments	Caryophyllen-e, Limonene, Transβ-ocimene	+++	Used to treat nerve pain, skin burns, epilepsy, and asthma
4.	*Callicarpa macrophylla*	Daiya	Verbenaceae	Assam, Arunanchal Pradesh, Bihar, and Jammu and Kashmir	1200–1500	Fruits are edible	Flavonoids, diterpenoids, and triterpenoids	+	Wine and jelly are derived from berries. Headaches can be treated with bark paste

5.	*Dendrocalamus strictus*	Bans	Poaceae	North East, Uttarakhand, Andhra Pradesh, and Karnataka	1200–1500	Bud and rhizome is used as vegetable	Flavonoids, tannins, and phenolics	++	Used for mouth and gum sores, inflammations, achin bones, and bloodshot eyes
6.	*Elaeagnus umbellata*	Autumn olive	Elaeagnaceae	Temperate parts of Himachal Pradesh	1200–3000	Raw fruits are edible	Phenolics, flavonoids, and alkaloids	++	Fruits are also processed into juice, jam and other preserves. In fresh berries, lycopene content is very high, which prevents heart diseases and cancer
7.	*Indigofera cassioides*	Sakena	Fabaceae	Assam, East Himalaya, Odisha, and Maharashtra	1000–2000	Foliage used as fodder	Phenolic compounds, terpenoids, and glucose esters of nitropropanoic acid	++	Powdered leaves are used to treat malaria. To alleviate chest pains, the dried root is ground into a powder and administered externally
8.	*Pyracantha crenulata*	Ghingaru	Rosaceae	Uttarakhand, Assam, Nepal, and Myanmar	1000–2600	Fruits are edible	Vitamins, minerals, flavonoids, and tannins	+	Branches that serve as tool handles and walking sticks. In apiculture, flowers are used as bee food, and plants serve as soil binder

(continued)

Table 13.2 (continued)

S. no.	Botanical name	Common name	Family	Local distribution	Altitudinal range	Plant part used	Major chemical constituents	Availability status	Ethnobotanical and medicinal uses
9.	*Rhus parviflora*	Tungla	Anacardiaceae	Garhwal Himalayas, Tamil Nadu, and Madhya Pradesh	1400–1800	Fruits are edible	Biflavonoids, flavonoids, beta-sitosterol, and lignoceric acid	++	Flour can be used with ground fruits. Tobacco combined with leaves
10.	*Rubus ellipticus*	Hisalu, Aakhe	Rosaceae	Himachal Pradesh, Uttarakhand, Uttar Pradesh, Tamil Nadu, Kerala, and Odisha	800–2000	Fruits are edible	Alkaloids, tannins, phenolic compounds, quinone, glycosides, carbohydrates, and proteins	++	Potential source of antifertility drugs Roots and young shoots are used in colic pains. Nectar is derived from the blossom and it is a valuable honey plant
11.	*Rubus niveus*	Kala Hinsalu	Rosaceae	Himachal Pradesh and Uttarakhand	1200–2200	Fruits are edible	Poluphenolics, alkanols, anthicyanins and tannins.	++	The berries are frequently used to make jam and a variety of sweets. Serve as antioxidants and may aid in the treatment of tumors, wounds, and skin conditions

12.	*Zanthoxylum armatum*	Timur	Rutaceae	Jammu and Kashmir, Himachal Pradesh, Assam, and Manipur	900–2200	Tender shoots as vegetable, ripe fruits are edible	Essential oils, flavonoids, alkaloids, and terpenoids	++	Fruits, twigs and seeds are used for toothache. It is used for low blood pressure, fever, and inflammation

Source: Bharti and Lata (2018), Chauhan et al. (2020), Agarwal and Chandra (2019), Radha et al. (2013), and Bhandari et al. (2021)
+ uncommon, ++ common, +++ abundant

Table 13.3 List of wild edible herbs of northwestern Himalayas

S. no.	Botanical name	Common name	Family	Local distribution	Altitudinal range	Plant part used	Major chemical constituents	Availability status	Ethnobotanical and medicinal uses
1.	*Amaranthus tricolor*	Chua	Amaranthaceae	Assam, Uttarakhand, and Himachal Pradesh	1300–2200	Chapatis are made from seed flour and fried leaves	Flavonoids, pigments, and amino acids	++	Used in external inflammation and treatment for bladder distress
2.	*Aloe barbadensis*	Aloevera	Liliaceae	Himachal Pradesh, Uttarakhand, Rajasthan, Karnataka, and Maharashtra	800–2000	Whole leaf, gel, and latex	Sugars, minerals, lignin, amino acids, saponins, and enzymes	+++	After being thoroughly boiled, fleshy leaves are used to wounds and sores. For instant relief from cuts and burns, the plant can be snapped off and applied
3.	*Angelica glauca*	Choru	Epiaceae	Himachal Pradesh, Uttarakhand, and Jammu and Kashmir	2800–3000	Root and dried seeds are used as spice	Volatile oils, essential oils, proteins, and saponins	+	Used in spice and flavoring. Used to treat constipation, ulcer, dysentery, asthma, cough, and colds

4.	*Berginiaciliata*	Silpari	Saxifragaceae	Uttarakhand, Himachal Pradesh, Sikkim, and Meghalaya	2200–2400	Dried leaves are used in tea	Bergenin, gallic acid, tannic acid, and catechin	+++	Used to treat kidney stones, skin infections, and gastrointestinal issues
5.	*Chenopodium album*	Baithu	Chenopodiaceae	Gujarat, Kerala, Odisha, Punjab, and Uttarakhand	1200–2300	Leaves used as pot vegetable	Phenols, flavonoids, fatty acids, and amino acids	++	Used as blood purifier and diuretic sedative.
6.	*Cleome viscosa*	Jakhiya	Cleomaceae	Uttarakhand, Odisha, Tamil Nadu, and Kerala	1200–1500	Seeds are used as condiments	Terpenes, flavonol, gallic acid, and linoleic acid	+	Used to treat fevers and wounds, jaundice
7.	*Duchesnea indica*	Kaphlia	Rosaceae	Kashmir, Manipur, Assam, Himachal Pradesh, and Uttarakhand	1400–2200	Ripened fruits are edible	Hydroxybenzoic acid, flavonols, and ellagic acid	++	Leaves used to treat skin conditions like boils, burns, snake bites, and insect bites
8.	*Mentha arvensis*	Paudina	Lamiaceae	Uttarakhand, Punjab, and Himachal Pradesh	1200–2300	Young leaves and twigs are used as flavorings and to make sauce	Menthol, limonene, isomenthone, and menthone	++	Used as a flavoring in medicines and in hair care

(continued)

Table 13.3 (continued)

S. no.	Botanical name	Common name	Family	Local distribution	Altitudinal range	Plant part used	Major chemical constituents	Availability status	Ethnobotanical and medicinal uses
9.	*Physalis peruviana*	Rashbhary	Solanaceae	Himachal Pradesh, Jammu and Kashmir, Uttar Pradesh, Kerala, and Odisha	Up to 1650	Raw fruits are edible	Alkaloids, phenols, saponins, steroids, starch, tannin, and terpenoids	++	The crushed leaves are used to treat itches and headaches. The plant's extracts have demonstrated anticancer properties.
10.	*Solanum nigrum*	Giwali, black nightshade	Solanaceae	Himachal Pradesh	1200–2400	Leaves and ripened fruits are edible	Glycosides, alkaloids, tannins, saponins, and steroids	++	The plant's paste is applied to burns to promote rapid healing. Beneficial for digestive system malignancies.

Source: Radha et al. (2013), Bharti and Lata (2018), Agarwal and Chandra (2019), Chauhan et al. (2020), and Bhandari et al. (2021)
+ uncommon, ++ common, +++ abundant

Table 13.4 Nutritional profile of some important species of northwestern Himalayas (value per 100 g)

S. no.	Species	Proteins	Carbohydrates	Phosphorous	Potassium	Calcium	Moisture %
1.	*Aegle marmelos*	2.0 g	–	25.0 mg	–	67.0 mg	90%
2.	*Bombax ceiba*	1.4 g	11.9 g	49.0 mg	–	92.2 mg	85.6%
3.	*Cornus capitata*	2.0 g	–	–	275.0 mg	55.0 mg	–
4	*Ficus palmata*	1.72 g	27.09 g	0.034 mg	0.296 mg	0.071 mg	80.5%
5.	*Juglans regia*	15.2 g	13.7 g	346.0 mg	0.8 mg	98.0 mg	–
6.	*Morus alba*	1.44 g	9.80 g	–	194.0 mg	39.0 mg	
7.	*Myrica esculenta*	1.30 g	16.13 g	–	1.98 mg	1.0 mg	76.60%
8.	*Phyllanthus emblica*	1.0 g	10.0 g	–	–	–	86.0%
9.	*Physalis peruviana*	1.9 g	11.2 g	40.0 mg	–	9.0 mg	–
10.	*Prunus armeniaca*	23.8 g	11.2 g	23.0 mg	259.0 mg	-	–
11.	*Punica granatum*	–	–	0.10 mg	0.48 mg	0.031 mg	–
12.	*Pyrus peshia*	3.70 g	–	0.03 mg	0.48 mg	0.06 mg	–
13.	*Rubus ellipticus*	3.7 g	27.1 g	0.2 mg	1.8 mg	0.9 mg	64.4%
14.	*Solanum nigrum*	5.5 g	–	5.1 g	–	–	60.3%
15.	*Ziziphus mauritiana*	1.34 g	9.4 g	–	-	–	84.39%

Source: Bharti and Lata (2018)

3 Market Potential of Wild Edible Plants

In recent years, increasing attention is being paid to medicinal plants value both due to their economic and conservation concerns (Dhar et al. 2002; Tewari et al. 2010). More than 90% of the medicinal plant raw materials used in manufacturing facilities come from natural forests; frequently, one or more of the species parts—such as the root, bark, stem, leaves, flowers, fruits, and seeds—make up the legally recognized produce. The entire plant is only occasionally utilized as a therapeutic produce. Ayurvedic medicine and other ancient sciences advise gathering various plant parts at various times of the year. This was perhaps done keeping in view the optimum activity of herbs when harvested at a specific season (Uttarakhand Open University).

Due to their distant location, inadequate economic and physical infrastructure, high transportation costs, low volume, lack of information, difficulty obtaining finance and other institutional services, and weak bargaining power, mountain farmers confront numerous obstacles when trying to market their agricultural products (Pandey et al. 2011; Choudhary et al. 2012; Hurni 2013).

Agricultural product's perishable nature and the lack of processing and storage facilities exacerbate marketing problems (Sanginga et al. 2004; Pokhrel and Thapa 2007; Tiwari et al. 2008; Shahbaz et al. 2010). There is a need to make the transportation route easier and accessible to the farmers so that the wild fruits of the hills that have huge market value reach different parts of country and world and benefit the farmers, which in turn will motivate them to add these wild edible species in their farm and practice agroforestry instead of relying solely on the farm produce.

There are several wild species that have a huge market value not only within the country but also globally. As people are shifting more toward the organic and farm-grown produce, it gives great opportunity to the wild edible species to get inside the market and earn huge profits.

The most treasured wild fruit of the Uttarakhand mountains is *Myrica esculenta*. In the hills of Uttarakhand, especially in the Kumaon region between 900 and 2100 msl, wild populations of *Myrica esculenta* are rather numerous and significantly contribute to the local community's livelihood and nutritional security (Bisht et al. 2018). It is sold to tourists for about Rs 45 per 100 gram. In large quantities, it is sold to tourists at prices ranging 350 to 450 per kg (by Jagran News Sun, 11 Jun 2023). Rich in vitamin C, *Phyllanthus emblica* fruits, whether raw or pickled, are a key component of "Trifala," a cure-all for various illnesses. There are numerous ecosystems in the warmer valleys and hills of Uttarakhand where aonla wild stands support local families' livelihood security (Herring 2015). *Diplazium esculentum*, often known as lingura, is an edible fern and leafy vegetable used all over the world. It is a familiar dish found in Uttarakhand's hillside local markets. *D. esculentum* has potential applications as a cytotoxic, antibacterial, and natural antioxidant (Bisht et al. 2018). Among the wild edibles, the most widely consumed fruits were *Myrica esculenta, Rubus ellipticus,* and *Berberis asiatica*. Furthermore, all three species were used to cure a range of ailments. They change their flavor and provide additional nutrients to promote general health; nevertheless, they were primarily consumed as seasonal fruits. Although the distribution range of some wild foods is limited, their high demand has created a market for them. In the past, those who were unable to gather and possess these natural delicacies in their community were compensated with paddy, wheat, or pulse. Currently, the ancient bartering system has been overtaken by the money-oriented economic structure. The inhabitants of Uttaranchal's lower hills buy *Myrica esculenta* from *Myrica esculenta*-growing regions because it is not found in the foothills or warmer valleys. The fruits of *Ziziphus mauritiana* and *Myrica esculenta* have emerged as viable income crops for the local population in warmer climates and colder regions, respectively (Chandra Prakash and Kala 2007).

Products from Chiuri are used in a number of industries, such as soap manufacture, candle making, vegetable ghee production, medicines (as a therapeutic salve for dry and cracked skin), and confectionery. Additionally, it has demonstrated efficacy in treating rheumatism, and it is occasionally added as an adulterant to animal ghee. After Chiuri is processed, the cake that is left over is used as fertilizer on banana and paddy fields. European consumers may find Chiuri butter to be a highly sought-after product. For example, Nepal exported about 20 tons of Chiuri butter per year to the European market between 2011 and 2016. Chiuri butter-containing products are already on the market in Europe and are listed among the raw materials for cosmetics that meet COSMOS standards. Numerous businesses have launched cosmetic products that use Chiuri butter as a base component in lotions, ointments, and creams. International businesses Officina Nature and Cime, as well as national businesses Alternative Herbal Products (P) Ltd., Deuti Herbal Industry, and Himalayan Bio-Trade Pvt. Ltd., are working together to produce herbal products using this butter (Rythkonen 2016). Additionally, Polygon Chemicals Pvt. Ltd. (India) develops medications and cosmetics (Nautiyal 2009).

The economically significant but underappreciated *Cornus capitata* is among the many wild edible fruit trees in the Uttarakhand Himalayan regions of Garhwal and Kumaun. The evergreen fruit tree *Cornus capitata* is referred to locally as Bhamora. The nutritional, therapeutic, and commercial worth of this wild edible fruit makes it imperative that extra care be taken to preserve and enhance this significant natural food source. To address this, it is also necessary to encourage a broader and longer-term recognition of wild fruits as essential nutritional components, especially in light of the COVID-19 pandemic (Purohit et al. 2020).

When used appropriately, wild edible plants have the potential to raise household incomes, according to a 2018 study by Bisht et al. on the sale of various wild-harvested food resources in the Nainital district. The study's findings are listed in the table, together with the market availability of wild species and the average annual household income of people engaged in market sales in the Nainital district in 2018 (Table 13.5).

4 Economic Development Through Wild Edible Plants

Harvesting and trading WEPs can provide economic opportunities for rural and indigenous communities. By creating local markets for WEPs, communities can reduce their dependence on global supply chains and foster economic self-sufficiency.

4.1 Fruits

Fruits are mostly consumed raw. *Phyllanthus emblica, Punica granatum, Cordia dichotoma, Ficus palmata, Berberis lycium, Syzygium cumini, Prunus armeniaca, Prunus persica, Pyrus pashia, Rubus ellipticus*, and others were among the most often consumed fruits by the inhabitants. According to Arya 2013, the people of the Garhwal Himalaya in India also consume *Ficus palmata, Carissa spinarum, Elaeagnus parvifolia, Rubus ellipticus*, and *Pyrus pashia* as wild edible fruits. In 2017, Thakur et al. also documented the use of raw fruits from the Western Himalayan tribal region, including *Prunus armeniaca, Pyrus pashia*, and *Rubus ellipticus*. Sharbat, a classic summertime beverage, is produced from *Aegle marmelos* pulp and the *Ocimum basilicum seeds*. In several regions of India and other nearby nations, *Aegle marmelos* and *Ocimum basilicum* are both utilized as beverages (Al-Shahib and Marshall 2003).

Table 13.5 ICAR, National Bureau of Plant Genetic Resources, Uttarakhand, India

S. no.	Plant species	Market availability	Average annual cash income per household for persons engaged in market sales (INR)
1.	*Asparagus adscendens* (Safed Musli)	July–August	1950
2	*Bauhinia variegata* (Kachnar)	April–May	3500
3	*Berbaris asiatica* (Kilmora)	April–June	600
4	*Chenopodium album* (Bathua)	March–November	1750
5	*Cleome viscose* (Jakhiya)	October	500
6	*Dioscorea glabra* (Bhatkando)	February–March	260
7	*Diplazium esculentum* (Linguda)	March–October	9000
8	*Ficus glomerata* (Indian fig tree)	May–June	750
9	*Ficus palmata* (Bedu)	April–June	400
10	*Girardinia diversifolia* (Himalayan nettle)	September–January	1500
11	*Indigofera heterantha* (Himalayan indigo)	March–April	600
12	*Myrica esculenta* (Kafal)	April–June	2725
13	*Phyllanthus emblica* (Amla)	September–November	650
14	*Punica granatum* (Anar)	September–October	300
15	*Rubus ellipticus* (Hisalu)	April–June	1300

For months or even years, some fruit species, like *Cydonia oblonga* and *Phyllanthus emblica*, can be preserved as murabba, or local jam, a regional specialty. It is prepared by boiling apples whole or cut into slices, let them to dry in the shade, and then by adding sugar or sugar syrup and storing them in airtight containers. *Citrus medica, Phyllanthus emblica*, and *Artocarpus lacucha* fruits are pickled in mustard oil with salt and spices. In addition to being a part of the local tradition, all of these methods of preserving fruits during times of scarcity are also used in the Himalayas and other regions of India (Thakur et al. 2017).

4.2 Vegetables

Zeghichi et al. in 2003 claim that vegetables collected from the wild contain a wider variety and possibly more nutrients than those grown for commercial purposes. The leaves of *Diplazium esculentum* with a high moisture level have an excellent nutritional value (3413.2 Kcal Kg^{-1}) and a crude protein content (143.8 g kg/ha) (Seal

2012). The protein content is even higher than that of several commercial fruits and green vegetables, such as cauliflower, apples, cabbage, and litchi (Gopalan et al. 2004). *Taraxacum campylodesa* is a vegetable that is high in dietary fiber, proteins, calcium, and phosphorus. In several Middle Eastern nations, yogurt and veggies are also frequently combined. A sizable portion of vegetables are pickled or sun-dried (10 species). Sun-dried veggies are intended for use throughout the winter months when resources are few, particularly in hilly areas. This is another traditional practice in the Himalayas; certain plants, such as juvenile fronds of *Diplazium esculentum* (Thakur et al. 2017; Singh et al. 2016), *Moringa oleifera* pods and roots (Verma and Chauhan 2005), *Agave americana* blooms (Arya 2013), and the entire plant of *Ranunculus arvensis*, are preserved in mustard oil and salt as pickles.

4.3 Food Medicine

It is often recognized that wild foods are used as medicines throughout the world, and these value-added products are generally referred to as nutraceuticals (Hook et al. 1993). In every way, including numbers (36 species) and citations (436), they rank third, after fruits and vegetables. These species, which can be wild fruits or vegetables, are eaten when they are available or on special occasions. For instance, natives boil the leaves of *Taraxacum campylodes* as a vegetable, especially to aid pregnant or postpartum women who are weak.

One multipurpose medicinal species is *Urtica dioica* (CI, 0.49) (Rao et al. 2015). The plant is a diuretic, and locals harvest and boil its tender leaves as a delicacy. *Cissampelos glaberrima* (CI, 0.40) is consumed only at the time of diarrhea (Bhatia et al. 2014). Rose petals are used to make jam (gulukand), which is then sweetened with sugar and used as a remedy for mouth ulcers, jaundice, and digestive issues.

4.4 Beverages

After adding sweeteners, four species—*Viola odorata* (flowers), *Persicaria amplexicaulis* (roots), *Flacourtia indica* (inner bark), and *Acacia catechu* (heartwood)—are boiled in hot water and used as tea substitutes (hot beverages) in tribal areas of Jammu and Kashmir (Bhatia et al. 2018). Tea made from *Viola odorata* flowers is used to treat jaundice, fever, colds, and coughs. *Viola odorata* is used to treat bronchitis, cancer, cold and cough, fever, kidney problems, liver diseases, rheumatism, sneezing, and urinary tract infections (Nadkarni 2009; Gwari et al. 2012). It also includes alkaloids, mucilage, and vitamin C, as well as diuretic and antioxidant characteristics.

4.5 Spices

In Jammu and Kashmir, three species—*Artemisia scoparia* (seeds, flowers, and leaves), *Carum carvi* (fruits, seeds), and *Murraya koenigii* (leaves)—are used as condiments and spices in pulses and vegetables because of their aromatic qualities (Bhatia et al. 2018). The important spice *carum carvi* is found in the wild throughout Europe, Russia, Siberia, and the Himalayas (Gupta et al. 2011). It has antibacterial, antiproliferative, antifungal, anticancer, and antihyperglycemic qualities.

4.6 Chutney

To prepare chutney, *Mentha longifolia* leaves are ground separately in a mortar and pestle, followed by the addition of salt and spices. Known generally as the poor man's food adjuster, it is a ready-to-eat food that can be consumed immediately (Ramchandran and Ali 1996).

Chutney is a staple of many meals, particularly in the summer, because it is a great appetizer, antispasmodic, and antigastric, and helps with digestion (Rao et al. 2015). Chutney made from the seeds of *Flemingia prostrata, Zanthoxylum armatum,* and *Punica granatum* is also consumed. Along with being a tasty appetizer, chutney helps with digestion (Bhatia et al. 2014).

5 Role of Wild Edibles in Nurturing Agroforestry System

An eco-friendly approach that can sustain and diversify production for greater social, economic, and environmental advantages is agroforestry. For smallholder farmers and other rural residents, agroforestry is especially important because it may boost food security, diversify production and revenue, and make farms more resilient to climate change. For farmers, incorporating trees that produce fruit, fodder, or nitrogen fixation can also have a number of direct and indirect advantages (FAO).

The native community in the Indian Himalayan region has domesticated wild edible species. Due to their numerous uses, *Ficus auriculata* and *Pyrus pashia* are important agroforestry species in Uttarakhand. *Pyrus pashia* and *Ficus auriculata* are also well-known for their applications as fuel wood and regional herbal remedies (maikhuri). Edible portions of many wild backyard plants and agroforestry system plants are frequently consumed and serve as essential sources during lean times. It is interesting to note that fruits and vegetables make up 70–80% of the wild plant resources in the hills of Uttarakhand. It is well recognized that eating fruits and vegetables lowers the risk of several oxidative stresses, including heart disease, stroke, and cancer. Phytochemicals like polyphenols, carotenoids, and vitamins are primarily responsible for these health advantages. Polyphenols are the most well-known of these phytochemicals for their anti-inflammatory, antiviral, antibacterial, and antioxidant properties (Pandey and Rizvi 2009). Agroforestry and wild plant resources derived from forests are used in Uttarakhand's traditional farming systems to enhance household food options, guarantee dietary diversity, and improve nutrition (Bisht et al. 2017). In the ancient farming systems of the Uttarakhand hills, the wild plant resources obtained from forestry and agroforestry offer nutritional diversity and better nutrition while also enhancing household food choices. For rural residents, wild-collected tubers, fruits, seeds, twigs, leaves, and flowers of several plants continue to be important yet minor food sources. Certain edible plants are very valuable economically and are closely related to the socioeconomic advancement of the state's tribal tribes. A few of these species might be added to agroforestry systems, which could provide genetic resources for tree breeding initiatives in other parts of the nation. They would also help communities by producing a photosynthetic pool to combat environmental degradation and by providing edible planter sources (Jasmine Therese).

To help safeguard such a significant wild resource, research elsewhere in the Indian Himalaya has suggested growing the wild edibles in traditional agroforestry systems and on marginal lands with otherwise low agricultural values (Sundariyal et al. 2004, Sundriyal and Sundriyal 2005). One way to restore marginal and degraded lands and turn such unproductive area into community nutrigardens or food gardens is to establish appropriate agroforestry models, particularly based on wild edible plants (Kumar and Purushotam 2023). Using sustainable land management techniques can lessen the harm that industrialization and agriculture inflict to WEPs. This covers methods, including organic farming, integrated pest management, and agroforestry (Shirsat and Koche 2004).

6 Threats to Wild Edible Plants (WEPs)

Both local food security and biodiversity depend heavily on wild edible plants. According to Shirsat et al. (2023), the diversity of wild edible plants is fast decreasing despite their significance. An autonomous organization called "The State of World Biodiversity for Food and Agriculture" (SOWBFA) was created by the Food and Agricultural Organization (FAO) in 2019 to address this problem. Its mission is to evaluate and protect the diversity of food and agriculture. The Royal Botanic Gardens and the FAO (2019) have reported that shifts in land management and use, such as the conversion of forest area to agricultural land, the development of infrastructure and industry, habitat destruction from forest fires, grazing, and overharvesting of forest products, are some of the main causes of the decline in plant diversity, including wild edible plants. These variables are responsible for almost 62% of the risks to wild edible plant, which impact those that grow outside of protected forest ranges (Mishra et al. 2021). In agroforestry systems, a large number of WEPs can be found growing as wild trees, hedges, weeds, and tiny forest patches.

(Powell et al. 2015). These biological resources may be adversely affected by modifications to agricultural methods, such as intensification, increasing use of pesticides, and the loss of native vegetation. SOWBFA used the Sampled Red List Index for Plants prepared by the International Union for Conservation of Nature (IUCN) (Kew 2016). Under this project, a total of 822 wild edible species evaluated in seven classifications, 73% are currently considered to be under low risk of extinction. But in the wild, certain classes face serious threats. Only 31% of all known wild edible plants have global conservation evaluations included in the IUCN Red List Index for Plants. For 1587 wild edible plant species, the International Center for Tropical Agriculture (CIAT) also assessed the comprehensiveness of conservation as part of a larger study to identify conservation gaps for valuable plants. According to the survey, 89.1% of WEPs urgently need off-site conservation measures, while only 3.3% are well conserved ex situ (Khoury et al. 2019). Of the WEP taxa that were examined, 11.1% need immediate conservation efforts, 46.7% have medium priority, and 42.1% are sufficiently conserved. The geographic distributions of many species can be changed by rapidly rising temperatures and habitat degradation, which can push them across artificially designated boundaries of many protected areas in search of ideal growing conditions (Shirsat et al. 2023). The conservation of wild food plants is further complicated by this dynamic.

7 Leading Factors Affecting Availability and Consumption of the Wild Edibles

7.1 Climate Change

However, exotic species usually lack the natural capacity to endure rapid climatic change, which may become more important during periods of low agricultural output connected to climate disasters. Wild food species offer a potentially crucial role in buffering against food stress caused by changes in the climate (Bharucha and Pretty 2010).

7.2 Local Ecological Knowledge (LEK)

LEK was shown to be higher and knowledge acquisition to be faster from an early age in low-income groups across all agro-ecologies of the Uttarakhand hills. This implies that knowledge becomes concentrated in fewer individuals with a sustained personal interest as communities grow wealthier. A significant decline in LEK was recorded where communities move away from land-based livelihoods and rely more on market-bought foods (Bharucha and Pretty 2010).

7.3 Unsustainable Harvesting Practices and Shifting Priorities

Unsustainable harvests have led to a reduction in wild food species in certain regions of the Uttarakhand highlands. The two important wild commodities that are taken from the alpine meadows/bugyals of the Uttarakhand hills are "Jambu or Faran" (*Allium* spp.) and "Gandhrain" (*Angelica glauca*). Unsustainable harvesting is a problem with these high-elevation wild food resources.

7.4 Degradation and Changes in Land Use, Especially Inadequate Resource Management

Wild food species have declined as a result of deforestation and inadequate resource management, and local communities' attempts to reverse this trend by domesticating significant species have mostly failed because many species do not live outside of their natural habitat. Changes in land use and agricultural expansion/urbanization have a significant impact on the availability of wild foods (Schmidt-Vogt 2001; Padoch et al. 2007).

7.5 Socioeconomic Change, Especially the Growth of Markets as a Result of Globalization

Public health issues in hill communities arise from the nutrition transition brought on by modern diets. When wild foods are replaced with products from the market, there is a loss in dietary diversity and a rise in the incidence of chronic lifestyle-related illnesses. In areas where children use wild plant resources, malnutrition rates were low (Bharucha and Pretty 2010).

8 Conservation of Wild Edible Plants

Since local communities and tribes are the primary users of wild edible plants, strong cooperation with them is essential to the conservation and sustainable management methods of wild edible plants (Shirsat et al. 2023). Conventional and scientific knowledge about wild edibles must be incorporated into participatory research approaches in order to maximize the benefits for these communities. Many publications emphasize the loss of wild edible plants from local markets, their natural habitats, and even from the typical meals of many local communities. As such, their preservation and sustainable management for future food security is a challenging undertaking that requires cooperation of all members of the human community. In order to solve issues with information, infrastructure, environmental circumstances, and sustainable practices, Borrelli et al. (2020) put out a good plan for the sustainable management of wild edible plants. Reportedly, there are over 20,000 species of wild edibles present globally (Shirsat et al. 2023). Because of carelessness and a lack of knowledge about these species globally, less than 10% of these WEPs are now being used, and the remaining ones are declining. The preservation of these species is essential for both ecological balance and future food security. To maintain WEP diversity, conservation strategies that are both in situ (on-site) and ex situ (off-site) must be implemented right away. In order to optimize the use of WEPs, we must create a particular knowledge system and offer the required infrastructure and assistance to bolster WEP conservation and management initiatives worldwide. In addition to its goals, particular strategies must be created to manage and preserve the WEP resources that are accessible in a particular location (Priyadarshini et al. 2024).

9 Conclusion

Wild edible plants are a precious gift from nature, and most people depend on them to survive on a daily basis. In times of food scarcity and rising prices, these wild edible plants and their products can be useful. These wild edibles may develop into a significant source of revenue. The ethnobotanic species of Uttarakhand are diverse, and the state has a long history of using wild food plant systems. The people of the Kumaun and Garhwal regions are responsible for maintaining these significant plants. Consuming wild edible fruits helps the region's rural population meet their protein, carbohydrate, fat, vitamin, and mineral needs. The health of the local rural communities depends on wild edible plants since they can be used to make money in addition to supplying extra nourishment, a nutritionally balanced diet, medications, fuel, and fodder. Many wild fruits, including *Prunus persica, Myrica esculenta, Fragaria × ananassa,* and *Rubus ellipticus*, are high in nutrients and are used by the locals to make a variety of food goods. Numerous other wild fruits have also been claimed to have good therapeutic qualities, including *Ficus auriculata, Berberis asiatica, Punica granatum,* and *Solanum nigram.* It should be a top priority to preserve traditional knowledge about wild edible plant resources. The public must be encouraged to support the cultivation of these wild edible plants. It is possible to guarantee food security and year-round availability by using wild fruits and their products. In terms of nutrition, wild fruits can be the ideal supplement. Even if they currently contribute significantly to the improvement of the rural economy, these natural resources should be properly commercialized with the help of appropriate marketing techniques. Over the past few decades, forests and many of these wild fruits have been disappearing. Therefore, it is crucial for their conservation and propagation plans to properly document their availability and use. The greatest way to use these neglected and underutilized Himalayan resources is to ensure food, nutrition, and livelihood security. We must develop these wild fruits into functional foods and nutraceuticals in addition to turning them into jams, pickles, and juices. For the sake of ecological balance and future food security, wild edible plants must be conserved and managed sustainably. Through the implementation of comprehensive conservation measures, integration of traditional and scientific knowledge, and collaboration with indigenous and local communities, we can protect the variety of wild edible plants. Furthermore, it is necessary to preserve the rapidly vanishing wild plants that are still widely distributed throughout many ecosystems.

References

Agarwal R, Chandra V (2019) Diversity of wild edible plants in the Mandal Chopta forest, Uttarakhand. J Plant Med Stud 7(1):89–92

Al-Shahib W, Marshall RJ (2003) The fruit of the date palm: its possible use as the best food for the future. Int J Food Sci Nutr 54:247–559

Arora RK, Anjula P (1996) Wild Edible Plants of India: diversity conservation and use. ICAR, New Delhi

Arya D (2013) Major wild edible fruits used by the locals of Garhwal Himalaya. Int J Adv Life Sci 6(3):145–149

Bandyopadhyay S, Mukherjee SK (2009) Wild edible plants of Koch Bihar district, West Bengal, 2009

Bhandari R, Sigh V, Uniyal AK (2021) Study on wild edible plant diversity in the parts of district Tehri Garhwal, Uttarakhand, India. Res J Agric For Sci 9(3):40–46

Bharti H, Lata K (2018) Wild Edible fruits of Himachal Pradesh, pp 1–61

Bharucha Z, Pretty J (2010) The roles and values of wild foods in agricultural systems. Philos Trans R Soc B Biol Sci 365:2913–2926

Bhatia H, Sharma YP, Manhas RK, Kumar K (2014) Ethnomedicinal plants used by the villagers of district Udhampur, J&K, India. J Ethnopharmacol 151(2):1005–1018

Bhatia H, Sharma YP, Manhas RK, Kumar K (2018) Traditionally used wild edible plants of District Udhampur, J&K, India. J Ethnobiol Ethnomed 14:73

Bhatt ID, Rawat RS, Dhar U (2000) The availability, fruit yield and harvest of *Myrica esculenta* Buch- Ham ex D. Don in Kumaun (West Himalaya), India. Mountain Res Dev 20(2):146–153

Bisht IS, Mehta PS, Negi KS, Rawat R, Singh R, Garkoti SC (2018) Wild plant food resources in agricultural systems of Uttarakhand Hills in India and its potential role in combating malnutrition and enhancing human health. J Food Sci Toxicol 2:3

Bisht IS, Mehta PS, Negi KS, Verma SK, Tyagi RK, Garkoti SC (2017) Farmers' rights local food systems and sustainable household dietary diversification: A case of Uttarakhand Himalaya in north-western India. Agroecol Sustain Food Syst 42(1):77–113. https://doi.org/10.1080/21683565.2017.1363118

Borrelli P, Robinson DA, Panagos P, Lugato E, Yang JE, Alewell C, Wuepper D, Montanarella L, Ballabio C (2020) Land use and climate change impacts on global soil erosion by water (2015–2070). Proc Natl Acad Sci USA 117(36):21994–22001

Chauhan PS, Parmar MPS, Tiwari I (2020) Ethno botanical and medicinal uses of some wild edible fruiting plants in hills region of Garhwal Himalaya. New York Sci J 13(3):64–68

Choudhary D, Kala SP, Todaria NP, Dasgupta S, Kollmair M (2012) Marketing of bay leaf in Nepal and Northern India: Lessons for improving terms of participation of small farmers in markets. Small Scale For 12(2):289–305

Dhar U, Manjkhola S, Joshi M, Bhatt A, Bisht AK, Joshi M (2002) Current status and future strategy for development of medicinal plants sector in Uttaranchal, India. Curr Sci 83(8):956–964

FAO (2019) The state of the world's biodiversity for food and agriculture. FAO, Rome

Food and Agricultural Organization of the United Nations, FAO (2004) The state of food insecurity in the world. Monitoring the progress towards the world food summit 2nd millennium development goals

Gopalan C, Ramasastry BV, Balasubramanian SC (2004) Nutritive value of Indian foods. Indian Council of Medical Research Hyderabad, Hyderabad

Gupta A, Dubey M, Parmar M, Mahajan S, Sharma R (2011) Evaluation of antimicrobial activity of Carum carvi (seeds) extract against E coli and *Aspergillus niger*. Drug Invent Today 3:211–213

Gwari G, Bhandari U, Andola HC, Lohani H, Chauhan N (2012) Aroma profile of s seeds of Carum carvi Linn cultivated in higher hills of Uttarakhand Himalaya. Indian J Natl Prod Resour 3:411–413

Herring RJ (2015) The Oxford handbook of food, politics and society. Oxford University Press

Hook I, McGee A, Henman M (1993) Evaluation of dandelion for diuretic activity and variation in potassium content. Int J Pharmacogn 31:29–34

Hurni H (2013) Global change and mountain livelihoods. In: Mountain Farming Is Family Farming: A Contribution from Mountain Areas to the International Year of Family Farming 2014. Food and Agriculture Organisation of the United Nations (FAO), Centre for Development and Environment (CDE), University of Bern, and University of Natural Resources and Life Sciences (BOKU), Vienna, Rome, pp 14–15

Jaarsveld AS, Biggs R, Scholes RJ, Bohensky E, Reyers B (2005) Measuring conditions and trends in ecosystem services at multiple scales: The Southern African Millenium Ecosystem Assessment (SAfMA). Philos Trans R Soc B 360:425–441

Kala CP (2005) Ethnomedicinal botany of the Apatani in the Eastern Himalayan region of India. J Ethnobiol Ethnomed 11(1):1746–4269

Kala CP (2007) Prioritization of cultivated and wild edibles by local people in the Uttaranchal hills of Indian Himalayas. Indian J Tradit Knowl 6(239):243

Kew (2016) Plants under pressure—a global assessment. Royal Botanic Gardens/Natural History Museum, London

Khoury CK, Amariles D, Soto JS, Diaz MV, Sotelo S, Sosa CC, Ramirez VJ, Achicanoy HA, Guarino L (2019) The comprehensiveness of conservation of useful wild plants: an operational indicator for biodiversity and sustainable development targets. Ecol Indic 98:420–429

Khyade MS, Kolhe SR, Deshmukh BS (2009) Wild edible plants used by the tribes of Akole Tehsil of Ahmednagar District (Ms) India. Ethnobot Leaflets 13(1328):1336

Kumar AB, Purushotam AC (2023) Securing wild edible plants for planetary healthy diet. 2:109–111

Maikhuri RK, Semwal RL, Singh A, Nautiyal MC (1994) Wild fruit as a contribution to sustainable rural development: a case study from the Garhwal Himalaya. Int J Sustain Dev:156–168

Medicinal & Aromatic Plants (MAPs) Department of Forestry and Environmental Science (SoEES) Uttarakhand Open University Haldwani, Nainital (U.K.)

Mishra A, Swamy SL, Thakur TK, Bhatt R, Bijalwan A, Kumar A (2021) Use of wild edible plants: can they meet the dietary and nutritional needs of indigenous communities in Central India. Foods 10:1453

Misra S, Maikhuri RK, Kala CP, Rao KS, Saxena KG (2008) Wild leafy vegetables: a study of their subsistence dietetic support to the inhabitants of Nanda Devi Biosphere Reserve, India. J Ethnobiol Ethnomed 4:15

Mohan R (2000) Plant resources of Indian Himalaya 9th G P Pant Memorial Lecture. G.B Pant Institute of Himalayan Development, Gangtok

Nadkarni KM (2009) The Indian Materia Medica. Popular Prakashan Publishers, Bombay

Nautiyal S (2009) Indian butter tree paves way for entrepreneurship despite economic slump

Negi KS, Gaur RD (1994) Principal wild food plants of western Himalaya, Uttar Pradesh, India. In: Higher plants of Indian sub continent, vol 3. Bishan Singh Mahendra Pal Singh, Dehradun, pp 1–78

Nordeide MB, Hatloy A, Folling M, Lied E, Oshoug A (1996) Nutrient composition and nutritional importance of green leaves and wild foods in an agricultural district, Koutiala. Int J Food Sci Nutr 47(6):455–468
Orech FO, Aagaard HJ, Friis H (2007) Ethnoecology of traditional leafy vegetables of the Luo people of Bondo district, western Kenya. Int J Food Sci Nutr 58(7):522–530
Padoch C, Coffey K, Mertz O, Leisz SJ, Fox J (2007) The demise of swidden in Southeast Asia? Danish J Geogr 107:29–41
Pandey KV, Rizvi SI (2009) Plant polyphenols as dietary antioxidants in human health and disease. Oxidative Med Cell Longev 2(5):270–278
Pandey D, Kumar A, Singh R (2011) Marketing of sweet orange (Malta) in Kumaon region of Uttarakhand. J Recent Adv Appl Sci 26:6–11
Pieroni A, Houlihan L, Ansari N, Hussain B, Aslam S (2007) Medicinal perceptions of vegetables traditionally consumed by south- Asian migrants living in Bradford, Northern England. J Ethnopharmacol 113:100–110
Pokhrel DM, Thapa GB (2007) Are marketing intermediaries exploiting mountain farmers in Nepal? A study based on market price, marketing margin and income distribution analyses. Agric Syst 94(2):151–164
Powell B, Thilsted SH, Ickowitz A, Termote C, Sunderland T, Herforth A (2015) Improving diets with wild and cultivated biodiversity from across the landscape. Food Secur 7:535–554
Prescott-Allen OC, Prescott-Allen R (1990) How many plants feed the world? Conserv Biol 4:365–374
Priyadarshini S, Tudu S, Dash SS (2024) Wild edible plants: diversity, use pattern a and livelihood linkage in Eastern India. Genetic Resour Crop Evol 71:3111–3133
Purohit VK, Chauhan J, Bhatt V, Negi MS (2020) Bhamora: an economically important but neglected wild edible fruit from higher Himalaya. ENVIS Newslett Himalayan Ecol 17(2):2020
Radha B, Singh RD, Tiwari JK, Tiwari P, Gairola A (2013) Wild Edible Plant Resources of the Lohba Range of Kedarnath Forest Division (KFD) Garhwal Himalaya, India. Int Res J Biol Sci 2(11):65–73
Ramchandran R, Ali M (1996) Chemical composition of fruit oil of *Zanthoxylum armatum*. Hamdard Med 42:51–54
Rao PK, Hasan SS, Bhellum BL, Manhas RK (2015) Ethnomedicinal plants of Kathua district, J&K, India. J Ethnopharmacol 171:12–27
Rythkonen A (2016) Forest-based value chains in Nepal. Published by the Multi Stakeholder Forestry Programme (MSFP) A Programme of the Government of Nepal (GoN) supported by theGovernments of Finland, Switzerland and the United Kingdom
Saklani S, Chandra S (2011) Evaluation of Nutritional profile, medicinal value and quantitative estimation in different parts of *Pyrus pashia, Ficus palmata and Pyracantha crenulata*. JGTPS 2(3):350–354
Sanginga PC, Best P, Chitsike C, Delve R, Kaaria S, Kirkby R (2004) Linking smallholder farmers to markets in East Africa: empowering mountain communities to identify market opportunities and develop rural agroenterprises. Mt Res Dev 24(4):288–291
Schmidt-Vogt D (2001) Secondary forests in swidden agriculture in the highlands of Thailand. J Trop Sci 13:748–767
Seal T (2012) Antioxidant activity of some wild edible plants of Meghalaya state of India: a comparison using two solvent extraction systems. Int J Nutr Metab 4(3):51–56
Shahbaz B, Ali T, Khan IA, Ahmad M (2010) An analysis of the problems faced by farmers in the mountains of northwest Pakistan: challenges for agri. extension. Pakistan J Agric Sci 47(4):417–420
Shirsat R, Koche D (2004) Edible plants (WEPs): a review on their importance, possible threats, and conservation. J Agric Ecol Res Int 978:93–48
Shirsat R, Jagtap T, Shirsat A, Rathod S, Koche D (2023) Current Scenario of Wild Edible Plants (WEPs), their Importance, possible threats, and conservation: A mini review. J Agric Ecol Res Int 24(5):18–27

Singh B, Sultan P, Hassan QP, Gairola S, Bedi YS (2016) Ethnobotany, traditional knowledge and diversity of wild edible plants and fungi: a case study in the Bandipora District of Kashmir Himalaya, India. J Herbs Spices Med Plants 22(3):247–278

Sundriyal M, Sundriyal RC (2001) Wild edible plants of the Sikkim Himalaya, nutritive values of selected species. Econ Bot 55:377–390

Sundriyal M, Sundriyal RC (2005) Seedling growth and survival of selected wild edible fruit species of the Sikkim Himalaya, India. Acta Oecologica 28:11–21

Sundriyal M, Sundriyal RC, Sharma E (2004) Dietary use of wild plant resources in the Sikkim Himalaya, India. Econ Bot 58:626–638

Tewari LM, Pangtey YPS, Tewari G (2010) Biodiversity potentials of the Himalaya. Jyanodaya Prakashan, Nainital

Thakur D, Sharma A, Uniyal SK (2017) Why they eat, what they eat: patterns of wild edible plants consumption in a tribal area of Western Himalaya. J Ethnobiol Ethnomed 13:70

Tiwari KR, Nyborg IL, Sitaula BK, Paudel GS (2008) Analysis of the sustainability of upland farming systems in the Middle Mountains region of Nepal. Int J Agric Sustain 6(4):289–306

Tiwari JK, Ballabha R, Tiwari P (2010a) Some promising wild edible plants of Srinagar and its adjacent area in Alaknanda Valley of Garhwal Himalaya, India. J Am Sci 6(4):167–174

Tiwari JK, Ballabha R, Tiwari P (2010b) Ethnopaediatrics in Garhwal Himalaya, Uttarakhand, India (Psychomedicine and Medicine). NY Sci J 3(4):123–126

Verma S, Chauhan NS (2005) Indigenous medicinal plants knowledge of Kunihar forest division. Indian J Tradit Knowl 6:494–497

Zeghichi S, Kallithraka S, Simopoulos AP, Kypriotakis Z (2003) Nutritional composition of selected wild plants in the diet of Crete. World Revolut Nutr Diet 91:22–40

Chapter 14
Integration of Medicinal Plants in Agroforestry Systems to Enhance the Farmer Income

Manish Jangid, Pooja Uniyal, Chandramohan Singh, Bhupendra Singh, Vinod Prasad Khanduri, Bijendra Lal, and Munesh Kumar

Abstract The northwestern Himalayas, encompassing Himachal Pradesh, Uttarakhand, and Jammu & Kashmir, represent a biodiversity-rich region with a long-standing tradition of agroforestry. Agroforestry systems (AFS), which integrate trees, shrubs, and crops, are vital for maintaining ecological balance while sustaining rural livelihoods. The incorporation of medicinal plants into agroforestry practices offers a dual advantage: enhancing farmer income through sustainable land use and contributing to biodiversity conservation. Owing to their wide-ranging applications in traditional and modern medicine, cosmetics, and the food industry, medicinal plants possess substantial economic value. Their cultivation within agroforestry frameworks not only provides an alternative source of income but also improves soil fertility, reduces erosion, and regulates microclimatic conditions. Furthermore, medicinal plants contribute to pest management, carbon sequestration, and the overall resilience of farming systems by strengthening on-farm biodiversity. In India and other developing countries, where access to modern healthcare remains limited in rural areas, traditional medicine continues to serve as a primary source of healthcare. Simultaneously, the rising global demand for herbal products creates lucrative opportunities for cultivating high-value medicinal species under AFS. By integrating medicinal plants, farmers can diversify agricultural production, optimize resource utilization, and enhance land productivity, while alleviating pressure on natural forests and conserving endangered species. This chapter examines the potential of medicinal plants in AFS of the northwestern Himalayas, emphasiz-

M. Jangid · P. Uniyal · C. Singh · B. Singh (✉) · V. P. Khanduri
Department of Silviculture and Agroforestry, College of Forestry, Veer Chandra Singh Garhwali Uttarakhand University of Horticulture and Forestry, Tehri Garhwal, Uttarakhand, India

B. Lal
Department of Forest Products and Utilization, College of Forestry, V.S.C.G. Uttarakhand University of Horticulture and Forestry, Ranichauri, Tehri Garhwal, Uttarakhand, India

M. Kumar
Department of Forestry and Natural Resources, Hemvati Nandan Bahuguna Garhwal University (A Central University), Srinagar Garhwal, Uttarakhand, India

B. Singh et al. (eds.), *Himalayan Agroforestry Systems*, Advances in Agroforestry 15,
https://doi.org/10.1007/978-3-032-25096-4_14

ing their contributions to rural economies, ecological sustainability, and traditional healthcare practices.

Keywords Biodiversity · Endangered species · Ethnomedicine · Ecological sustainability · Climate change mitigation

1 Introduction

The northwestern Himalayas, encompassing Himachal Pradesh, Uttarakhand, and Jammu and Kashmir, are known for their rich biodiversity and long-standing agroforestry traditions (Samant et al. 1998). This region harbors a rich repository of medicinal and aromatic plants (MAPs), which have sustained indigenous healthcare systems, including Ayurveda, Unani, and Amchi medicine, while simultaneously supplying raw material to modern pharmaceutical industries (Kala 2003). An estimated 70–80% of the rural population in the Himalayan belt continues to rely on herbal remedies for primary healthcare (WHO 2013). Ethnobotanical surveys and regional reviews document hundreds of species used locally for ailments ranging from fever, respiratory and gastrointestinal disorders to skin diseases and musculoskeletal problems (Bhat et al. 2021). The generations of empirical knowledge have refined plant identification, selection of therapeutic parts, and preparation methods—such as decoctions, pastes, and infusions—forming the foundation of both household-level care and organized community healthcare systems. Increasingly, modern phytochemical and pharmacological investigations validate these traditional claims, uncovering bioactive compounds with potential for drug development and nutraceutical applications (Joshi et al. 2016). This intersection of conventional ethnomedicine and scientific research presents a critical opportunity to integrate MAPs into contemporary healthcare strategies, bridging cultural heritage and modern innovation. The biodiversity and socioeconomic value of these plants are considerable: regional inventories report several hundred to over 600 medicinal species in particular states, many of them endemic or of high market value, which provide income through collection, cultivation, and value-added products for rural households and small enterprises. However, these resources face escalating threats from habitat degradation, unsustainable harvesting, and urban expansion (Ved and Goraya 2008), while smallholder farmers confront challenges of fragmented landholdings, low productivity, and limited income diversification (Maikhuri et al. 2001). Addressing these challenges is essential to conserve Himalayan biodiversity, sustain traditional healthcare practices, and harness the full socioeconomic potential of medicinal plants.

Agroforestry, the intentional integration of trees, shrubs, and crops, plays a pivotal role in maintaining ecological balance while supporting local livelihoods. Beyond its ecological functions, agroforestry addresses diverse human needs by providing economic benefits and essential environmental services, including carbon sequestration, watershed protection, and climate change mitigation and adaptation. Tree-based systems influence local economies by stabilizing income, diversifying products, ensuring food and fuel security, and improving the natural environment (Dar et al. 2018). Within these systems, MAPs are particularly valuable, contributing to traditional healthcare, biodiversity conservation, and rural economic development (Sharma et al. 2019). As a scientific and management practice, agroforestry offers significant potential to enhance rural livelihoods. Its diverse forms create multiple opportunities for farmers to increase productivity and income while safeguarding the agricultural environment (Butola et al. 2012). By combining trees and crops, farmers can diversify agricultural practices, improve land productivity, and strengthen resilience. Integrating medicinal plants into agroforestry—through alley cropping, home gardens, silvopastoral strips, or understory cultivation beneath tree canopies—provides a practical approach to reconcile conservation objectives with rural development goals. On-farm cultivation of high-value MAPs reduces pressure on wild populations, diversifies household portfolios against climate and market shocks, and enhances biodiversity, soil conservation, and microclimate buffering in fragile mountain ecosystems. Empirical and review studies at regional and global scales show that including MAPs in agroforestry or smallholder systems often stabilizes and increases farm income compared to exclusive reliance on annual crops or unmanaged wild harvest (Schippmann et al. 2002).

The economic viability of MAP agroforestry in the northwestern Himalayas is promising but context-dependent. Key factors such as species selection (e.g., perennial aromatics versus threatened root or rhizome species), availability of quality planting material, postharvest handling, value-chain development, and market access determine whether farmers achieve tangible income gains. Recent field studies and programmatic initiatives in the region highlight both the potential and practical challenges: successful schemes integrate technical training, nursery development, certification support, and market aggregation to translate conservation-friendly cultivation into sustainable livelihoods (Singh et al. 2024a). At the policy level, national and international frameworks increasingly recognize agroforestry and MAP cultivation as tools for rural development, biodiversity conservation, and climate-smart agriculture. The National Agroforestry policy and FAO-supported programs emphasize certified planting material, nursery protocols, and farmer capacity building—measures that are directly relevant to scaling medicinal-plant agroforestry in the Himalayan hills. Embedding MAP cultivation within agroforestry therefore aligns with broader objectives of income enhancement, landscape restoration, and carbon-sensitive rural growth. This chapter synthesizes current knowledge on medicinal-plant diversity and traditional uses in the northwestern Himalayas, reviews agroforestry models suitable for mountain smallholders, evaluates socioeconomic and ecological outcomes from case studies and pilot projects, and proposes pragmatic, evidence-based recommendations for researchers, extension practitioners, and policymakers to scale farmer-centric MAP-agroforestry systems that enhance livelihoods while conserving Himalayan plant heritage.

2 Presence of Medicinal Plants in Agroforestry Systems of Northwestern Himalayas

The northwestern Himalayas are recognized as one of the world's most vital reservoirs of medicinal plant diversity (Samant et al. 1998). Its complex topography, expansive altitudinal gradient (~300 m to >7000 m), pronounced climatic variability, and heterogeneous soils generate a mosaic of habitats that sustain exceptionally rich and often highly endemic floras. The region alone harbors over 1700 plant species used in traditional medicine (Gupta et al. 2025), with state-level assessments reporting hundreds of medicinal taxa—approximately 643 species in Himachal Pradesh and over 1100 species across Jammu, Kashmir, and Ladakh. This unparalleled biodiversity not only underpins local healthcare traditions but also represents a globally significant genetic resource with potential applications in pharmaceuticals, nutraceuticals, and sustainable agroforestry. AFS offer an optimal framework for the sustainable cultivation and conservation of these medicinal plants, simultaneously promoting ecological stability, enhancing rural livelihoods, and strengthening traditional healthcare practices (Negi et al. 2011).

2.1 Agroforestry Systems in the Northwestern Himalayas

The northwestern Himalayas encompass a wide range of agro-climatic zones, spanning from temperate to subtropical regions, and hosts a rich mosaic of traditional AFS that have evolved over centuries. These systems integrate trees, crops, and livestock in a balanced, synergistic manner, enhancing agricultural productivity while simultaneously contributing to biodiversity conservation, soil fertility, and carbon sequestration. These include agrisilvicultural, agrihorticultural, silvopastoral, horti-silvipastoral, and other integrated systems, which collectively optimize resource use, enhance ecosystem services, and support rural livelihoods (Sharma et al. 2024). For instance, in the Garhwal Himalaya, traditional agroforestry has been shown to improve carbon sequestration and soil quality, albeit sometimes with trade-offs in crop yields (Maurya et al. 2025; Kumar et al. 2025). The inclusion of multipurpose tree species enhances soil structure, reduces erosion, and provides diverse products, including timber, fodder, and medicinal plants, thereby supporting rural livelihoods. The adaptability of these systems across different altitudes and climatic conditions underscores their resilience and long-term sustainability (Mahato et al. 2016). A particularly notable feature of northwestern Himalayan agroforestry is the incorporation of medicinal and aromatic plants, which hold both economic and cultural significance. Species such as *Saussurea costus*, *Inula racemosa*, *Sinopodophyllum hexandrum,* and *Bunium persicum* are highly valued for their therapeutic properties and strong demand in the pharmaceutical industry (Thakur et al. 2025). The major agroforestry systems in this region include the following.

2.1.1 Agrisilviculture System

In the face of escalating environmental degradation and growing demand for medicinal resources, agrisilvicultural systems have emerged as a strategic approach for sustainable land management. By integrating multipurpose tree species, such as *Melia composita* and *Eucalyptus*, with high-value medicinal plants, including *Withania somnifera*, *Ocimum sanctum*, and *Curcuma longa*, these systems simultaneously enhance ecosystem resilience and rural livelihoods. Empirical studies highlight that such integration enhances soil fertility, mitigates erosion, and increases carbon sequestration, thereby contributing to both ecological restoration and climate mitigation (Sahu et al. 2025). Beyond environmental benefits, agrisilvicultural systems generate diversified income streams, reduce dependence on monocultures, and facilitate the conservation of threatened medicinal species, such as *Rheum australe* (Bhardwaj and Sood 2022). Moreover, shade-tolerant medicinal plants, including *Swertia chirayita* and *Picrorhiza kurroa*, thrive under the canopy of trees such as *Cedrus deodara* and *Pinus roxburghii*, exemplifying the synergistic interactions between tree and understory crops. By linking biodiversity conservation, climate adaptation, and socioeconomic resilience, agrisilvicultural systems exemplify an integrated land-use model that addresses multiple global sustainability challenges (Kala 2005).

2.1.2 Silvipastoral System

In the northwestern Himalayas, the silvipastoral system represents a traditional land-use practice wherein farmers integrate multipurpose tree species with forage crops and livestock to optimize both productivity and ecological balance. Trees such as *Grewia optiva*, *Bauhinia variegata*, *Terminalia* spp., *Phyllanthus emblica*, *Myrica esculenta*, *Rhododendron arboreum*, *Juniperus polycarpos*, *Salix* spp., *Prunus armeniaca*, *Prunus mira*, *Pinus gerardiana*, *Bamboo* spp., *Bauhinia vahlii*, *Vitex negundo*, *Punica granatum*, *Hippophae* spp., and *Berberis* spp. Provide multiple benefits such as timber, fuelwood, fodder, and other ecosystem services. The integration of medicinal plant cultivation further diversifies farmers' incomes while conserving biodiversity (Lata et al. 2024). This synergistic approach exemplifies sustainable agriculture, balancing ecological integrity with socioeconomic resilience in rural Himalayan communities (Table 14.1).

2.1.3 Agrihortisilviculture System

The multilayered system in the northwestern Himalayas offers a sustainable and integrated approach to land use, combining agriculture, horticulture, and forestry to enhance productivity, conserve biodiversity, and improve the livelihoods of local communities (Atul et al. 2017). Incorporating MAPs into the AHS system provides

Table 14.1 List of medicinal plants found in Uttarakhand, Himachal Pradesh, and Jammu and Kashmir

		Representative medicinal plants			References
Altitudinal zone	Altitude (m)	Uttarakhand	Himachal Pradesh	Jammu and Kashmir	
Tropical and subtropical zone	300–1800	*Aloe barbadensis*, *Withania somnifera*, *Bacopa monnieri*, *Aegle marmelos*, *Terminalia arjuna*, *Tinospora cordifolia*, Chirayata, *Swertia chirayita*, *Terminalia chebula*, *Azadirachta indica*, *Ocimum sanctum*	*Zizypus mauritiana*, *Rauwolfia serpentina*, *Mentha piperata*, *Cyperus rotundus*, *Ocimum canum*, *Achyranthes aspera*, *Atropa belladonna*, *Tinospora cordifolia*, *Dioscorea bulbifera*, *Curculigo orchioides*, *Justicia adhatoda*, *Azardirachta indica*, *Terminalia arjuna*, *Rubia cordifolia*, *Aloe barbadensis*, *Urtica parviflora*, *Bauhinia variegata*, *Asparagus adscendens*, *Salvia lanata*, *Swertia angustifolia*, *Hypericum perforatum*, *Calotropis gigantea*, *Withania somnifera*, *Tagetes minuta*, *Cinnamomum tamala*, *Emblica officinalis*, *Apium graveolens*, *Prinsepia utilis*, *Berberis asiatica*, *Rauwolfia serpentina*, *Terminalia chebula*, *Woodfordia fruticosa*, *Zanthoxylum armatum*, *Terminalia atata*, *Bergenia ligulata*, *Ocimum sanctum*, *Curculigo orchioides*, *Mangifera indica*, *Cassia fistula*, *Bacopa monnierii*, *Hedychium spicatum*, *Rorippa indica*, *Syzygium cuminii*, *A. racemosus*, *A. japonica*, *Malaxis acuminata*, *Elaeagnus conferta*, *M. arvensis*, *C. decapitala*, *A. barbadensis*, *C. tamala*, *Salvia plebei*	*Aloe barbadensis*, *Withania somnifera*, *Bacopa monnieri*, *Ocimum sanctum*, *Tinospora cordifolia*, *Azadirachta indica*, *Terminalia bellirica*, *Phyllanthus emblica*, *Swertia chirayita*, *Asparagus racemosus*, *Andrographis paniculata*	Kumar et al. (2021a), Sharma et al. (2024), Tiwari et al. (2020), Singh et al. (2017), Prakash (2015), Gairola et al. (2022), Kumar et al. (2021b)
Temperate zone	1800–3000	*Picrorhiza kurroa*, *Aconitum heterophyllum*, *Zanthoxylum armatum*, *Viola odorata*, *Commiphora wightii*	*Ferula jaeschkeana*, *Heracleum candicans*, *Asparagus filicinus*, *Ainsliaea aptera*, *Berberis aristata*, *Betulaalnoides*, *Sagina saginoides*, *Corylus jacquemontii*, *Rosularia rosulata*, *Gentiana kurroo*, *Skimmia laureola*,*Geranium nepalense*, *Elsholtzia fruticosa*, *Polygala sibirica*, *Taxus baccata subsp. wallichiana*, *Oxalis corniculata*, *Malva verticillata*, *Rhododendron arboreum*, *Phytolacca acinosa*.	*Picrorhiza kurroa*, *Saussurea costus*, *Jurinea dolomiaea*, *Viola odorata*, *Valeriana jatamansi*, *Arnebia benthamii*, *Pinus roxburghii*, *Selinum tenuifolium*, *Podophyllum hexandrum*, *Rhododendron arboreum*	Dwivedi et al. (2019), Upadhyay et al. (2024), Singh et al. (2017, 2020), Gairola et al. (2021)

Sub-alpine zone	Above 3000	*Nardostachys jatamansi, Rhododendronarboretum, Jurinea macrocephala, Saussurea costus*	*Allium humile, Bunium persicum, Malaxis muscifera, Carum carvii, Geranium wallichianum, Angelicaglauca, Archangelica himalaica, Bupleurum falcatum, Heracleum lanatum, Rhododendron campanulatum, Arisaema flavum, Pedicularis pectinata, Impatiens gladulifera, Polygonatum verticillatum, Aconitum ferox, Tanacetum tenuifolium, Arnebia benthamii, Saussurea auriculata, Picrorhiza kurrooa, Dactylorhiza hatagirea, Polygonatum multiflorum, Eritrichium canum, Plantago depressa, Tanacetum gracile, Aconitum heterophyllum, Aconitum falconeri, Ribes orientale, S. costus, A. leave, T. tomentosum*	*Aconitum heterophyllum, Bergenia ciliate, Inula racemose, Picrorhiza kurroa, Valeriana jatamansi, Delphinium denudatum, Geranium wallichianum, Angelica glauca, Viola serpens, Paris polyphylla, Dioscorea deltoidei, Thalictrum foliolosum, Berberis aristate, Polygonatum verticillatum*	Pandey et al. (2019), Haq et al. (2023), Chauhan et al. (2020), Devi et al. (2023), Singh et al. (2016, 2023), Jishtu et al. (2023)
Alpine zone	Above 3500	*Ophiocordyceps sinensis, Dactylorhiza hatagirea, Meconopsis aculeate, Saussurea obvallata, Ephedra gerardiana*	*Cortia depressa, Selinum tenuifolium, Heracleum wallichii, Inula royleana, Saussurea graminifolia,S. obvallata, Rheum moorcroftianum, Delphinium vestitum, Picrorhiza kurrooa, S. gossypiphora, Dactylorhiza hatagirea, Arnebia euchroma, Iriskumaonensis, Nardostachys grandiflora, Aconitum rotundifolium, S. simsoniana, Rheum australe, Fritillaria roylei, Aconitum violaceum, Rhododendron anthopogon, Corydalis govaniana, Aconitum spicatum, Aconitum heterophyllum, Corydalis meifolia, Delphinium cashmerianum.*	*Dactylorhiza hatagirea, Ophiocordyceps sinensis, Rhodiola imbricate, Nardostachys jatamansi, Gentiana kurroo, Saussurea obvallata, Aconitum violaceum, rnebia benthamii, Dactylorhiza hatagirea, Podophyllum hexandrum, Rheum emodi, Saussurea costus, Meconopsis aculeate, Swertia chirayita, Gentiana kurroo, Lagotis cashmeriana, Fritillaria roylei, Ephedra gerardiana, Jurinea macrocephala*	Thapliyal et al. (2024), Verma and Tewari (2016), Chauhan et al. (2020), Haq et al. (2023), Devi et al. (2023), Singh et al. (2016, 2023), Jishtu et al. (2023)

multiple ecological and economic benefits. Cultivating MAPs like *Withania somnifera*, *Ocimum sanctum*, and *Andrographis paniculata* enhances species diversity and supports ecosystem resilience. Moreover, some other medicinal plants like Aloe vera, *Tinospora cordifolia*, and *Ocimum sanctum* are often intercropped with fruit trees such as apple, walnut, and apricot (Maikhuri et al. 2001).

2.1.4 Home Gardens

Home gardens in the northwestern Himalayas serve as vital agro-ecosystems that integrate agricultural practices with the cultivation of medicinal plants, offering a sustainable model for biodiversity conservation and livelihood enhancement. These gardens not only provide essential food and income but also play a pivotal role in preserving the region's rich ethnobotanical heritage. Typically, traditional home gardens support a diverse assemblage of fruit trees, vegetables, and medicinal species (Negi and Maikhuri 2018), fostering ecological resilience and ensuring the sustainable use of biological resources. Key medicinal plants cultivated include *Azadirachta indica*, *Terminalia chebula*, *Aegle marmelos*, and *Tinospora cordifolia*, reflecting both ecological and cultural significance (Gaur 1999). In Uttarakhand's Nainital district, surveys documented 111 medicinal plant species within home gardens, highlighting their essential role in in situ conservation and the perpetuation of traditional medicinal knowledge (Tiwari et al. 2017). Collectively, these observations underscore the potential of home gardens as strategic platforms for integrating conservation, sustainable agriculture, and ethnobotanical knowledge, which is critical for advancing resilient agro-ecosystems in the Himalayan region.

3 Key Medicinal Plants in Agroforestry Systems

The northwestern Himalayan region of India is renowned for its rich biodiversity and a long tradition of agroforestry practices. In this region, AFS often integrate medicinal plants alongside forest trees, crops, and livestock, thereby promoting ecological balance while simultaneously supporting local livelihoods. A number of high-value medicinal plants are cultivated within these agroforestry frameworks. *Withania somnifera*, for example, is a widely used adaptive in traditional and modern herbal medicine and is often intercropped with cereals such as wheat and maize. *Ocimum sanctum*, revered for its antimicrobial, antioxidant, and immune-enhancing properties, is commonly grown in homestead agroforestry plots due to its spiritual as well as medicinal importance. At higher altitudes, *Picrorhiza kurroa* represents a critical yet vulnerable resource, extensively harvested for its hepatoprotective and anti-inflammatory uses but increasingly threatened by overexploitation and habitat degradation. Similarly, *Zanthoxylum armatum* holds dual importance, serving both culinary and medicinal roles, and is frequently cultivated in boundary plantations and mixed cropping systems (Table 14.2).

Table 14.2 Medicinal plants and preferred AFS in northwestern Himalayas

Altitude zone	Botanical name	Common name	Family	Plant part used	Medicinal uses	Role in agroforestry systems
High-altitude (above 3000 m)	*Aconitum heterophyllum*	Atis	Ranunculaceae	Roots	Used in Ayurveda for fever, diarrhea, and as a tonic; roots are highly valued	Cultivated under conservation-focused high-altitude agroforestry
	Aconitum balfourii	–	Ranunculaceae	Roots	Analgesic and anti-inflammatory; traditionally used in controlled doses	Cultivated in high-altitude alpine agroforestry for conservation and medicinal purposes
	Saussurea costus	Kuth	Asteraceae	Roots and leaves	Treats respiratory ailments, digestive disorders, and skin diseases	Promoted in community-based alpine agroforestry initiatives
	Dactylorhiza hatagirea	Salam Panja	Orchidaceae	Roots and shoot	Aphrodisiac, tonic, and restorative herb; used in Ayurveda and Unani medicine	Grown in high-altitude agroforestry plots under conservation programs
	Rheum emodi	Revand Chini	Polygonaceae	Rhzimoe, stalk, and lLeaves	Laxative, digestive aid, and blood purifier	Integrated into high-altitude mixed agroforestry systems
	Arnebia benthamii	Ratanjot	Boraginaceae	Roots, aerial parts, flowers, and leaves	Used for skin ailments, wound healing, and as a natural dye	Cultivated in temperate agroforestry niches; often grown on field margins

(continued)

Table 14.2 (continued)

Altitude zone	Botanical name	Common name	Family	Plant part used	Medicinal uses	Role in agroforestry systems
Mid-altitude (1500–3000 m)	*Berberis aristata*	Daruharidra	Berberidaceae	Roots, leaves, stem, bark, and fruits	Antimicrobial, anti-inflammatory, liver tonic	Boundary plantations and hedgerows in agroforestry system
	Valeriana jatamansi	Tagar	Valerianaceae	Roots and leaves	Source of essential oil; sedative, anti-anxiety, used in perfumery	Grown in moist, shady niches within mid- to high-altitude agroforestry
	Picrorhiza kurroa	Kutki	Plantaginaceae	Roots, leaves, and bark	Hepatoprotective, anti-inflammatory; used for liver disorders	Integrated into high-altitude community agroforestry systems
	Podophyllum hexandrum	Himalayan Mayapple	Berberidaceae	Roots, leaves, and fruit	Cancer treatment	Cultivated under alpine agroforestry and conservation-based systems
	Hippophae rhamnoides	Sea buckthorn	Elaeagnaceae	Berries, seeds, and leaves	Rich in vitamins; used for skin, cardiovascular, and digestive health	Planted in cold arid agroforestry systems for soil conservation and fruit harvesting

Lower-altitude (below 1500 m)	*Withania somnifera*	Ashwagandha	Solanaceae	Leaves, flowers, and stem	Adaptogen; reduces stress; boosts vitality and immunity	Intercropped with wheat and maize in lower- to mid-altitude fields
	Tinospora cordifolia	Giloy	Menispermaceae	Stem and leaves	Immunomodulator; antipyretic; used for diabetes and liver disorders	Grown as a climber on trees in homestead and boundary agroforestry systems
	Asparagus racemosus	Shatavari	Asparagaceae	Roots	Rejuvenate tonic, supports female reproductive health	Cultivated in homestead and field-margin agroforestry
	Aloe vera	–	Liliaceae	Leaf	Skin care, wound healing, digestive aid	Widely cultivated in homestead agroforestry and intercropped in arid and semi-arid regions
	Ocimum sanctum	Tulsi	Lamiaceae	Leaves, flowers, and seeds	Antimicrobial, antioxidant, immune-enhancing; household medicine	Cultivated in homestead agroforestry and boundary plantations
	Curcuma longa	Turmeric	Zingiberaceae	Rhizome, leaves	Anti-inflammatory, antioxidant, antimicrobial; widely used in food and medicine	Intercropped with cereals, pulses, and horticultural crops in agroforestry systems

The integration of these species within AFS offers multiple ecological benefits, including enhanced biodiversity, improved soil fertility, increased water retention, and the stabilization of fragile mountain slopes. At the same time, medicinal plants provide additional economic opportunities through value-added processing, market linkages, and growing demand in pharmaceutical, nutraceutical, and cosmetic industries (Kholiya et al. 2025). Despite these advantages, the sustainable cultivation of medicinal plants remains constrained by overexploitation, limited market access, and insufficient technical knowledge among farmers. The region's favorable climatic conditions further support the growth of diverse medicinal species (Samant and Dhar 1997). This ecological heterogeneity supports a wide range of species with significant therapeutic potential, many of which form the backbone of both Ayurvedic and modern herbal medicine. An overview of key medicinal species integrated into AFS of the northwestern Himalayas is presented below.

4 Land Preparation and Plantation Techniques for Medicinal Plants in Agroforestry Systems

Sustainable land-use strategies that simultaneously enhance biodiversity, strengthen rural livelihoods, and build climate resilience are increasingly vital in the face of global environmental challenges. Agroforestry, defined as the deliberate integration of trees, crops, and sometimes livestock, has emerged as a multifunctional approach that addresses these needs by combining ecological sustainability with economic development. Within this context, medicinal plants have emerged as particularly valuable components owing to their high economic potential, ecological adaptability, and cultural significance (Rao et al. 2004). Their integration into AFS not only augments land-use efficiency and farm productivity but also plays a critical role in conserving indigenous species and safeguarding traditional knowledge systems that are rapidly eroding in the face of modernization. Nevertheless, the successful incorporation of medicinal plants requires informed site selection, systematic land preparation, and scientifically grounded plantation techniques to ensure optimal growth and sustainable yields. When strategically embedded within agroforestry landscapes, medicinal plants strengthen ecosystem services, diversify farm income, and advance both conservation and livelihood security, positioning them as key drivers of sustainable agricultural transformation in the twenty-first century (ICRAF 2019).

4.1 Land Preparation

4.1.1 Site Selection and Soil Assessment

The success of medicinal plant agroforestry fundamentally relies on the careful selection of suitable sites, which form the foundation for sustainable cultivation and high-quality yield. Optimal sites are characterized by well-drained soils with adequate fertility, as highlighted by Kumar et al. (2020), and require comprehensive evaluation of key soil properties, including pH, nutrient availability, and organic matter content. Medicinal plants exhibit highly species-specific ecological preferences; for example, *Withania somnifera* thrives in well-drained sandy loam, whereas *Ocimum sanctum* performs best in loamy soils with moderate moisture (Li et al. 2025). Beyond soil characteristics, microclimatic factors such as light intensity, temperature, and relative humidity must be carefully considered to align with the physiological and phenological requirements of each species. Strategic site selection not only enhances plant growth and yield quality but also contributes to ecosystem health by promoting soil conservation, biodiversity, and resilience to climate variability. Furthermore, well-planned site selection has direct socioeconomic implications, supporting local livelihoods, sustainable resource use, and the long-term viability of medicinal plant-based agroforestry systems.

4.1.2 Soil Fertility Management

Enhancing soil fertility is fundamental for the sustainable cultivation of medicinal plants in the northwestern Himalayas, where fragile ecosystems and variable climatic conditions pose significant challenges. The application of organic amendments—such as compost, green manure, and biofertilizers—not only supplies essential nutrients but also stimulates beneficial soil microbial activity, which is critical for nutrient cycling and plant health (FAO 2019). The use of organic mulches further supports soil health by conserving moisture, moderating temperature fluctuations, and suppressing weed proliferation. Incorporating leguminous cover crops into cropping systems can substantially improve soil nitrogen content, reduce dependence on synthetic fertilizers, and contribute to long-term soil fertility (Kholiya et al. 2025). Collectively, integrating these practices within traditional agroforestry systems enhances soil resilience, supports biodiversity, and strengthens the ecological and economic sustainability of medicinal plant production in the region.

4.1.3 Water Management

Effective water management is critical for the sustainable growth and productivity of medicinal plants. Precision irrigation techniques, such as drip and sprinkler systems, ensure efficient, targeted water delivery, particularly during periods of water scarcity. Supplementary strategies, including rainwater harvesting and small check dams, enhance water availability and strengthen resilience to climatic variability. Maintaining proper drainage is essential, as water logging can compromise root development, nutrient uptake, and overall plant health. Integrating these measures supports both consistent plant growth and climate-resilient, sustainable cultivation practices (ICAR 2020).

4.1.4 Erosion Control and Land Structuring

In hilly and sloped terrains, soil erosion is a major constraint to sustainable agriculture. Strategic land preparation and plantation techniques—such as contour planting, terracing, and bunding—effectively reduce runoff, enhance soil conservation, and improve water infiltration. In the northwestern Himalayan AFS, where undulating landscapes are prevalent, these measures are essential for the successful establishment and productivity of medicinal plants, ensuring both soil stability and optimal growth conditions. Building on these soil and water conservation strategies, the selection of appropriate land preparation methods and plantation techniques becomes critical to maximize the growth, yield, and ecological integration of medicinal species within agroforestry system.

4.2 Plantation Techniques

4.2.1 Agroforestry System Design

Medicinal plants can be strategically incorporated into AFS through carefully designed plantation techniques that maximize both productivity and ecological benefits. In multi-strata systems, these species are cultivated as understory crops beneath taller trees, optimizing light interception, soil moisture retention, and spatial resource use. Intercropping with compatible annual or perennial crops, such as cereals and legumes, further enhances land-use efficiency while mitigating pest and disease pressures through diversified cropping patterns. Effective plantation practices include site-specific planning, soil preparation, selection of suitable propagation methods (seed sowing, vegetative cuttings, or tissue culture), and appropriate spacing to ensure healthy growth and sustainable yields. Species such as *Withania somnifera*, *Ocimum sanctum*, *Centella asiatica*, and *Aloe vera* have demonstrated high adaptability in AFS, offering economic returns alongside ecosystem services such as soil stabilization, biodiversity conservation, and microclimate regulation (Rao et al. 2004). By integrating medicinal plants into agroforestry, practitioners can achieve a balance between productivity, sustainability, and conservation, making these systems both economically and ecologically resilient.

4.2.2 Nursery Practices

Raising healthy seedlings is essential for the successful establishment of plantations, particularly in AFS that incorporate medicinal plants. Seeds should be sown in well-prepared nursery beds or containers using sterilized soil media to ensure uniform germination and reduce disease incidence. Seedlings must be nurtured under controlled conditions with adequate light, moisture, and nutrient supply to promote vigorous growth. Hardening off—the gradual acclimatization of seedlings to field conditions—minimizes transplant shock and improves survival rates. In agroforestry systems, these propagation practices are critical for integrating medicinal plants alongside timber, fruit, or other cash crops. Strategic spatial arrangement and intercropping optimize light, nutrient, and water availability, enhancing both plant growth and the accumulation of bioactive compounds. Complementary practices such as soil fertility management, mulching, and microclimate regulation further support productivity and resilience against biotic and abiotic stresses. By combining rigorous nursery management with site-specific agroforestry techniques, medicinal plants can thrive, contributing simultaneously to biodiversity conservation, ecosystem services, and the economic sustainability of rural farming systems.

4.2.3 Field Planting

The establishment and productivity of medicinal plants within AFS critically depend on proper planting techniques. Optimal spacing and planting density vary with species and growth habit; for instance, *Withania somnifera* is typically cultivated at 45–60 cm between plants, whereas *Ocimum sanctum* requires closer spacing of 30–40 cm to maximize canopy development and light capture. The planting depth should align with the seedling root ball to prevent desiccation or root rot, while immediate irrigation following transplantation promotes effective soil–root contact, reduces transplant stress, and facilitates early establishment (Li et al. 2025). Within agroforestry systems, these practices gain added significance, as medicinal plants are integrated with trees, shrubs, and other crops, necessitating careful attention to spatial arrangement, light availability, and soil moisture dynamics. When executed effectively, such plantation strategies not only enhance plant growth and yield but also contribute to broader ecosystem functions, including biodiversity conservation, soil fertility improvement, and system resilience. Therefore, meticulous attention to species-specific planting protocols forms the cornerstone of sustainable medicinal plant production in multifunctional agroforestry landscapes.

4.2.4 Canopy and Crop Management

Effective management of the tree canopy is crucial to ensure adequate light availability for understory medicinal plants in agroforestry systems. Practices such as selective pruning and thinning of trees enhance light penetration and reduce competition for soil nutrients, thereby promoting optimal growth of medicinal species. Complementary interventions—including regular weeding, soil aeration, and periodic replenishment of organic mulches—help maintain soil fertility and moisture, further supporting plant health (Kholiya et al. 2025). These canopy and soil management strategies can be integrated with targeted plantation techniques to maximize productivity and sustainability. For instance, careful selection of shade-tolerant versus light-demanding medicinal species, appropriate spatial arrangements, optimal planting densities, and intercropping with complementary crops can improve resource use efficiency, enhance biodiversity, and reduce pest and disease pressures.

By combining these approaches, agroforestry systems can provide a conducive microenvironment for medicinal plants while maintaining overall ecosystem resilience and productivity.

4.2.5 Harvesting and Postharvest Care

The sustainable cultivation of medicinal plants requires an integrated approach that combines optimized plantation techniques with careful postharvest management. Proper harvesting practices are essential to maintain the therapeutic and pharmacological value of plants, while drying, processing, and storage under controlled conditions are critical for preserving bioactive compounds (WHO 2020). In AFS, plantation strategies—such as selecting appropriate sites, arranging species for complementary growth, and implementing intercropping or multi-strata designs—play a pivotal role in enhancing plant health, yield, and the concentration of medicinal compounds. By linking cultivation with standardized harvesting and processing practices, agroforestry not only ensures the sustainable production of high-quality medicinal plants but also contributes to biodiversity conservation, soil fertility improvement, and socioeconomic benefits for local communities.

5 Value Addition in Medicinal Plant Production

Value addition in medicinal plant production refers to the processes that enhance the economic, functional, and market potential of plant products beyond their raw form. Despite the rapidly growing global demand for natural remedies, nutraceuticals, and herbal pharmaceuticals, the potential of medicinal plants remains underexploited when marketed solely as raw materials. Value addition involves multiple stages, including cleaning, sorting, drying, grinding, extraction of bioactive compounds, formulation into diverse dosage forms (such as tablets, capsules, teas, syrups, and ointments), and standardization to ensure therapeutic consistency. Complementary practices such as packaging, branding, and accurate labeling further enhance consumer confidence, product appeal, and market competitiveness. By transforming raw plant materials into higher-value products, producers can substantially increase profitability, minimize postharvest losses, and gain greater access to both domestic and international markets (Fig. 14.1). From an economic perspective, value addition fosters rural entrepreneurship, supports small-scale industries, and ensures continuous income generation for farming communities, while promoting sustainability through eco-friendly harvesting practices, organic cultivation, and reduced reliance on synthetic chemicals.

Nonetheless, several challenges constrain the full potential of value addition, including inadequate technical expertise, limited processing infrastructure, stringent regulatory frameworks, and increasing market competition. Addressing these barriers requires the integration of traditional knowledge with modern scientific advancements, the development of innovative herbal formulations, and the adoption

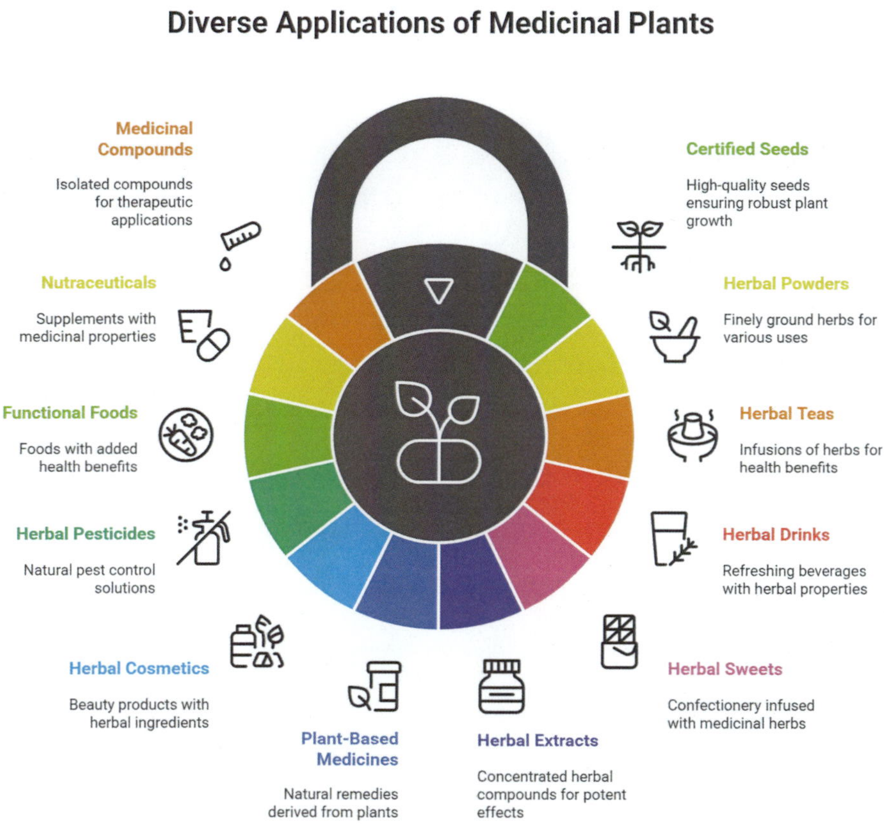

Fig. 14.1 Different value-added products from medicinal plants

of digital marketing strategies and export-oriented business models. In essence, value addition not only strengthens the financial viability of medicinal plant production but also enhances product quality, efficacy, and global reach, thereby contributing simultaneously to rural livelihood security and the expansion of the herbal product industry (Table 14.3).

5.1 *Processing Techniques for Medicinal Plants*

Effective processing techniques play a critical role in preserving the therapeutic properties of medicinal plants and in ensuring the production of high-quality, standardized products. Broadly, these techniques can be classified into primary processing methods, which involve basic handling, cleaning, drying, and storage, and advanced processing methods, which encompass modern technologies aimed at enhancing bioactive compound stability, extraction efficiency, and product safety (Fig. 14.2). Together, these approaches form the foundation of quality assurance in herbal drug development and commercialization.

Table 14.3 Value addition in the final product from raw medicinal plant (Kumar et al. 2021a, b)

Species	Part used	Uses/properties	End product	Raw material (price in Kg)	Product after value addition	Price of the final product in the market (eg.)	Value added (%)
Picrorhiza kurroa	Root	Dyspepsia, chronic diarrhea, and scorpion sting	Rohitkyadi churna (powder), Arogyavardini vati (tablet), Dardnasak oil (syrup), Heptogen capsule (capsule)	1500–2000	Kutki Livertone, spices, Kutki powder, Kutki suprime	100 gm pack Rs. 512	192%
Aconitum heterophyllum	Root	Expectorant, anthelmintic, anti-diarrheal, anti-emetic, febrifuge, and anti-inflammatory	Ashtvardhak, atis shirin	4500–5000	[a]	[a]	[a]
Rheum australe	Leaves	Gastritis, stomach problems, blood purification, menstrual problems, and liver diseases	Kaulo bokra powder	[a]	[a]	Bread, Chatani, cosmetics	[a]
Bunium persicum	Seed	Antispasmodic, carminative, antiobesity, and lactogage	Spice and flavoring agent in foods and beverages such as bread cooking, rice, yoghurt, cheese, and in confectionery products	2200–2500	Baked into breads, added to curry pastes and blends, or steamed with rice to give it a distinctive fragrance, especially in northern Indian meat and rice dishes, Zeera biskit, Zeera laddu, Zeera Namkeen	Zeera biskit Rs. 35/10 gm.	49%

Polygonatum verticillatum	–	Cardiotonic, antiperiodic, demulcent, diuretic, energizer, aphrodisiac, and sedative	Chyavanprasha, Sudarshana Churna, Taila (oil), Ghritam (medicated clarified butter), and Churana (powder)	[a]	Rumi mastagi, Salab	10 gm Rs. 350	[a]
Fritillaria cirrhosa	Bulb	Antiasthmatic, hemostatic, ophthalmic, and oxytocic	Astavarga, Chyavanprash	12,000–18,000	[a]	[a]	[a]
Dactylorhiza hatagirea	Root	Root is used as sweet, cooling, emollient, astringent, aphrodisiac, demulcent, rejuvenating and nervine tonic, and also as a nutritional supplement	Patanjali Youvan Chourna	[a]	[a]	[a]	[a]
Sinopodophyllum hexandrum	Root	Typhoid fever, jaundice, rheumatism, skin diseases, tumorous growth, and kidney and bladder problems	Anticancer drug etoposide and teniposide	200–400	Chatani	[a]	[a]
Rheum webbianum	Leaves	Renal function disorders, hyperlipidemia, cancer, and improves the memory in senile patients	[a]	[a]	Bread, Chatani, cosmetics	[a]	[a]

[a]Lack of information on the final product

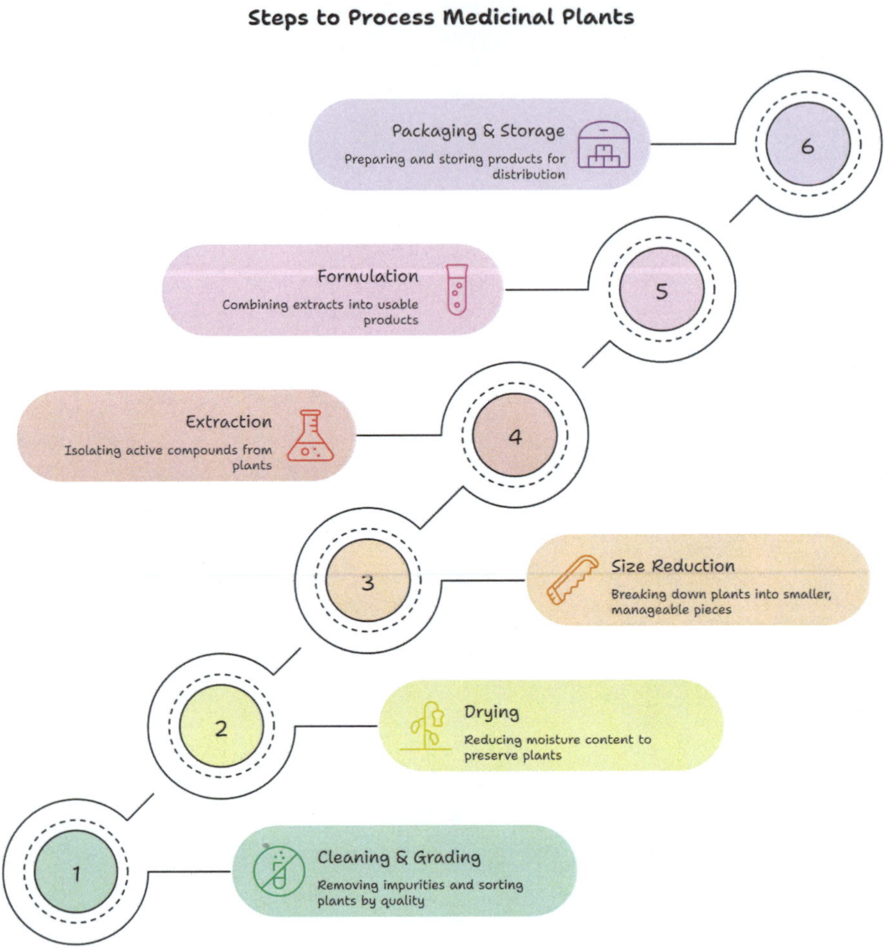

Fig. 14.2 Different processing techniques for medicinal plants

5.1.1 Primary Processing

1. *Cleaning and Grading*: Removing dirt, debris, and damaged plant parts to ensure uniformity and quality.
2. *Drying*: Reducing moisture content to prevent microbial growth and degradation. Methods include sun drying, air drying, and controlled drying using mechanical dryers.
3. *Size Reduction*: Grinding or cutting plant materials into smaller sizes to facilitate extraction and formulation. This can be achieved using crushers, grinders, or cutting machines.

5.1.2 Advanced Processing

1. *Extraction*: Isolating active compounds from plant materials using solvents or other methods. Common extraction techniques include maceration, percolation, Soxhlet extraction, and ultrasound-assisted extraction.
2. *Formulation*: Developing final products such as capsules, tablets, creams, or tinctures by combining extracted compounds with other ingredients. This step requires adherence to quality standards to ensure product efficacy and safety.
3. *Packaging and Storage*: Protecting processed products from contamination and degradation through appropriate packaging materials and storage conditions. This ensures the shelf life and stability of the products.

6 Significance of Integration of Medicinal Plants in Agroforestry Systems of the Northwestern Himalayas

6.1 Ecological Benefits

The integration of medicinal plants into AFS in the northwestern Himalayas offers a wide spectrum of ecological benefits that extend well beyond livelihood security and income diversification. A key advantage lies in biodiversity conservation, as the

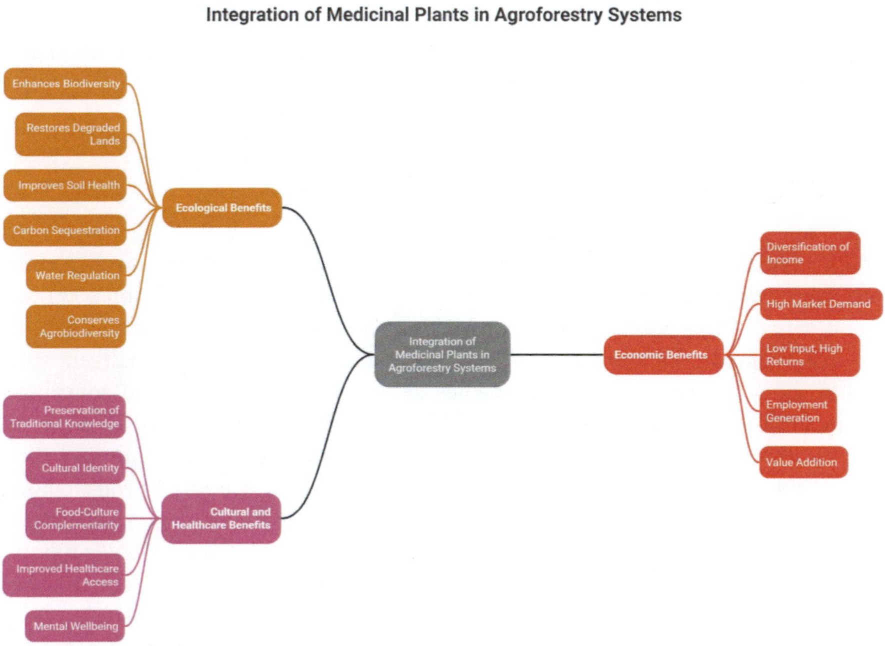

Fig. 14.3 Significance of integration of medicinal plants in agroforestry systems

on-farm cultivation of medicinal plants functions as an effective in situ strategy for maintaining genetic resources while simultaneously reducing extraction pressure on vulnerable wild populations (Shinwari et al. 2024). These enriched AFS also enhance soil health and mitigate erosion, particularly on steep Himalayan slopes where monocultural farming is prone to severe degradation (Fig. 14.3). The presence of tree cover alongside perennial medicinal species improves litter deposition, organic matter content, and soil structure, thereby facilitating water infiltration and reducing nutrient losses (Kc et al. 2022). These systems are equally important in terms of climate change mitigation, as empirical studies from the northwestern Himalayas demonstrate higher above- and belowground carbon storage in agroforestry plots compared with adjacent croplands, highlighting their role as carbon sinks with potential for inclusion in carbon crediting mechanisms (Singh et al. 2024b; Gupta et al. 2024; Saleem et al. 2023).

In addition to carbon and soil-related functions, medicinal plants contribute to the ecological regulation of pollination and pest control. The staggered flowering of diverse medicinal plants provides a continuous supply of nectar and pollen for pollinators, while also sustaining natural enemies of crop pests, thereby improving ecosystem resilience and supporting crop productivity in mixed farming landscapes (Joshi et al. 2016). Additionally, medicinal plant-based agroforestry systems improve hydrological functions and microclimate stability by reducing surface runoff, promoting infiltration, and maintaining water availability—an essential service for fragile Himalayan watersheds (Kc et al. 2022). They also play a critical role in land restoration, as several species are well suited for rehabilitating degraded or marginal lands, stabilizing soils, and re-establishing vegetation cover, while simultaneously generating marketable products for local communities. Such interventions have proven effective in the Western Himalayas for curbing outmigration and revitalizing abandoned agricultural terraces (Rawat et al. 2022).

6.2 Economic Benefits

MAPs play a pivotal role in sustaining rural livelihoods in the northwestern Himalayas, where exceptional species diversity and deep-rooted traditional knowledge create a strong foundation for their integration into AFS. Their incorporation into tree–crop combinations not only enhances biodiversity but also delivers substantial economic gains. Empirical studies demonstrate that MAPs can contribute significantly to household cash income, accounting for as much as 40–64% of annual earnings in forest-fringe communities. This contribution helps mitigate economic vulnerability and reduces dependence on seasonal agriculture or wage labour (Asigbaase et al. 2023). High-value species such as *Withania somnifera*, *Ocimum* spp., and Aloe vera have proven particularly profitable, with farm-level trials reporting favorable benefit–cost ratios and attractive net returns per hectare. The benefits of MAP cultivation extend beyond direct income generation. By diversifying production systems, MAPs enhance household resilience through staggered harvests, longer storage potential (e.g., dried herbs, essential oils, and extracts), and market diversification. Such features enable farmers to buffer against crop failures and price fluctuations, a particularly critical function in the fragile mountain economies of the Himalayas (Khan et al. 2022). Moreover, the growing domestic and international demand for herbal medicines, nutraceuticals, and AYUSH formulations has created expanding market opportunities. India's exports of AYUSH and herbal products alone reached approximately USD 612–628 million in recent years, underlining the lucrative potential for smallholders who integrate MAPs into their AFS and participate in organized value chains (Dadhich et al. 2024; IBEF 2024).

Beyond cultivation, the processing and marketing of MAPs stimulate local employment and enterprise development. Value addition activities such as drying, powdering, distillation, packaging, and transport generate rural jobs, foster micro-enterprises, and retain greater economic value within communities (Singh et al. 2024a, b). However, constraints such as price volatility, weak market channels, quality standardization issues, limited access to credit, and uncertain tenure rights often reduce farmers' ability to capture full economic benefits (Shinwari et al. 2024). National and state-level interventions, particularly through the National Medicinal Plants Board, have sought to address these barriers by promoting good cultivation practices, supplying quality planting material, establishing processing facilities, and facilitating linkages with AYUSH and pharmaceutical markets (Mbelebele et al. 2024).

6.3 Cultural and Healthcare Benefits

The northwestern Himalayas are globally renowned for their exceptional biodiversity and rich cultural heritage. Indigenous communities in this region have long relied on medicinal plants, integrating them into healthcare, rituals, and daily cultural practices. AFS, characterized by the deliberate integration of trees, crops, and livestock, play a central role in the cultivation and conservation of these species. Medicinal plants are deeply embedded in local traditions, serving not only as primary healthcare resources but also as integral components of traditional healing practices, religious ceremonies, and cultural rituals. Species such as *Nardostachys jatamansi* and *Valeriana wallichii* are valued for both therapeutic and spiritual purposes (Joshi et al. 2016), while *Aconitum heterophyllum* and *Berberis aristata* are widely used to treat fever and infections, respectively (Kumar et al. 2021b). Sacred species, including *Ficus religiosa* (Peepal) and *Tulsi*, further highlight the cultural significance of plant conservation (Sundriyal and Sharma 2018). Agroforestry provides a sustainable framework for the in situ conservation and cultivation of medicinal plants, incorporating them into home gardens, field margins, and understorey layers. This approach reduces pressure on wild populations while ensuring a consistent supply for healthcare and supplementary income. Ethnobotanical studies report over 100 medicinal plant species within agroforestry plots, with diversity shaped by altitude, ethnicity, and local cultural practices (Parihaar et al. 2014).

7 Market Opportunities for Medicinal Plants in the Northwestern Himalayas

The northwestern Himalayas offer significant opportunities for the cultivation and trade of medicinal plants, underpinned by its rich biodiversity, traditional knowledge, and growing global demand for natural health products. The region hosts approximately 1748 medicinal plant species, supplying nearly 90% of the raw material for India's herbal industries and exports (Bhat et al. 2013). Many endemic and rare species, such as *Aconitum heterophyllum* and *Picrorhiza kurroa*, are highly valued for their therapeutic properties. The WHO estimates that India's demand for medicinal plants will increase from the current $14 billion per year to $5 trillion by 2050. A study found that 92% of 71 rare medicinal plant species assessed in the northwestern Himalayas are actively traded, with 74% traded nationally and 35% internationally (Kala et al. 2006). In response to this growing demand, initiatives have been launched to promote cultivation and sustainable utilization of medicinal plants. In Uttarakhand's Almora district, a cultivation project aims to mitigate rural migration and revitalize hill agriculture by focusing on high-value species such as rosemary, chamomile, timur, bay leaf, and kapoor kachri (TOI 2025). Similarly, the Himachal Pradesh government has established herbal gardens in areas including Joginder Nagar and Neri, encouraging farmer clusters to participate in cultivation with a minimum landholding of two hectares to access financial incentives (The Tribune 2021). The global medicinal plant market is expanding, with increasing consumer preference for natural and herbal products. This trend presents opportunities for producers in the northwestern Himalayas to tap into international markets. There is a growing emphasis on value-added products such as essential oils, herbal teas, and nutraceuticals, which can enhance the economic returns for local communities (Ritika Bora et al. 2024).

7.1 *Current Market Trends and Demand*

The global herbal medicine market is experiencing rapid growth, projected to reach USD 550 billion by 2030 (WHO 2022). In India, the demand for medicinal plants has surged by 50%, while their availability has declined by 26% due to overexploitation and habitat destruction (TOI 2025). The Indian Himalayan region contributes approximately 90% of the raw materials used in the country's herbal industries and for export, making it second only to China in the global supply of medicinal plants. This growing demand, however, poses sustainability challenges, as overharvesting threatens certain species and the ecological balance. To address this, conservation efforts prioritize 43 out of 152 important medicinal plant species in the Western Himalayas to promote sustainable cultivation. The sector also offers opportunities for farmers and entrepreneurs by providing alternative, high-value crops, diversified income streams, and access to technical support and market linkages through institutions such as the Centre for Aromatic Plants (CAP). Himachal Pradesh alone exports approximately 2500 tonnes of medicinal plants and their derivatives annually, with the legal trade valued at around 10 crore (Kumar et al. 2021a, b). With the integration of traditional knowledge, scientific support, and sustainable practices, the cultivation and trade of medicinal and aromatic plants in the northwestern Himalayas have the potential to significantly contribute to regional economies and global markets.

7.2 Strategies to Enhance Market Opportunities

The following strategies can be adopted to fully realize the market potential of medicinal plants in the northwestern Himalayas.

7.2.1 Sustainable Harvesting and Conservation

- Establishing community-managed conservation areas and medicinal plant nurseries.
- Promoting nondestructive harvesting techniques.
- Implementing participatory forest management with local communities.

7.2.2 Promoting Large-Scale Cultivation

- Encouraging farmers to cultivate high-demand species like Kutki, Jatamansis, and Chirayata under agroforestry models.
- Providing financial incentives and subsidies for medicinal plant farming.
- Developing tissue culture and nursery programs for mass propagation.

7.2.3 Strengthening Market Linkages

- Setting up direct procurement mechanisms between farmers and herbal industries.
- Establishing herbal mandis and processing units to reduce middlemen exploitation.
- Encouraging public–private partnerships for value addition and branding.

7.2.4 Capacity Building and Policy Support

- Training local communities in sustainable collection, processing, and quality control.
- Implementing traceability mechanisms for herbal products to meet global standards.
- Revising legal frameworks to promote sustainable trade while ensuring conservation.

8 Key Challenges in Integrating Medicinal Plants into Agroforestry in the Northwestern Himalayas

8.1 Ecological and Climatic Constraints

The northwestern Himalayan region encompasses a wide spectrum of climatic zones, ranging from temperate valleys to alpine slopes, resulting in a complex mosaic of microclimates. This environmental heterogeneity poses significant challenges for integrating medicinal plants into agroforestry systems, as species must be carefully matched to specific ecological niches. For instance, *Saussurea costus* and *Valeriana jatamansi* are highly sensitive to altitude and soil characteristics, which limit their adaptability and complicate their cultivation alongside other crops (Sahu et al. 2025).

8.2 Overexploitation and Conservation Concerns

Medicinal plants are vital components of both biodiversity and traditional healthcare systems, offering significant ecological, economic, and cultural value. Many high-value species, including *Sinopodophyllum hexandrum, Saussurea costus, Valeriana jatamansi,* and *Taxus wallichiana*, are increasingly threatened due to overharvesting from the wild. Integrating these high-value plants into agroforestry systems provides a strategic pathway for their conservation and sustainable utilization. In the northwestern Himalayas, however, this integration is constrained by key challenges, including the implementation of sustainable harvesting, preservation of genetic diversity, and adaptation of agroforestry models to complex ecological and socioeconomic contexts. Effectively addressing these challenges is crucial not only for conserving medicinal plant resources but also for enhancing the livelihoods and resilience of local communities dependent on them.

8.3 Economic Viability and Market Access

Medicinal plants are a cornerstone of AFS, offering substantial economic and ecological benefits. However, farmers often face barriers such as limited market access, inadequate infrastructure, and a lack of value-added processing facilities. Strengthening local processing capabilities and establishing direct market linkages are therefore essential to enhance profitability, support sustainable livelihoods, and promote rural development while conserving biodiversity (Sahu et al. 2025).

8.4 Technical Knowledge and Capacity Building

The effective integration of medicinal plants into AFS requires advanced expertise in plant propagation, integrated pest and disease management, and postharvest processing. Strengthening the capacity of farmers and local communities through targeted training programs is therefore critical. Moreover, synergistic collaborations with agricultural universities and research institutions can accelerate the transfer of scientific knowledge and foster the adoption of best management practices (Kholiya et al. 2025).

8.5 Policy and Institutional Support

Medicinal plants are increasingly recognized for their critical ecological, economic, and healthcare roles; however, policy frameworks often remain insufficiently equipped to support their systematic integration into AFS. Advancing the inclusion of medicinal plants within agroforestry policies, coupled with dedicated investment in research and development, is essential to overcome institutional barriers and promote sustainable.

8.6 Climate Change and Environmental Stress

Climate change presents an escalating threat to the distribution, growth, and productivity of medicinal plants, which are integral to healthcare systems, biodiversity conservation, and rural livelihoods. Rising temperatures, altered precipitation patterns, and increased frequency of extreme climatic events can disrupt plant phenology, constrain growth, and significantly reduce yields. By mitigating climate-induced stressors, these interventions not only safeguard the ecological and economic value of medicinal plants but also support sustainable livelihood frameworks and the broader objectives of environmental sustainability (Rana et al. 2020).

9 Future Prospects and Recommendations

Integrating medicinal plants into AFS in the northwestern Himalayas presents a strategic approach to promote ecological sustainability, biodiversity conservation, and socioeconomic development. These systems not only increase species richness and biodiversity indices but also improve soil fertility and facilitate the restoration of degraded landscapes. By generating heterogeneous microhabitats, they support the regeneration of native flora and bolster ecosystem resilience, thereby enhancing long-term ecological stability. For instance, Balkrishna et al. (2020) documented 29 medicinal plant species—including herbs, shrubs, and trees—within agroforestry systems of the Govind Pashu Vihar Wildlife Sanctuary, Uttarakhand, illustrating the ecological potential of such integrated land-use practices. From an economic perspective, the global herbal medicine market, projected to exceed USD 150 billion by 2025, underscores the capacity of high-value medicinal plants to provide diversified and sustainable income streams for local communities. Species such as *Withania somnifera, Ocimum sanctum, Phyllanthus emblica,* and *Tinospora cordifolia* are particularly suited to the region due to their adaptability to local agro-climatic conditions and strong market demand. Integrating these species with traditional crops allows farmers to diversify livelihoods, mitigate economic risk, and achieve greater financial stability. Furthermore, systematic cultivation of medicinal plants can reduce pressure on wild populations, thereby contributing to the conservation of rare and endangered species, including *Coptis teeta* (Gupta et al. 2025).

The effective integration of medicinal plants into AFS requires the careful selection of species that are ecologically compatible, economically valuable, and culturally significant. Incorporating traditional agroforestry practices, such as multi-tiered cropping and pollarding, can further enhance plant growth while maintaining ecological balance. Capacity-building initiatives such as farmer training programs, demonstration plots, and knowledge-sharing platforms are crucial for equipping local communities with sustainable cultivation techniques, integrated pest and disease management strategies, and postharvest processing skills. Additionally, the establishment of robust market linkages and promotion of value-added products can significantly improve profitability, while government support in the form of subsidies, incentives, and research facilitation can further catalyze adoption. Collaborative efforts among research institutions, agricultural departments, and local communities are essential to optimize species-specific cultivation strategies and align production with market trends. Overall, integrating medicinal plants into agroforestry systems in the northwestern Himalayas offers a multifaceted solution that simultaneously addresses ecological, economic, and conservation objectives. By leveraging indigenous knowledge, promoting sustainable practices, and fostering effective market mechanisms, such integration can contribute to biodiversity conservation, sustainable land management, and improved livelihoods for local communities, establishing a model for resilient and profitable agroforestry systems in the region.

10 Conclusion

Incorporating medicinal plants into AFS of the northwestern Himalayas offers significant ecological, economic, and health benefits. These plants play a crucial role in supporting biodiversity, enhancing rural livelihoods, and preserving traditional healthcare practices. For the long-term success of medicinal plant-based agroforestry, sustainable harvesting, farmer training, and supportive policies are crucial. With increasing global demand and government initiatives promoting the AYUSH sector, strategic investments in cultivation, conservation, and trade can contribute to economic growth and biodiversity conservation. Ensuring certification and legal compliance is vital for maintaining product quality and sustainability. By adopting sustainable practices and market reforms, the northwestern Himalayas can emerge as a key hub for medicinal plants, benefiting local communities and the pharmaceutical industry alike. Future efforts should focus on expanding research, improving market access, and fostering collaborations between farmers, policymakers, and industries to maximize the potential of medicinal plant agroforestry in the region. Raising public awareness and conducting educational campaigns can promote sustainable harvesting practices, helping to preserve these valuable plant species for the long term. Moreover, leveraging technological advancements such as biotechnology and digital marketplaces can enhance productivity, streamline supply chains, and improve farmer incomes. By creating a robust policy framework and fostering sustainable business models, the northwestern Himalayas can set an example for medicinal plant conservation and agroforestry, contributing to both environmental sustainability and economic prosperity.

References

Asigbaase M, Adusu D, Anaba L, Abugre S, Kang-Milung S, Acheamfour SA, Ackah DK (2023) Conservation and economic benefits of medicinal plants: insights from forest-fringe communities of Southwestern Ghana. Trees For People 14:100462

Atul K, Sharma N, Kumar R (2017) Agri-horti-silviculture systems for sustainable land use and productivity enhancement. IJAF 12(3):45–52

Balkrishna A, Saxena P, Patel S, Vashistha RK, Gond DK, Butola JS, Rawat B (2020) Diversity, ethnomedicinal knowledge and conservation status of plant species in a part of North West Himalaya. Med Plants Int J Phytomed Relat Ind 12(3):356–370

Bhardwaj R, Sood M (2022) Rheum australe: an endangered high-value medicinal herb of North Western Himalaya. IJEP 9(2):106–110

Bhat JA, Kumar M, Bussmann RW (2013) Ecological status and traditional knowledge of medicinal plants in Kedarnath Wildlife Sanctuary of Garhwal Himalaya, India. J Ethnobiol Ethnomed 9(1):1

Bhat MN, Singh B, Surmal O, Singh B, Shivgotra V, Musarella CM (2021) Ethnobotany of the Himalayas: safeguarding medical practices and traditional uses of Kashmir regions. Biology 10(9):851

Butola JS, Singh R, Mehta A (2012) Agroforestry: a sustainable approach for rural livelihoods and carbon sequestration. Int J Sustain Agric 10(4):89–102

Chauhan A, Jishtu V, Thakur L, Dolma T (2020) Medicinal plants of the trans-Himalayan Cold Desert of Ladakh-a review. Int J Sci Environ 9(2):239–253

Dadhich MK, Sanwal CS, Mahajon B, Vaishya JK, Dutt S, Rath C, Sharma RK (2024) Contributions of the national medicinal plants board (NMPB), Ministry of Ayush in the medicinal plants sector-an appraisal. Int J Ayurveda Res 5(2):62–75

Dar M, Qaisar K, Ahmad S, Wani A (2018) Inventory and composition of prevalent agroforestry systems of Kashmir Himalaya. Adv Res 14(1):1–9

Devi U, Sharma P, Ahmad A, Khan AS (2023) Wild plant resources used as food in Spiti Valley: a cold arid zone of Himachal Pradesh, India. J Med Plant Stud 11:106–114

Dwivedi T, Kanta C, Singh LR, Prakash I (2019) A list of some important medicinal plants with their medicinal uses from Himalayan State Uttarakhand, India. J Med Plant Stud 7(2):106–116
FAO (2019) Medicinal plants production systems in rural Indonesia. 15(1):75–88
Gairola S, Singh K, Kumar P, Kumar B, Sharma P (2021) The wild edible plants of Paddar Valley, Jammu division, Jammu and Kashmir, India. Ethnobot Res Appl 22:1–21
Gairola S, Singh K, Sharma J (2022) Plants used for socio-magico-religious purposes by the indigenous communities of sub-Himalayan tract, Uttarakhand. Ethnobot Res Appl 23:1–19
Gaur RD (1999) Flora of the North-West Himalaya. Bishen Singh Mahendra Pal Singh, New Delhi
Gupta H, Janju S, Mahajan A, Singh C, Sharma S, Prajapati A (2024) Forests and agroforestry: nature-based solutions for climate change mitigation. In Forests and climate change: biological perspectives on impact, adaptation, and mitigation strategies. Singapore: Springer Nature Singapore, pp 421–443
Gupta AK, Ravikumar K, Gopal V, Begum N, Sangeetha S, Narendran P, Shankar D (2025) A trans-disciplinary agro-ecology strategy to grow medicinal plants. J Ayurveda Integr Med 16(1):100985
Haq SM, Waheed M, Khoja AA, Amjad MS, Bussmann RW, Ali K (2023) A cross-cultural study of high-altitude botanical resources among diverse ethnic groups in Kashmir Himalaya, India. J Ethnobiol Ethnomed 19(1):12
IBEF (2024) Ayurveda and herbal products export performance 2021–2023. India Brand Equity Foundation Report
ICAR (2020) Advancements, challenges, and policy interventions in agriculture. 22(4):110–125
ICRAF (2019) Medicinal plant production system management in rural Java, Indonesia: Views of local actors from a participatory rural appraisal approach. 9(3):98–113
Jishtu V, Chauhan A, Bano H (2023) Medicinal flora of the trans-Himalayan Cold Desert in Ladakh, India. In: Biosynthesis of bioactive compounds in medicinal and aromatic plants: manipulation by conventional and biotechnological approaches. Springer Nature Switzerland, Cham, pp 257–286
Joshi RK, Satyal P, Setzer WN (2016) Himalayan aromatic medicinal plants: a review of their ethnopharmacology, volatile phytochemistry, and biological activities. Medicine 3(1):6
Kala CP (2003) Medicinal plants of Indian trans-Himalaya: focus on Tibetan use of medicinal resources. Bishen Singh Mahendra Pal Singh
Kala CP (2005) Indigenous uses, population density, and conservation of threatened medicinal plants in protected areas of the Indian Himalayas. Conserv Biol 19(2):368–378
Kala CP, Dhyani PP, Sajwan BS (2006) Developing the medicinal plants sector in northern India: challenges and opportunities. J Ethnobiol Ethnomed 2(1):32
Kc KB, Tzadok E, Pant L (2022) Himalayan ecosystem services and climate change driven agricultural frontiers: a scoping review. Discov Sustain 3(1):35
Khan S, Masoodi TH, Islam MA, Wani AA, Gattoo AA (2022) Ethnomedicinal study of wild plants used by fringe communities in temperate forests of Himalayan Kashmir, India. Phytomed Plus 2(2):100251
Kholiya D, Uttam LD, Rajan P, Sharma S, Manjeet M, Jain G, Bhandari A, Kaur D (2025) Integrating medicinal plant cultivation into agroforestry systems for biodiversity and income generation. J Neonatal Surg 14(22):944–953
Kumar S, Sharma P, Meena DC (2020) Linking farmers to markets: schemes/programmes, progress and impact. 12(3):99–113
Kumar A, Sathyakumar S, Goraya GS, Gupta AK, Adhikari BS, Rawat GS (2021a) Assessment of medicinal and aromatic plant species on their collection, usage, demand, markets, price trends and life cycle in Lahaul and Pangi landscape, Himachal Pradesh. A report submitted to Himachal Pradesh Forest Department and United Nations Development Programme, 141
Kumar M, Radha Devi H, Prakash S, Rathore S, Thakur M, Mekhemar M (2021b) Ethnomedicinal plants used in the health care system: survey of the mid hills of Solan district, Himachal Pradesh, India. Plants 10(9):1842

Kumar S, Kumar S, Khanduri VP, Singh B, Joshi R, Riyal MK, Kumar KS (2025) Tree diversity, carbon sequestration and production potential of *Oryza sativa* L. in traditional agroforestry systems of Garhwal Himalaya, India. Carbon Res 4(1):6

Lata S, Kamakshi Paul S, Barman T, Sharma D (2024) Traditional silvipastoral systems: an important sustainable agroforestry system for greening, restoration of wasteland and livelihood security in Himachal Pradesh, India. Indian For 150(9):861–868

Li X, Liang H, Liang J, Ma J, Huang Y, Liang Y (2025) Understory cultivation of medicinal plants promotes sustainable forestry development. Agrofor Syst 99(2):39

Mahato S, Dasgupta S, Todaria NP, Singh VP (2016) Agroforestry mapping and characterization in four districts of Garhwal Himalaya. Energy Ecol Environ 1(2):86–97

Maikhuri RK, Nautiyal S, Rao KS, Saxena KG (2001) Conservation policy–people conflicts: a case study from Nanda Devi Biosphere Reserve (a world heritage site), India. For Policy Econ 2(3–4):355–365

Maurya R, Kumar N, Gupta M, Sood KK, Singh C (2025) Agroforestry systems in the North-Western Himalayas: a nexus of biodiversity conservation, livelihood enhancement, and ecosystem services. Ind J Eco 52(5):1281–1287

Mbelebele Z, Mdoda L, Ntlanga SS (2024) Growing wealth from nature: analyzing the profitability and determinants of high-value medicinal plants in smallholder production systems. Cogent Econ Finance 12(1):2426536

Negi VS, Maikhuri RK (2018) Traditional homegardens for sustainable livelihood and biodiversity conservation in the Indian Himalayas. Agrofor Syst 92(3):635–648

Negi VS, Maikhuri RK, Rana TS (2011) Agroforestry systems: a sustainable land use option for medicinal plant conservation in the North-West Himalaya. J Sustain For 30(7):659–673

Pandey A, Chandra Sekar K, Joshi B, Rawal RS (2019) Threat assessment of high-value medicinal plants of cold desert areas in Johar valley, Kailash Sacred Landscape, India. Plant Biosyst 153(1):39–47

Parihaar RS, Bargali K, Bargali SS (2014) Diversity and uses of Ethno-medicinal plants associated with traditional agroforestry systems in Kumaun Himalaya. Indian J Agric Sci 84(12):1470–1476

Prakash R (2015) Medicinal plants used by tribal communities: a study of Uttarakhand Himalayan region. IJHSSI 4(2):55–61

Rana SK, Rana HK, Ranjitkar S, Ghimire SK, Gurmachhan CM, O'Neill AR, Sun H (2020) Climate-change threats to distribution, habitats, sustainability and conservation of highly traded medicinal and aromatic plants in Nepal. Ecol Indic 115:106435

Rao MR, Palada MC, Becker BN (2004) Medicinal and aromatic plants in agroforestry systems. Agrofor Syst 61(1):107–122

Rawat N, Purohit S, Painuly V, Negi GS, Bisht MPS (2022) Habitat distribution modelling of endangered medicinal plant *Picrorhiza kurroa* (Royle ex Benth) under climate change scenarios in Uttarakhand Himalaya, India. Ecol Inform 68:101550

Ritika Bora B, Ismail BB, Garba U, Mishra S, Jha AK, Gupta AK (2024) Himalayan fruit and circular economy: nutraceutical potential, traditional uses, challenges and opportunities. Food Prod Process Nutr 6(1):71

Sahu S, Rathod A, Rathod CB (2025) Medicinal and aromatic agroforestry: current scenario, practices, and future prospects in the context of Western and Central Himalayas. J Sci Res Rep 31(2):101–121

Saleem I, Mugloo JA, Pala NA, Bhat GM, Masoodi TH, Mughal AH, Mehraj B (2023) Biomass production, carbon stock and sequestration potential of prominent agroforestry systems in north-western Himalaya, India. Front For Glob Change 6:1192382

Samant SS, Dhar U (1997) Diversity, endemism and economic potential of wild edible plants of Indian Himalaya. Int J Sustain Dev World Ecol 4(3):179–191

Samant SS, Dhar U, Palni LMS (1998) Medicinal plants of Indian Himalaya: diversity, distribution, and potential values. Gyanodaya Prakashan

Schippmann U, Leaman DJ, Cunningham A B (2002) Impact of cultivation and gathering of medicinal plants on biodiversity. Global trends and Issues. FAO, Rome

Sharma J, Gairola S, Gaur RD, Painuli RM, Siddiqi TO (2013) Ethnomedicinal plants used for treating epilepsy by indigenous communities of sub-Himalayan region of Uttarakhand, India. J Ethnopharmacol 150(1):353–370

Sharma A, Verma R, Kumar P (2019) Agroforestry and biodiversity conservation in the Northwestern Himalayas. J Agrofor Environ 25(3):120–135

Sharma H, Pant KS, Bishist R, Prakash P, Gautam KL (2024) An insight into prevalent agroforestry land use systems of North Western Himalayan Region, India: challenges and future prospects. In: Agroforestry to combat global challenges: current prospects and future challenges, pp 419–439

Shinwari AK, Ahmad M, Alsahli AA, Israr A, Shinwari ZK, Shinwari S, Yilmaz ME (2024) Himalayan medicinal plants wealth, threats and conservation for securing the future of biodiversity. Pak J Bot 56(6):2327–2350

Singh KN, Lal B, Todaria NP (2016) Ecological status and conservation of rare plants in high altitude landscapes of Indian Western Himalaya. Glob J Med Plant Res 5(1):1

Singh A, Nautiyal MC, Kunwar RM, Bussmann RW (2017) Ethnomedicinal plants used by local inhabitants of Jakholi block, Rudraprayag district, Western Himalaya. India J Ethnobiol Ethnomed 13(1):49

Singh K, Kumar B, Sharma J, Gairola S (2020) Medicinal plants used to manage CNS and memory-related problems by indigenous communities of Jammu & Kashmir (UT) and Ladakh (UT). India Plant Arch 20(2):1959–1974

Singh S, Thakur S, Mehta JP, Dutt HC (2023) Documentation of ethnobotany: a source for conservation plan of Medicinal plant species in alpine region of Madhmaheshwar, Uttarakhand, India. Vegetos 36(3):989–998

Singh N, Riyal MK, Singh B, Khanduri VP, Rawat D, Singh C, Kumar M (2024a) Carbon sequestration potential of agroforestry versus adjoining forests at different altitudes in the Garhwal Himalayas. Atmos 15(3):313

Singh P, Bargali H, Pant S, Negi GCS (2024b) Diversity, yield potential and economic value of medicinal herbs in dominant forests of western Himalaya. Trees For People 17:100625

Sundriyal M, Sharma E (2018) Cultivation and conservation of medicinal plants in the Sikkim Himalaya. Mt Res Dev 16(2):108–116

Thakur N, Kumar N, Kumar S, Sharma A, Ahlawat YK, Malik A, Ahmed M (2025) Ethnobiological survey on medicinal plants used by gaddi and Gujjar tribes of riparian region of Beas River of Himachal Pradesh in North Western Himalayas, India. Clin Phytosci 11(1):4

Thapliyal N, Bhojak P, Arya D, Sekar KC (2024) Ethnobotanical assessment of medicinal plants used in alpine regions of Uttarakhand, west Himalaya. Ethnobot Res Appl 27:1–24

The Tribune (2021) Himachal farmers get financial aid to cultivate medicinal plants. https://www.tribuneindia.com/news/himachal/farmers-get-financial-aid-to-cultivate-medicinal-plants-274938/. Accessed 12 July 2021

Tiwari V, Negi KS, Rawat R, Mehta PS (2017) In-situ conservation and traditional uses of medicinal plants: a case study of home gardens in Nainital, Uttarakhand. Asian Agri Hist 21(1):25–30

Tiwari D, Sah AN, Bawari S, Bussmann RW (2020) Ethnobotanical investigations on plants used in folk medicine by native people of Kumaun Himalayan region of India. Ethnobot Res Appl 20:1–35

TOI (2025) Medicinal plant project in Almora to curb migration and revive hill agriculture. https://timesofindia.indiatimes.com/city/dehradun/medicinal-plant-project-in-almora-to-curb-migration-and-revive-hill-agriculture/articleshow/121384422.cms. Accessed 25 Aug 2025

Upadhyay G, Pandey NC, Tewari LM, Upreti D, Lodhiyal V, Koranga S (2024) Status, threats, diversity and distribution of threatened medicinal plants of Kumaun Himalaya, Uttarakhand, India. In: Threatened medicinal plants in the Indian Himalayan region: sustainability challenges and conservation strategies. Springer Nature Switzerland, Cham, pp 89–108

Ved DK, Goraya GS (2008) Demand and supply of medicinal plants in India. FRLHT, Bangalore
Verma RK, Tewari VP (2016) Some important medicinal plants of cold desert regions of District Kinnaur of Himachal Pradesh state in India: their uses and chemical ingredients. J Plant Chem Ecophysiol 1:1009
WHO (2013) WHO traditional medicine strategy 2014–2023. World Health Organization, Geneva
WHO (2020) Medicinal plants production systems in rural Indonesia
WHO (2022) Traditional medicine. World Health Organization, Geneva

FSC
www.fsc.org
MIX
Papier | Fördert
gute Waldnutzung
FSC® C083411

Kolibri 360 GmbH
Ferdinand-Jühlke-Straße 7
99095 Erfurt, Deutschland
produktsicherheit@kolibri360.de